Processing of Polymer Matrix Composites

Processing of Polymer Matrix Composites

P. K. Mallick
William E. Stirton Professor of Mechanical Engineering
University of Michigan-Dearborn

CRC Press
Taylor & Francis Group
Boca Raton London New York

CRC Press is an imprint of the
Taylor & Francis Group, an **informa** business

CRC Press
Taylor & Francis Group
6000 Broken Sound Parkway NW, Suite 300
Boca Raton, FL 33487-2742

First issued in paperback 2021

ISBN-13: 978-1-4665-7822-7 (hbk)
ISBN-13: 978-1-03-217894-3 (pbk)
DOI: 10.1201/9781315157252

This book contains information obtained from authentic and highly regarded sources. Reasonable efforts have been made to publish reliable data and information, but the author and publisher cannot assume responsibility for the validity of all materials or the consequences of their use. The authors and publishers have attempted to trace the copyright holders of all material reproduced in this publication and apologize to copyright holders if permission to publish in this form has not been obtained. If any copyright material has not been acknowledged please write and let us know so we may rectify in any future reprint.

Publisher's Note

The publisher has gone to great lengths to ensure the quality of this reprint but points out that some imperfections in the original copies may be apparent.

Visit the Taylor & Francis Web site at
http://www.taylorandfrancis.com

and the CRC Press Web site at
http://www.crcpress.com

In memory of my parents.

Contents

Preface

Polymer matrix composites are finding an increasing number of applications in aerospace, automotive, sporting goods, biomedical, construction, and many other industries due to their high weight-saving potential as well as many unique characteristics, such as high strength-to-density ratio, fatigue resistance, high damping factor, and freedom from corrosion. By using a variety of fiber architectures and fiber–matrix combinations, it is possible to tailor their mechanical and thermal properties to suit the product design and performance requirements. Since polymer matrix composites are nonisotropic in nature and their response characteristics to applied loads are different from that of metals, the mechanics and design methodology for polymer matrix composites are different from the ones used for metals. Similarly, the processes used to manufacture polymer matrix composites are also different from the conventional manufacturing processes, such as casting and forming, used for metals.

In undergraduate engineering curricula, manufacturing process courses are taught from textbooks that are mostly based on manufacturing processes used for metals and contain a short chapter on composite manufacturing. While many textbooks are available on the mechanics of polymer matrix composites, there are very few textbooks on their processing. The current book, *Processing of Polymer Matrix Composites*, is written to fill the need in this area. It contains chapters on the major manufacturing processes used for polymer matrix composites. It describes the process details, how they are practiced, the process parameters, and their effects on properties and process-induced defects. It is my expectation that this book will be useful to both undergraduate and graduate students as well as practicing engineers who are beginning to learn about composite materials and their processing.

The book starts with an introduction to polymer matrix composites and their applications, the roles of fibers, matrix and fiber–matrix interface, and key concepts related to the selection and design of polymer matrix composites. The next two chapters describe the fiber forms, fiber architectures, and matrix materials commonly used in polymer matrix composites. Process fundamentals are described next. Thus, the first four chapters in the book give the essential knowledge required to understand the material and manufacturing process parameters that affect the properties, performance, and quality of composite parts. There are problems at the end of these chapters, which are included to reinforce the fundamental knowledge needed for understanding the processing characteristics of polymer matrix composites. The primary manufacturing processes are covered in the subsequent chapters. The processes include bag molding, compression molding, liquid composite molding, filament winding, pultrusion, and forming. The book then examines the joining and repair processes used for polymer matrix composites. Quality inspection methods, tooling, and manufacturing cost considerations are also considered. The health and safety issues related to the processing of polymer matrix composites are described in the Appendix.

The concept of this book has been with me since the second edition of my first book, *Fiber Reinforced Polymers*, was published in 1998. For a variety of reasons, the project never took off, and it remained in the planning-to-write stages all these years until I was contacted by Allison Shatkin of CRC Press to consider writing this book. I want to thank her for not only asking me to write this book, but also giving me several extensions to finish it.

I also want to sincerely thank my wife, Sunanda, who has given me encouragement and support during the writing of the book. The last few weeks of trying to finish the manuscript were particularly tough for me, and her support during this period was crucial.

P. K. Mallick

1 Introduction

In the last few decades, polymer matrix composites (PMCs) have become an important and useful class of structural materials. They are used in many aerospace, automotive, industrial and consumer applications, often competing with other structural materials, such as steels, aluminum alloys, and titanium alloys. In a PMC, high-strength and high-modulus fibers are used as reinforcement, and a polymer serves as the matrix material. The combined material has a much higher modulus and strength than the polymer matrix itself, and hence, it is often called a fiber-reinforced polymer (FRP).

PMCs have found much greater use than either metal matrix composites or ceramic matrix composites in aerospace, automotive, and other industries, primarily because of their lower density, high strength-to-density ratio, high modulus-to-density ratio, and relative ease of processing. The lower density of PMCs compared to metal matrix composites and ceramic matrix composites is due to the lower density of polymers compared to metals and ceramics. The density of polymers commonly used for PMCs is in the range of 0.9–1.4 g/cm^3. The density of reinforcing fibers is also very low, typically between 1.4 and 2.6 g/cm^3. Depending on the fiber type and fiber volume fraction, the density of PMCs is between 1.2 and 2 g/cm^3. Because of such a low density, the strength-to-density ratios and, in many cases, modulus-to-density ratios of PMCs are comparatively higher than those of metals (Table 1.1). This gives a great weight saving opportunity in many applications if a PMC is selected instead of metals.

In a PMC, the matrix can be either a thermoplastic polymer, such as polyether ether ketone (PEEK) and polypropylene (PP) or a thermosetting polymer, such as an epoxy and a polyester. Fibers commonly used as reinforcement are glass, carbon, or aramid (e.g., Kevlar 49). A variety of fibers in each of these types is commercially available. They can be either continuous or discontinuous in length and arranged in unidirectional, bidirectional, multidirectional, or completely random orientation (Figure 1.1). A combination of these arrangements is also possible. A common form of composite structure is a laminate in which very thin layers of continuous fibers embedded in a polymer matrix are stacked and laminated together. Fiber orientations and fiber arrangements can be different in different layers of the laminate. Two or more fiber types can also be combined to create a hybrid construction.

The mechanical properties, such as tensile strength and modulus, of a PMC can be altered relatively easily by changing the fiber type, fiber length, and fiber arrangement. They can also be combined with aluminum honeycomb, structural plastic foam, or wood to produce sandwich structures that are stiff and very lightweight. Considering the variety of options available in each of these categories, it is easy to see that there are many alternative ways of designing PMCs to suit the design and performance requirements of a structure or a component. This design flexibility is one of the principal advantages of PMCs.

1

TABLE 1.1

Comparison of Properties of a Few Selected Metals and Composites

Material	Density (g/cm³)	Tensile Modulus (GPa)	Tensile Strength (MPa)	Tensile Modulus-to-Weight Ratio (10⁶ m)	Tensile Strength-to-Weight Ratio (10³ m)
SAE 1010 steel (cold worked)	7.87	207	365	2.68	4.72
AISI 4340 steel (quenched and tempered)	7.87	207	1722	2.68	22.3
6061-T6 aluminum alloy	2.70	69	310	2.60	11.7
7178-T6 aluminum alloy	2.70	69	606	2.60	22.9
Ti-6Al-4V titanium alloy	4.43	110	1171	2.53	26.9
High-strength carbon fiber–epoxy composite (unidirectional)	1.55	137.8	1550	9.06	101.9
High-modulus carbon fiber–epoxy composite (unidirectional)	1.63	215	1240	13.44	77.5
E-glass fiber–epoxy composite (unidirectional)	1.85	39.3	965	2.16	53.2
Kevlar 49 fiber–epoxy composite (unidirectional)	1.38	75.8	1378	5.60	101.8
Carbon fiber–epoxy composite (quasi-isotropic)	1.55	45.5	579	2.99	38
SMC composite (isotropic)	1.87	15.8	164	0.86	8.9

Notes: (1) For unidirectional composites, the fibers are unidirectional, and the reported tensile properties are in the direction of fibers. Properties in the direction normal to the fibers are much lower. (2) Tensile modulus-to-weight ratio is calculated by dividing the tensile modulus with specific weight, which is defined as the weight per unit volume and is obtained by multiplying density with the acceleration due to gravity (9.81 m/s^2). Tensile strength-to-weight ratio is calculated in a similar way.

There are several other unique attributes of PMCs that make them desirable in many applications. One of these attributes is their nonisotropic nature, which allows one to tailor their properties according to the design requirement. For example, fibers can be selected and oriented purposely to strengthen or stiffen a structure in a preferred direction, produce zero coefficient of thermal expansion, or fabricate curved panels without using any secondary forming operations. Another attribute of PMCs is their high damping factor, which is an order of magnitude greater than that of metals. High damping factor is useful in reducing structure-borne vibration and noise transmission. The third attribute is that, unlike metals, they do not corrode, and therefore, no anticorrosion measures need to be applied to them. However, it should be noted that some metals in direct contact with some PMCs, for example, aluminum fasteners in contact with carbon fiber-reinforced epoxy, may corrode due to

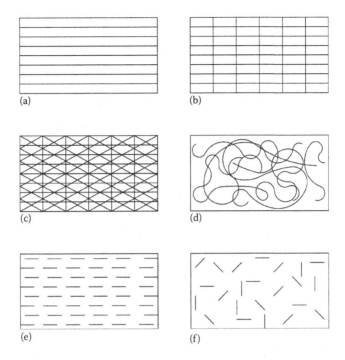

FIGURE 1.1 Schematic representation of fiber orientations in PMCs. (a) Unidirectional, continuous; (b) bidirectional, continuous; (c) multidirectional, continuous; (d) random, continuous; (e) unidirectional, discontinuous; and (f) random, discontinuous.

galvanic potential difference between the two and, therefore, must be protected by using paints or coatings acting as an anticorrosive barrier between them.

This chapter gives a brief overview of the constituents used in PMCs and examines the important parameters that control or influence the properties of PMCs. A brief introduction to the important processing techniques used for making structural PMC parts is also given. More details on the materials, mechanics, design, and processing of PMCs are available in several references [1–3] listed at the end of this chapter.

1.1 CONSTITUENTS

The major constituents in a PMC are the fibers and the polymer matrix. Fibers are the reinforcing phase and are dispersed in the polymer matrix, which is the continuous phase surrounding the fibers (Figure 1.2). The matrix participates in load sharing with the fibers and transfers loads between the fibers. Some PMCs may also contain fillers (such as calcium carbonate) mainly to reduce cost and control shrinkage, impact modifiers (such as rubber particles) to reduce notch sensitivity and increase impact energy absorption, and ultraviolet (UV) absorbers to reduce deterioration of properties under long exposure to sunlight, etc. The fiber surface may be coated with a chemical coupling agent or modified by oxidation to help improve fiber surface wetting with the polymer matrix and create good bonding at the interface between

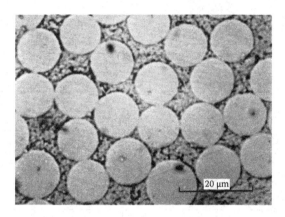

FIGURE 1.2   Cross section of a continuous fiber composite.

the fibers and the polymer matrix. Both fiber surface wetting and good interfacial bonding are important in enhancing the properties of PMCs.

1.1.1   FIBERS

Fibers are the principal load-carrying members in a fiber-reinforced composite material. Mechanical properties, such as modulus and strength, and thermal properties, such as coefficient of thermal expansion, of a composite depend on the following parameters related to fibers:

1. Fiber type and properties
2. Fiber volume fraction
3. Fiber length
4. Fiber orientation
5. Fiber architecture (arrangement of fibers in the composite)

In most high-performance PMCs, fibers are used in continuous lengths, and fiber volume fraction is in the range of 50–65% of the composite. Discontinuous fibers are used in random fiber PMCs. The fiber volume fraction in these composites is usually in the range of 20–40%.

The three most commonly used fibers used in PMCs are glass, carbon, and aramid fibers. A selected list of these fibers and their properties are given in Table 1.2. Other fibers available in the market, but used in lower quantities, include boron fibers, ceramic fibers (such as silicon carbide), extended chain polyethylene fibers, and natural fibers, which include jute, hemp, sisal, etc. The general characteristics of glass, carbon, and aramid fibers are briefly discussed in the following.

1. Fibers are produced in the form of very thin continuous filaments ranging from 7 to 15 μm in average diameter. In practice, they are supplied and used as bundles of filaments. These bundles are called *strands* in the

TABLE 1.2
Properties of a Few Selected Fibers Used in PMCs

Fiber	Filament Diameter (μm)	Density (g/cm³)	Tensile Modulus (GPa)	Tensile Strength (MPa)	Strain at Failure (%)
E-glass	10	2.54	72.4	3450	4.8
S-glass	10	2.49	86.9	4300	5
T-300 (PAN carbon)	7	1.76	230	3650	1.5
AS-4 (PAN carbon)	7.1	1.79	231	4410	1.7
IM-7 (PAN carbon)	5.2	1.78	276	5516	1.9
M-55J (PAN carbon)	–	1.91	540	4020	0.8
P120 (pitch carbon)	–	2.16	823	2200	0.3
Kevlar 49 (aramid)	11.9	1.45	131	3620	2.8

case of glass fibers and *tows* in the case of carbon and aramid fibers. The number of filaments in a strand or a tow can be varied. In both strands and tows, filaments are parallel and continuous. Glass fiber strands are bundled together to form either *rovings* (parallel strands) or *yarns* (twisted strands). The number of strands in rovings and yarns can be varied. Carbon filaments are bundled together to form tows of parallel filaments without any twist and yarns in which some filaments are twisted. The fiber packages in which glass fiber rovings and carbon fiber tows are supplied by the fiber manufacturers are shown in Figure 1.3. Similar fiber packages are used for aramid fibers that are also available as tows and yarns. The larger the number of filaments in a strand or a tow, the higher the yield in processing (in terms of production rate). However, composite properties, such as strength, may be lower if the strand or the tow is too thick, since the polymer matrix may not be able to penetrate into the inside of a thick strand or tow and wet out the interior filaments.

2. Fibers have very high tensile modulus and tensile strength; however, their tensile strain at failure is very low. Their tensile stress–strain diagrams (Figure 1.4) are nearly linear until failure. They behave like a classic brittle material when subjected to tensile load. Their tensile strength depends on fiber length and diameter and, in general, shows a large statistical variation.

3. Carbon fibers have modulus values that can range from 220 to 800 GPa. The lower number in this range is close to the modulus of steel. Glass and aramid fibers have a much lower modulus, but their tensile strength is comparable to that of some of the carbon fibers (see Table 1.2). Because of a very low strain-at-failure, high-modulus carbon fibers have a tendency to break easily during processing. This can cause process shutdowns unless proper care is taken in handling these fibers during processing. For example, in many composite manufacturing processes, fiber bundles are pulled over small-diameter rollers to create tension in them. If the roller diameter is very small, the tensile strain in the fibers may exceed the failure strain of

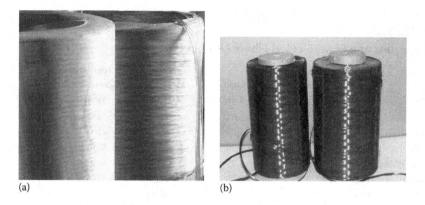

(a)   (b)

FIGURE 1.3   Fiber packages of (a) glass fiber rovings and (b) carbon fiber tows.

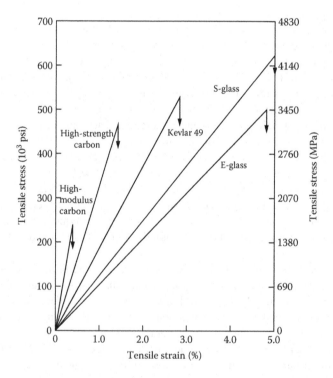

FIGURE 1.4   Tensile stress–strain diagrams of reinforcement fibers used in PMCs.

the fibers, and they may break. This will cause interruption of the processing line and reduce the production rate.

4. Carbon and aramid fibers have negative coefficient of thermal expansion in their longitudinal direction, but a relatively high positive coefficient of thermal expansion in the radial direction. Glass fibers, on the other hand,

have equal coefficients of thermal expansion in both longitudinal and radial directions. It should also be noted that the coefficient of thermal expansion (and contraction) of fibers is generally much lower than that of the polymer matrix. Due to the large difference in thermal contractions between the fibers and the polymer matrix, thermal residual stresses may develop at the fiber–matrix interface when the polymer matrix cools down from the processing temperature to room temperature.

1.1.1.1 Glass Fibers

E-glass fibers are the least expensive of all commercially available fibers and are used in much larger quantity than any of the other fibers in PMCs. As can be seen in Table 1.2, their modulus is not as high as that of carbon or aramid fibers, but their tensile strength is very high. *S-glass fibers* are slightly higher in modulus as well as strength, but they are more expensive. It should be noted that the strength of glass fibers is affected by abrasion and moisture absorption and is often much lower than the strength of freshly manufactured glass fibers.

Glass fibers have low thermal and electrical conductivities, and they are very good insulators. In some applications (such as, in a radar), their low electrical conductivity may pose problems by causing static charge accumulation and, thereby, creating electromagnetic interference. Glass fibers are very hard and abrasive on cutting tools. Glass FRPs should be cut or machined with carbide-tipped or diamond-tipped cutters. Another point to note about glass fibers is that during the making of glass fiber strands, the filaments are coated with a mix of chemicals, called *sizing*. It is a mixture of lubricants (which prevent abrasion between the filaments), antistatic agents (which reduce static electricity between the filaments), and a binder (which packs the filaments into a strand and promotes adhesion between the filaments and the polymer matrix). It also protects the filaments from abrasion due to rubbing against each other and, in many instances, promotes chemical bonding between the glass fibers and the polymer matrix.

Glass fibers are available in a variety of lengths and forms. Glass fiber strands are a collection of continuous parallel filaments numbering 204 or more. A glass fiber roving is a group of untwisted parallel strands (called *ends*) wound around a cylindrical drum, called a fiber package (Figure 1.3). Glass fiber rovings are specified by tex number (weight in grams per 1000 m) or by yield (length in yards per pound). Common tex numbers are 600, 1200, 2400, etc., while the common yields are 225, 450, 900, etc. *Chopped strands* are produced by cutting continuous strands into small lengths ranging from 3.2 to 12.7 mm and are used in injection molding. Longer chopped strands, ranging in lengths of up to 50 mm are mixed with a resinous binder and are spread in the form of a two-dimensional mat to make *chopped strand mats*.

1.1.1.2 Carbon Fibers

Carbon fibers are made from either polyacrylonitrile (PAN) precursor or pitch precursor. Carbon fibers are more expensive than glass fibers, but they have lower density and higher tensile modulus, which makes them very attractive in stiffness-critical applications. PAN-based carbon fibers are available in four different varieties—high strength, high modulus, intermediate modulus, and ultrahigh modulus. In general, density and price are higher for higher modulus carbon fibers. Pitch-based carbon fibers have tensile modulus values greater than 350 GPa, but their tensile strength is lower than that of PAN-based carbon fibers. The thermal conductivity of pitch-based carbon fibers is also higher than that of PAN-based carbon fibers.

Carbon fibers are available in a variety of tow sizes—ranging from 1000 filaments per tow (i.e., 1K tow) to 50,000 filaments per tow (i.e., 50K tow). The cost of carbon fibers is lower for larger tows than for smaller tows; however, there may be processing difficulty with very large tows, particularly in terms of resin penetration into the interior of the tow.

Carbon fibers are currently used primarily for aerospace and sporting good applications. The reason for selecting carbon fibers for these applications is their exceptionally high modulus-to-density ratio. Their strength-to-density ratio is also very high. Other advantages of carbon fibers are their high electrical conductivity, high thermal conductivity, and low coefficient of thermal expansion. Their high electrical conductivity, however, possesses a problem in processing plants, since carbon dust or fine particles released from carbon fibers during processing may short-circuit nearby electric motors and other electrical machines; thus, they have to be properly protected (carbon-proofed).

1.1.1.3 Aramid Fibers

Aramid fibers, such as Kevlar 49 and Kevlar 149, are polymeric fibers, with modulus values nearly twice that of E-glass fibers. Their tensile strength-to-density ratio is the highest among the reinforcing fibers used. However, their compressive strength is low, which is the reason these fibers are not selected for applications involving high compressive stresses. They are not very thermally stable above 160°C. They tend to absorb moisture when exposed to humid environments. Their properties are also adversely affected by long exposure to sunlight. From the processing standpoint, aramid fiber-reinforced composites are difficult to machine, since aramid fibers tend to mushroom at the cutting point. Special cutting tools are used for machining (e.g., drilling) aramid FRPs.

1.1.2 Matrix

The role of the matrix in a PMC is to keep the fibers in place, transfer load among the fibers, and protect them from moisture, chemicals, and other environments. Compressive and shear properties of the composite are strongly influenced by the matrix properties and any changes in the matrix properties due to increased temperature or moisture absorption are reflected in these matrix-controlled properties of the composite. Other properties that are affected by the matrix characteristics are flammability, resistance to weathering (e.g., long exposure to sunlight), and environmental effects (e.g., moisture absorption from the surrounding atmosphere).

Matrix plays a very important role in the processing of PMCs. For example, the viscosity of the polymer at the time of processing is a major factor contributing to the wetting of the fibers, filling the mold with liquid resin, expelling the air out of the mold, and making a good quality part. Processing parameters, such as temperature, pressure, and processing time, which control production rate, depend on the choice of matrix. For example, epoxy resins, in general, have a longer processing time than polyester resins and are not the first choice for matrix material in mass production applications.

Polymers commonly used in PMCs and their key characteristics are listed in Table 1.3. Both thermosetting and thermoplastic polymers are used in PMCs. Processing method and processing time for PMCs depend on whether the matrix is a thermosetting or a thermoplastic polymer. In general, thermoset matrix composites require longer processing time than thermoplastic matrix composites. Other distinctions between thermosetting and thermoplastic polymers are listed in Table 1.4.

Both thermosetting and thermoplastic polymers contain long chain molecules of many repeating chemical units. On a molecular level, the principal difference between the two is that in thermosetting polymers, the molecules are chemically connected or cross-linked, whereas in thermoplastic polymers, there are no chemical connections between the molecules (Figure 1.5). In thermoplastic polymers, the molecules are entangled with each other and can exist either in random orientation or in a combination of orderly and random orientations. The former is called an amorphous polymer, and the latter is called a semicrystalline polymer. The crystalline

TABLE 1.3
Properties of a Few Selected Polymers Used in PMCs

Polymer	Density (g/cm³)	Tensile Modulus (GPa)	Tensile Strength (MPa)	Strain at Failure (%)	Glass Transition Temperature, T_g (°C)	Maximum Service Temperature (°C)
Thermosetting Polymers[a]						
Epoxies	1.2–1.3	2.75–4.1	55–130	1.5–8	150–260	125
Polyesters	1.1–1.4	2.1–3.5	35–104	1–7		60–150
Vinyl esters	1.12–1.32	3–3.5	73–81	3–8		60–150
Bismaleimides	1.2–1.32	3.2–3.5	48–110	1.5–3.3	230–290	232
Polyurethanes	1.21	0.7	30–40	400	120–167	
Thermoplastic Polymers						
PEEK	1.30	3.7	110	25	143	260
PPS	1.35	3.45	93	15	85	218
Polyether-imide	1.28	3.45	117	60	210	171
Polyamide-imide	1.42	4.5	150	8	275	
PP	0.90	1.5	36	>100%	−10	50–75
Polyamide-6	1.13	3.10	80	50	60	

[a] Properties of thermosetting polymers are given in ranges, since they depend on the chemical structure of the polymer and the curing condition.

TABLE 1.4
General Characteristics of Thermoplastic and Thermosetting Polymers

Thermoplastic Polymers

- The polymer molecules are not chemically joined; however, there are entanglements between them
- Are processed and formed by heat softening and/or melting
- High viscosity during processing
- Are directly recyclable; however, the polymer properties may deteriorate with repeated recycling
- Have lower heat and chemical resistance than thermosetting polymers
- Have lower hardness than thermosetting polymers
- Are usually more ductile

Thermosetting Polymers

- The polymer molecules are chemically joined (crosslinked) at the processing stage
- Cannot be reprocessed by heat softening or melting
- Have low viscosity during processing
- Cannot be directly recycled
- Have higher heat and chemical resistance than thermoplastic polymers
- Have higher hardness than thermoplastic polymers
- Are usually more brittle

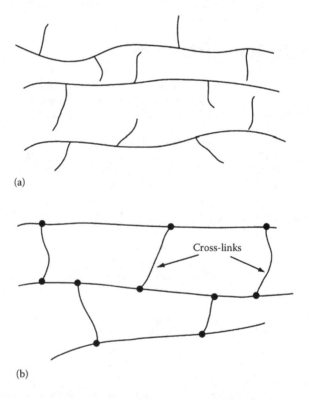

(a)

(b)

FIGURE 1.5 Schematic representations of (a) thermoplastic and (b) thermosetting polymer molecules.

phase of a semicrystalline polymer contains molecules that are orderly oriented in a folded chain configuration, and the amorphous phase contains molecules that are randomly oriented. The microstructure of a semicrystalline polymer usually contains a large number of *spherulites* (Figure 1.6). The center of each spherulite is the nucleation site from which the crystalline phase grows radially outward, thus roughly creating a spherical shape. The amorphous phase is located in between the crystalline phase. Spherulites are nucleated at numerous sites as the liquid polymer cools down from the processing temperature to room temperature and eventually stops growing as they impinge on each other. The number of nuclei and size of the spherulites depend on the crystallization condition, including the cooling rate from the liquid state to the solid state, the crystallization temperature, and the presence of nucleating agents, if any. Properties of semicrystalline polymers depend on the amount of crystallinity (indicated by the weight percentage of crystallinity) and the average size of spherulites.

Thermosetting polymers or simply *thermosets*, such as epoxies, are the traditional matrix materials in many high-performance composites, particularly for aerospace applications. The starting materials for many thermosets are low-viscosity liquid prepolymers. Fiber surface can be easily coated and wetted by the prepolymer before starting the chemical reaction, called *curing*, that changes it from a low-viscosity liquid to a solid thermosetting polymer. Curing, in most cases, requires elevated temperature, called the *cure temperature*, which varies from polymer to polymer. The cure temperature for many epoxy systems is in the range of 120–180°C. Once the curing reaction is complete, the thermosetting polymer cannot be changed to a liquid state by the application of heat. Properties

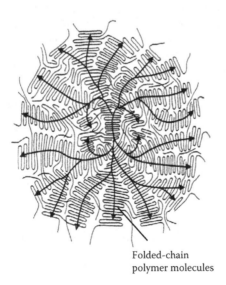

Folded-chain
polymer molecules

FIGURE 1.6 Schematic of a spherulite in a semicrystalline thermoplastic. (Note: Arrows indicate the directions of folded chain growth.) (From Ehrenstein, G. W., *Polymeric Materials*, Hanser, Munich, 2001.)

of a thermosetting polymer depend on the number of cross-links formed and their closeness in the cured polymer.

Thermoplastic polymers or thermoplastics, on the other hand, have very high viscosity in the liquid state. The coating and wetting of the fiber surface with liquid thermoplastic polymers are much more difficult and require more expensive processes, such as melt impregnation and powder coating, when continuous fibers are used. Short fibers are more commonly used with thermoplastic polymers, since they can be compounded (mixed) with liquid thermoplastic polymers using extruders and then injection molded into the desired part shape and size at a relatively low cost. Note that thermoplastics can be softened and melted repeatedly by the application of heat— a major advantage over thermosets in terms of their reusability and recyclability. However, it should be noted that repeated recycling may deteriorate their properties.

Thermoplastic polymers, when heated, change from a hard solid to a soft solid at a temperature called the *glass transition temperature*, denoted by T_g. At this temperature, the modulus of the thermoplastic sharply reduces to a much lower value as shown in Figure 1.7. If the temperature is raised further, the thermoplastic transforms into a viscous liquid. Amorphous thermoplastics will melt as the temperature is raised well above their T_g, but will not show a melting temperature (Figure 1.7a). Semicrystalline thermoplastics will show a sharp *melting temperature* on solid-to-liquid transition, which is denoted by T_m in Figure 1.7b. The melting temperature T_m corresponds to the melting of the crystalline phase in the solid polymer. On cooling to room temperature, both amorphous and semicrystalline polymers revert to their solid state and exhibit hardness, modulus, and other mechanical properties that are their room temperature characteristics.

Thermosetting polymers, when heated, will soften and exhibit a glass transition temperature, but will not melt and show a melting temperature. As shown in Figure 1.7c, the glass transition temperature increases with increasing numbers of cross-links; however, for a very highly cross-linked thermoset, the glass transition and associated softening may not be clearly observed. If the temperature is raised much higher than the glass transition temperature, thermosetting polymers will start to char and ultimately burn and disintegrate.

A majority of continuous FRP matrix composites used today contain a thermoset as the matrix. The principal reason for this is their very low viscosity at the time the fibers are incorporated into them. Other reasons for selecting thermosets

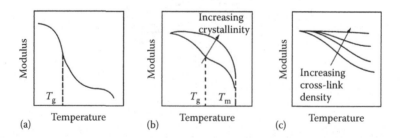

FIGURE 1.7 Modulus vs. temperature diagrams of polymers (a) amorphous thermoplastic, (b) semicrystalline thermoplastic, and (c) thermoset.

as matrix are their higher thermal stability, higher creep resistance, and better chemical resistance than thermoplastic polymers. The glass transition temperature of thermosetting polymers is, in general, higher than that of thermoplastic polymers. A few thermoplastics, such as PEEK and polyphenylene sulfide (PPS), have high glass transition temperatures, but they are also very expensive materials compared to most thermosets, such as epoxies and polyesters. The processing temperature for these thermoplastics is also very high compared to that of most thermosetting polymers. The advantages of thermoplastics and the reasons they are often considered in high-performance applications despite the difficulty of processing are as follows:

1. Have higher tensile strain-at-failure
2. Have higher fracture toughness and greater resistance to crack propagation
3. Have higher impact energy absorption
4. Have unlimited storage (shelf) life
5. Have no special requirements for storage (such as in a freezer for most thermoset matrix composites) before final processing
6. Have shorter processing time
7. Can be postformed and postshaped by heating
8. Can be welded using vibration welding, ultrasonic welding, etc.
9. Can be reprocessed and recycled

1.1.3 Fillers and Other Additives

Fillers are added in some PMCs to reduce cost and control molding shrinkage. Among the inorganic fillers used in the composite industry are low-cost minerals, such as calcium carbonate, talc, and mica [5]. Hollow glass spheres, solid glass spheres, and milled fibers are also used, but they are generally more expensive than the mineral fillers. Fillers can be surface treated with coupling agents to improve their bonding with the polymer matrix. A list of fillers commonly used with PMCs is given in Table 1.5.

TABLE 1.5
Properties of Commonly Used Fillers in PMCs

Filler	Color	Density (g/cm³)	Modulus (GPa)	CLTE (10⁻⁶/°C)	Thermal Conductivity (W/m °C)
Calcium carbonate	White	2.63	179	10	2.34
Anhydrous kaolin	White or brown	2.58	138	8	1.97
Talc	White	2.8	138	8	2.09
Mica	Gray or white	2.82	172	8	2.51
Solid glass microsphere	Clear	2.48	–	8.47	1.05
Hollow glass microsphere	Clear	0.08–0.64	–	8.8	–
Alumina trihydrate	White	2.42	–	–	–

Fillers are usually mechanically mixed with thermosetting prepolymers and melt-mixed with thermoplastic polymers. In general, the addition of fillers will increase the viscosity of the liquid polymer and, if added in significant amounts, will make the processing difficult. With the exception of hollow glass spheres, inorganic fillers increase the density of the polymer. Hollow glass spheres, because of their low density, reduce the density of the polymer.

Some of the other additives used with polymer matrix and their primary functions are listed in the following:

1. Pigment or colorant to impart color to the polymer
2. UV stabilizer to prevent degradation of the polymer due to long exposure to UV energy from the sun
3. Flame retardant to reduce flammability of the polymer
4. Impact modifier to increase impact strength of the polymer
5. Antioxidant to reduce the possibility of oxidation of the polymer in an oxidizing atmosphere
6. Heat stabilizer to reduce the possibility of degradation of the polymer on long exposures to heat
7. Mold release agent to easily make the molded part release from the mold surface

1.2 FIBER–MATRIX INTERFACE

An important aspect in promoting load transfer between the fibers and the matrix and, therefore, developing the load-sharing *composite* action, is good bonding between them at the interface. When a fiber breaks during loading, a part of the load carried by the fiber before breaking is transferred to the neighboring fibers through the matrix, and for the load transfer to occur, the matrix must be bonded to the fibers. In general, good bonding is necessary for high transverse strength, shear strength, and fatigue durability. However, if the bonding is very strong, both fracture toughness and impact damage tolerance are reduced.

The reinforcing fibers in a PMC have a very large surface area-to-volume ratio, and therefore, the total interfacial area over which the matrix is bonded to the fibers is quite large. Good bonding between the fibers and the matrix requires (1) wetting of the fibers by the matrix, (2) good mechanical bonding, and (3) good chemical bonding. Fiber surface wetting by the matrix requires that the fiber surface energy is higher than the matrix surface energy. In general, the fiber surface energy is higher than the surface energy of many polymers used as matrix in PMCs. The other requirement for fiber surface wetting is the matrix having low enough viscosity to flow and coat the fibers during processing. The polymer viscosity depends on the polymer type and the processing conditions.

The mechanical bonding at the fiber–matrix interface is created during the time of processing. Since the polymer matrix has a higher coefficient of thermal contraction than the fibers, a mechanical bonding is formed between them as they cool down from the processing temperature to room temperature. However, the mechanical bonding is not always very strong and may easily break when a small external load is applied to the composite, causing *debonding* or separation of the fibers from the matrix at the fiber–matrix interface (Figure 1.8). In general, debonding is undesirable, since it not only diminishes the load transfer mechanism, but also exposes the fiber surface to environmental attack from moisture and chemicals that may diffuse through the matrix. If the fibers are sensitive to moisture or chemicals, their properties will also deteriorate and the composite may fail easily.

Increased bond strength is achieved by fiber surface treatment that helps in forming a chemical linkage between the fiber surface and the polymer molecules. Chemical bonding increases the stress level at which debonding is initiated and helps improve the strength of the PMC, particularly in adverse environmental conditions, such as a high-humidity environment. The two strength properties that are affected the most are the transverse tensile strength and interlaminar shear strength. Damage development in the composite is also influenced by the bonding between the fibers and the matrix.

The most common surface treatment for glass fibers is a group of chemicals called *silane coupling agents* [6], which have the general chemical formula $R'–Si(OR)_3$. To be an effective coupling agent, the functional group R' must be compatible with the polymer matrix. Thus, the type of silane selected for an application depends on the polymer matrix being used. To use a silane coupling agent, glass fibers are first heated to burn off any residual sizing from their surface, which is applied on the filament surface at the time of the fiber manufacturing stage. The process of burning off the sizing is called *heat cleaning* and is done at 320–350°C for 15–20 hours. The heat cleaned glass fibers are then immersed in an aqueous solution of the silane for 15–20 minutes to coat their surface with a very thin film of $R'–Si(OH)_3$ molecule, which is produced by the reaction of the silane and the water molecules. When the surface-treated glass fibers are processed with a polymer matrix, the functional group represented by R' in this thin surface film reacts with the polymer molecules to form chemical bridging between the fibers and the polymer matrix.

FIGURE 1.8   Debonding at the fiber–matrix interface.

Several different fiber surface treatments are available for carbon fibers. They include gaseous oxidation (in the presence of an oxygen atmosphere), liquid phase oxidation (in an acid environment), plasma oxidation, and electrolytic oxidation [7,8]. The oxidation process removes the weak surface layers (which will otherwise prevent strong bonding with the matrix), roughen the filament surface (which improves mechanical interlocking), and introduces polar groups that improves bonding with the matrix. In addition, a protective coating, also called sizing, is applied on the filaments before gathering them into a tow. The sizing is used to prevent damage during fiber handling in subsequent processing steps, such as prepregging, weaving, and filament winding. It also prevents the fiber tows from fuzzing and fragmentation. The sizing material can be based on epoxy, polyvinyl alcohol, or other polymers. Both oxidation and sizing application are done in line with the fiber manufacturing process.

1.3 FIBER VOLUME FRACTION

The fiber volume fraction v_f in a fiber-reinforced composite is calculated using the following equation:

$$v_f = \frac{\dfrac{w_f}{\rho_f}}{\dfrac{w_f}{\rho_f} + \dfrac{w_m}{\rho_m}} = \frac{\rho_m w_f}{\rho_m w_f + \rho_f w_m}, \tag{1.1}$$

where ρ_f is the density of the fiber, ρ_m is the density of the matrix, w_f is the fiber weight fraction, and w_m is the matrix weight fraction. Note that the matrix volume fraction is $v_m = (1 - v_f)$ and the matrix weight fraction is $w_m = (1 - w_f)$. The fiber weight fraction w_f is the weight of the fibers W_f in a composite divided by the weight of the composite W_c. The weight of the matrix in the composite is W_m, which is equal to $(W_c - W_f)$.

The average volume fraction in a PMC is experimentally determined using the resin burn-off method. In this method, the polymer matrix in a small sample of the composite is burnt off by heating it in a ventilated muffle furnace at 500–600°C for several hours. The fiber weight W_f is first calculated by comparing the weights of the sample before and after burning off the polymer matrix. After determining the fiber weight in the sample, the fiber weight fraction is calculated and then Equation 1.1 is used to calculate the fiber volume fraction.

It should be noted that all PMCs contain voids due to the entrapment of air or volatiles during the manufacturing of composite parts. Voids may exist in the matrix, at the fiber–matrix interface or within the fiber bundles. One of the challenges in manufacturing composite parts is to reduce the void content as much as possible. In high-performance aerospace composites, void volume fraction higher than 2% is not considered acceptable. Assuming that the composite part is void free, the theoretical density ρ_c of the composite part can be calculated using the following equation.

$$\rho_c = \rho_f v_f + \rho_m v_m. \tag{1.2}$$

However, the actual density of the composite part will be lower than the theoretical density due to the presence of voids.

Example 1.1

A hybrid composite contains intermingled T-300 carbon fibers and E-glass fibers in an epoxy matrix. The carbon fiber weight is 500 g, the glass fiber weight is 600 g, and the epoxy weight is 900 g. Determine the density of the composite. The fiber densities are given in Table 1.2. The density of the epoxy matrix is 1.2 g/cm³.

Solution:

Step 1: Calculate weight fractions w_{fc}, w_{fg}, and w_m.
Total weight of the composite = 500 g + 600 g + 900 g = 2000 g

$$w_{fc} = \frac{500\,g}{2000\,g} = 0.25, \quad w_{fg} = \frac{600\,g}{2000\,g} = 0.3, \quad w_m = 1 - 0.25 - 0.3 = 0.45$$

Step 2: Calculate volume fractions v_{fc}, v_{fg}, and v_m.
From Table 1.2, $\rho_{fc} = 1.76$ g/cm³ and $\rho_{fg} = 2.54$ g/cm³.
The density of the epoxy matrix $\rho_m = 1.2$ g/cm³.

$$v_{fc} = \frac{(0.25/1.76)}{(0.25/1.76)+(0.3/2.54)+(0.45/1.2)} = 0.224$$

$$v_{fg} = \frac{(0.3/2.54)}{(0.25/1.76)+(0.3/2.54)+(0.45/1.2)} = 0.186$$

$$v_m = 1 - 0.224 - 0.186 = 0.59$$

Step 3: Calculate the composite density ρ_c.

$$\rho_c = \rho_{fc}v_{fc} + \rho_{fg}v_{fg} + \rho_m v_m = (1.76)(0.224)+(2.54)(0.186)+(1.2)(0.59) = 1.575\,g/cm^3$$

1.4   FIBER ORIENTATION ANGLE

Fiber orientation angle in a thin unidirectional fiber composite lamina (ply or layer) is the angle between the direction of fibers and the x-direction, which is considered the principal loading direction. In Figure 1.9, the fiber direction and normal to the fiber direction are denoted by the 1–2 axes, and the loading directions are denoted by the x–y axes. The z-axis represents the thickness direction. The fiber orientation angle θ is measured from the positive x-direction. It is considered positive in the counterclockwise direction and negative in the clockwise direction.

For randomly oriented fiber-reinforced composites, the three-dimensional orientation of each fiber can be specified in terms of angles θ and ϕ with respect to a reference coordinate system as shown in Figure 1.10. Angles θ and ϕ will have random values

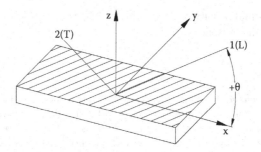

FIGURE 1.9   Fiber orientation angle in a thin lamina of a unidirectional fiber composite.

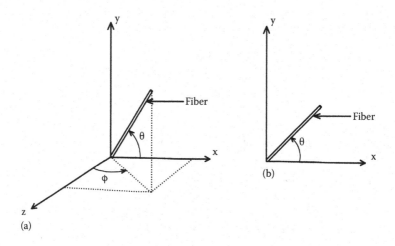

FIGURE 1.10   (a) Three-dimensional orientation of a fiber and (b) two-dimensional orientation of a fiber.

between 0° and 360°. If the composite part is thin and the fibers are long, the fibers may be constrained to lie in one plane, giving rise to a two-dimensional planar orientation of fibers. If the planar orientation is on the 1–2 plane, $\phi = 90°$ and $0° \leq \theta \leq 360°$.

In a molded composite part with random fiber orientation, the fibers will exhibit a statistical distribution of orientation angles, such as the one shown in Figure 1.11. The average fiber orientation in the plane of the thin composite part can be described by the orientation parameters f and g, which are defined as follows:

$$f = \frac{1}{2}\left[3\langle\cos^2\theta\rangle - 1\right],$$ (1.3)

$$g = \frac{1}{4}\left[5\langle\cos^4\theta\rangle - 1\right],$$ (1.4)

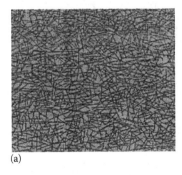
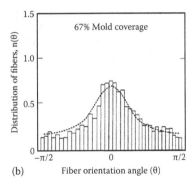

(a)

(b) Fiber orientation angle (θ)

FIGURE 1.11 Distribution of fiber orientations in a compression-molded randomly oriented short fiber composite. (a) Orientations of short fibers with 67% mold coverage and (b) statistical distribution of fiber orientations. (From Advani, S. G., and Sozer, E. M., *Process Modeling in Composites Manufacturing*, CRC Press, Boca Raton, FL, 2003.)

where

$$\left\langle \cos^m \theta \right\rangle = \int_0^{\pi/2} n(\theta) \cos^m \theta \sin \theta \, d\theta,$$

and $n(\theta)$ is the distribution of fibers with fiber orientation angle θ.

The fiber orientation angle θ is usually measured with respect to the principal loading direction. If the fibers are completely aligned in the principal loading direction, $\theta = 0°$ and $f = g = 1$. For a complete random orientation, $f = g = 0$. For most random fiber composites, both f and g lie between 0 and 1. Also to be noted is that f and g can not only vary from part to part made with the same batch of material and using the same manufacturing conditions, but they can also vary from location to location on the same part.

Another representation of fiber orientation is a second-order orientation tensor defined as

$$a_{ij} = \int p_i p_j \psi(p) \, dp, \tag{1.5}$$

where p is a unit vector in the fiber direction, with components p_i that are related to the angles θ and ϕ by the following equations.

$$\begin{aligned} p_1 &= p_x = \cos \theta \\ p_2 &= p_y = \sin \theta \sin \phi \\ p_3 &= p_z = \sin \theta \cos \phi \end{aligned} \tag{1.6}$$

$\psi(p)$ in Equation 1.5 is the probability of finding a fiber between angles θ and $(\theta + d\theta)$ and angles ϕ and $(\phi + d\phi)$. Note that a_{ij} is a (3×3) symmetric matrix with $a_{ij} = a_{ji}$ and

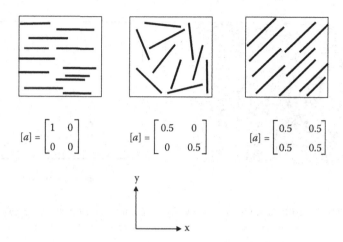

FIGURE 1.12   Two-dimensional fiber orientation matrix.

$a_{11} + a_{22} + a_{33} = 1$. For planar orientation, $\phi = 90°$ so that $\sin\phi = 1$ and $\cos\phi = 0$, and the orientation tensor becomes a (2 × 2) matrix. If all the fibers are aligned parallel to the x-axis, $a_{11} = 1$ and $a_{22} = a_{12} = 0$. If the fibers are randomly oriented, $a_{11} = a_{22} = 0.5$ and $a_{12} = 0$ (Figure 1.12).

The components of the orientation tensor are determined by examining a polished cross-sectional cut from the composite sample under a microscope and then measuring the orientation of N fibers. The weight average value of a_{ij} is calculated using the following equation.

$$a_{ij} = \frac{\sum\limits_{k=1}^{N} p_i^k p_j^k w^k}{\sum\limits_{k=1}^{N} w^k}, \tag{1.7}$$

where p_i^k represents the orientation of the kth fiber, and w^k is the weighting factor accounting for the noncircularity of the fiber cross section created by fibers not being perpendicular to the cutting plane. If the fiber cross section is circular, $w^k = 1$ and Equation 1.7 becomes

$$a_{ij} = \frac{1}{N} \sum\limits_{k=1}^{N} p_i^k p_j^k. \tag{1.8}$$

1.5   MECHANICAL PROPERTIES: TENSILE MODULUS AND STRENGTH

In this section, we will consider the effects of fiber volume fraction, fiber orientation, and fiber length on the tensile modulus and strength of unidirectional fiber-reinforced composites. To start with, let us consider a unidirectional continuous FRP matrix composite (Figure 1.13a) subjected to a tensile force in the fiber direction, also referred to as the longitudinal direction or L-direction. The fiber orientation angle θ in this case is 0°. The transverse direction or T-direction is normal to the fiber direction. Note that the L- and T-directions are also the 1- and 2-directions shown in Figure 1.9.

The micromechanics models used for predicting the tensile modulus and strength for a unidirectional continuous fiber composite are based on the following assumptions:

1. The matrix is free of voids and/or microcracks.
2. Fibers are all parallel, and there is no misalignment between the fiber direction and the loading direction.
3. Fibers are uniformly distributed in the matrix, i.e., fibers are not concentrated only in certain areas, leaving the other areas of the composite fiber starved or resin rich.
4. Residual thermal stresses arising from cooling the matrix from the processing temperature to room temperature are negligible.
5. Perfect bonding exists at the interface of the fibers and the matrix.

Some or all these assumptions may not be valid for real composite parts, and the deviations from these assumptions occur mostly due to manufacturing process-induced defects. For example, it is difficult, if not impossible, to produce a PMC without any voids. Through proper manufacturing process control, voids are kept to a minimum. Similarly, fiber orientation may not be so precisely controlled that there are no misalignments with the loading direction. Fiber distribution in the manufactured composite parts may not be completely uniform, and there may be resin-rich areas where there are fewer fibers than desired or fiber-rich areas where fibers are more concentrated with very little matrix between them. In some cases, the residual thermal stresses arising due to the difference in thermal contractions of the fibers and the matrix may be high enough to cause fine microcracks in the matrix. Thus, appropriate adjustments must be made to the theoretical micromechanics equations to take these deviations into account. Nevertheless, the micromechanics equations are useful in understanding the fiber and matrix selection criteria for a PMC.

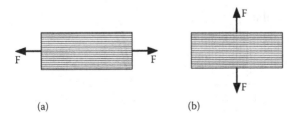

(a)   (b)

FIGURE 1.13 Unidirectional continuous fiber composite with (a) load acting in the fiber direction (longitudinal loading) and (b) load acting normal to the fiber direction (transverse loading).

The derivations of these micromechanics equations are available in many textbooks on composite materials [10–13].

The *longitudinal modulus* of a unidirectional continuous FRP matrix composite is given by the following *rule of mixture* equation.

$$E_L = E_f v_f + E_m v_m,$$ (1.9)

where E_L is the longitudinal modulus of the composite; E_f is the fiber modulus in its longitudinal direction; E_m is the matrix modulus; v_f is the fiber volume fraction; and v_m is the matrix volume fraction, which is equal to $(1 - v_f)$.

Since in PMCs $E_f \gg E_m$ (Figure 1.14), the contribution of the matrix to E_L is relatively small, and for all practical purposes, we can write

$$E_L \cong E_f v_f.$$ (1.10)

Equation 1.10 indicates that to obtain a high longitudinal modulus for the composite, a high-modulus fiber should be selected. Carbon fibers have a much higher modulus than glass or aramid fibers and are therefore selected for stiffness-critical applications where high longitudinal modulus of the composite is desired to obtain high stiffness of the structure.

Since the tensile strength of polymers is much lower than that of reinforcing fibers, the longitudinal tensile strength S_{Lt} of a unidirectional continuous fiber composite can be written as

$$S_{Lt} \cong S_f v_f,$$ (1.11)

where S_f is the tensile strength of the fiber. As Equation 1.11 indicates, a high-strength fiber should be selected if a high longitudinal tensile strength of the composite is

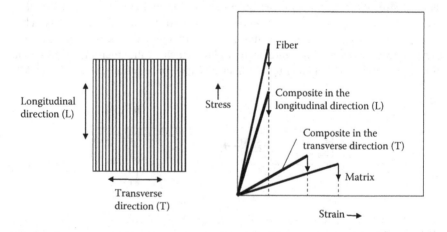

FIGURE 1.14 Tensile stress–strain diagrams of fiber, matrix, and unidirectional continuous fiber composite in the longitudinal and transverse directions of fibers.

desired. Some of the carbon fibers have higher tensile strength than glass or aramid fibers and are preferred for strength-critical applications as well. This is true for not only static load applications, but also fatigue load applications. Although the tensile strength of freshly drawn glass fibers is higher than that of many carbon fibers, their strength is affected by moisture absorption (e.g., during storage) and abrasion (e.g., due to rubbing against each other during processing). In practice, the tensile strength of glass fibers is assumed to be approximately one-half the strength of the freshly drawn glass fibers. Aramid fibers are selected in tensile load applications if tensile strength-to-density ratio is an important design consideration; however, they are not selected for compressive load applications, since aramid fibers are inherently weak in compression.

Both Equations 1.10 and 1.11 demonstrate that an important factor in obtaining either high longitudinal modulus or high longitudinal tensile strength is the *fiber volume fraction* v_f, which should be as high as possible. The fiber volume fraction in most high performance composites is between 0.5 and 0.65 (i.e., 50–65%). If the fibers were all arranged in a regular, repeating array throughout the matrix, the composite volume can be divided into unit cells, and the maximum fiber volume fraction can be theoretically calculated using the fiber diameter and the unit cell dimensions (Table 1.6). Thus, for example, if the fibers are arranged in a body-centered square array shown in Figure 1.15, the theoretical maximum fiber volume fraction will be 78.5%. However, in practice, the fibers are usually not uniformly distributed and the processing technique used

---

TABLE 1.6

Theoretical Maximum Fiber Volume Fraction of Unidirectional Fibers Based on Unit Cell Arrangements

Unit Cell Type	Fiber Arrangement in the Unit Cell	Maximum Fiber Volume Fraction
Hexagonal array	One fiber at each corner of a hexagon and one fiber at the center of the hexagon	0.907
Square array	One fiber at each corner of a square	0.785
Body-centered square array	One fiber at each corner of a square and one fiber at the center of the square	0.785
Random packing	Randomly distributed fibers	0.85

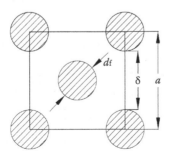

FIGURE 1.15   Body-centered square array of fiber arrangement.

can produce areas that are either *fiber rich* or *matrix rich* (i.e., *fiber-starved*). Thus although the theoretical fiber volume fraction has very little use in the calculation of actual composite properties, it can be used for determining processing limits.

Example 1.2

1. Calculate fiber volume fraction if the fibers are arranged in a body-centered square array as shown in Figure 1.15. Assume that the unit cell dimension is twice the fiber diameter.
2. Verify that the theoretical maximum fiber volume fraction is 78.5%.

Solution:

Part 1

There is one fiber at each corner of the unit cell, and one fiber is at the center. Since each corner fiber is shared by four unit cells, the number of fibers in each

unit cell $= (4)(\frac{1}{4})+1 = 2$.

Therefore, the total fiber cross-sectional area in each unit cell $=$

$(2)\left(\frac{\pi}{4}\right)(d_f^2) = \left(\frac{\pi}{2}\right)d_f^2$.

Area of each unit cell $= a^2$.

Therefore, the theoretical fiber volume fraction is

$v_f = \dfrac{\text{Fiber cross-sectional area}}{\text{Unit cell area}} = \dfrac{\pi d_f^2}{2a^2}$.

Since $a = 2d_f$, $v_f = 0.3925$ or 39.25%.

Part 2

For the maximum theoretical fiber volume fraction, the interfiber spacing (R) between the central fiber and each corner fiber is calculated first.

From part a, $a = \left(\dfrac{\pi}{2v_f}\right)^{1/2} d_f$.

Therefore, $R = \dfrac{a}{\sqrt{2}} - d_f = d_f\left[\dfrac{\pi^{1/2}}{2v_f^{1/2}} - 1\right]$.

For the maximum theoretical fiber volume fraction, $R = 0$, which gives

$v_{f,max} = \dfrac{\pi}{4} = 0.785$ or 78.5%.

The matrix properties play a much larger role in the transverse modulus and strength of a unidirectional continuous fiber composite than the fiber properties. The transverse tensile loading condition is shown in Figure 1.13b. In this case, the fiber orientation angle θ is 0°, and the loading direction is the T-direction or 2-direction. The *transverse modulus* E_T can be approximated by the following simplified equation:

$$E_T = \frac{E_m E_f}{E_f(1 - v_f) + E_m v_f} \cong \frac{E_m}{1 - v_f}. \tag{1.12}$$

Equation 1.12 shows that the high matrix modulus as well as the high fiber volume fraction are desirable for obtaining high transverse modulus.

It can be seen by comparing Equations 1.10 and 1.12 that the longitudinal modulus of a unidirectional continuous fiber composite is much higher than its transverse modulus. This is shown in Figure 1.16. The difference in the longitudinal and transverse modulus values is an indication of the anisotropic behavior of unidirectional composites, and their ratio E_L/E_T is often referred to as the degree of anisotropy. The degree of anisotropy of unidirectional ultrahigh modulus carbon fiber-reinforced epoxy is in the range of 30. The degree of anisotropy of unidirectional glass fiber-reinforced epoxy, on the other hand, is about 8.

The modulus and strength of unidirectional fiber composites depend on the fiber orientation angle θ with respect to the loading direction (Figure 1.17). It shows that both tensile modulus and tensile strength have their highest values when fibers are oriented in the same direction as the load. Thus, in a composite bar subjected to an axial tensile load, the ideal fiber orientation will be 0°, i.e., the fibers should be oriented along its length so that the fiber direction and the tensile load direction coincide. This will give the highest tensile load carrying capacity and the highest axial stiffness to the composite bar. Similarly, fibers in a thin composite beam should be oriented along its length (0° direction), since both tensile and compressive stresses in the beam are along its length direction. This will give the highest bending moment carrying capacity and the highest bending stiffness to the composite beam. Although the ideal fiber orientation in both the bar and the beam is in the 0° direction, it is always a good idea to include a layer or two of 90° fibers near the surfaces to prevent cracking in the matrix between the 0° fibers.

Figure 1.17 also indicates the effect of *fiber misalignment* from the 0° orientation on the modulus and strength of a unidirectional composite. The fiber misalignment may be caused by transverse flow of unidirectional fibers during processing. Even with a small misalignment of 2–5°, there is a significant decrease in both modulus and strength.

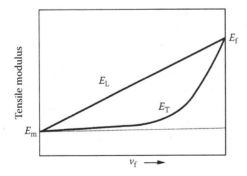

FIGURE 1.16   Variation of longitudinal modulus (E_L) and transverse modulus (E_T) of a unidirectional continuous fiber composite with fiber volume fraction (v_f).

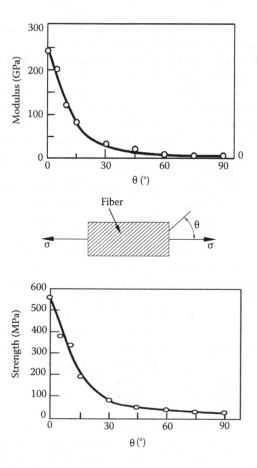

FIGURE 1.17 Influence of fiber orientation angle on tensile modulus and strength of a unidirectional carbon fiber/epoxy composite.

For discontinuous fiber composites, which are also called short fiber composites, strength depends not only on the fiber volume fraction, but also on the fiber length (Figure 1.18). Similarly, their modulus depends on both fiber volume fraction and the *fiber aspect ratio*, which is defined as the ratio of fiber length and fiber diameter. The effect of fiber aspect ratio on the longitudinal modulus is shown in Figure 1.19. Both Figures 1.18 and 1.19 show that the higher the fiber length, the higher the modulus and tensile strength. On the other hand, processing becomes more difficult with longer fibers. For effective reinforcement, the rule of thumb is that the fiber length l_f should be at least 10 times greater than the critical fiber length l_c, which is given by the following equation.

$$l_c = \frac{S_f d_f}{2S_i},$$

(1.13)

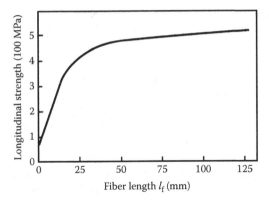

FIGURE 1.18 Longitudinal strength variation of a unidirectional discontinuous fiber composite as a function of fiber length (l_f).

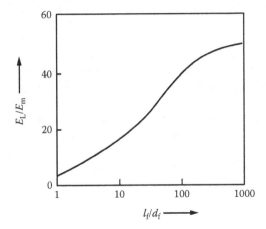

FIGURE 1.19 Longitudinal modulus variation of a unidirectional discontinuous fiber composite as a function of fiber aspect ratio, defined as the ratio of fiber length (l_f) to fiber diameter (d_f).

where l_c is the critical fiber length; d_f is the fiber diameter; S_f is the tensile strength of fibers; and S_i is the fiber–matrix interfacial shear strength, which depends on the bonding between the fibers and the polymer matrix.

For a unidirectional discontinuous fiber composite containing fibers of length l_f greater than the critical fiber length l_c, the tensile strength in the longitudinal direction can be calculated using the following equation:

$$S_{Lt} \cong S_f \left(1 - \frac{l_c}{2l_f} \right) v_f, \tag{1.14}$$

Note that Equation 1.14 applies if $l_f \geq l_c$. For $l_f = l_c$, $S_{Lt} \cong 0.5\, S_f v_f$.

Comparing Equations 1.11 and 1.14, it can be observed that the longitudinal tensile strength of discontinuous fiber composites is lower than the longitudinal tensile strength of continuous fiber composites. To achieve greater than 90% of the longitudinal tensile strength of continuous fiber composites, the fiber length l_f in a discontinuous fiber composite with the same fiber volume fraction must be greater than five times the critical fiber length l_c.

Equation 1.13 shows the importance of improving the fiber–matrix interfacial shear strength S_i. If the fiber length l_f is lower than the critical fiber length l_c, the maximum stress in the fiber does not reach the tensile strength of the fiber, and the failure in the composite occurs either by matrix failure or by interfacial bond failure. To achieve the reinforcement effect of fibers in a discontinuous fiber composite, the fiber length must be greater than the critical fiber length. This can be accomplished by increasing the fiber–matrix interfacial shear strength S_i, which, according to Equation 1.13, will reduce the critical fiber length. In the case of glass fibers, this is done by treating the fiber surface with a chemical coupling agent that improves their bonding with the polymer matrix. For carbon fibers, the oxidation of the fiber surface is used to improve the fiber–matrix interfacial bond strength. With proper surface treatment, the critical fiber length can be made smaller than the fiber length being used in the composite.

Note that fiber orientation in many discontinuous fiber composites is either two-dimensional random or three-dimensional random. While random fiber orientation will produce isotropic behavior, the modulus and strength will both be much lower than the longitudinal modulus and strength of the unidirectional discontinuous fiber composite. For example, if there is a two-dimensional orientation of fibers in a thin composite part, it is planar isotropic, meaning that its tensile modulus and strength will be independent of the direction in which they are measured. But its modulus and strength will be about one-third the values that can be obtained in the longitudinal direction of a unidirectional discontinuous fiber composite with the same fiber volume fraction.

Example 1.3

A unidirectional short fiber composite contains 40 vol.% of T-300 carbon fibers in an epoxy matrix. The fiber length and fiber bundle diameter are 5 mm and 20 μm, respectively. The tensile strength of the composite was determined as 803 MPa. Determine the fiber–matrix interfacial shear strength of the composite.

Solution:

Equation 1.14 is applicable only if l_f is greater than the critical fiber length l_c. Since l_c is not known, we will assume that l_f is greater than the critical fiber length l_c and see if this assumption is valid by calculating l_c using Equation 1.14.

From Table 1.2, the S_f for T-300 carbon fibers is 3650 MPa. Since it is stated that $S_{Lt} = 803$ MPa, $l_f = 5$ mm, and $v_f = 0.4$, let us use Equation 1.14 to calculate l_c.

$$803 = 3650 \left(1 - \frac{l_c}{(2)(5)} \right)(0.4).$$

This gives l_c = 4.5 mm, which is smaller than l_f. Thus, our assumption is valid. We can now use Equation 1.13 to calculate S_i.

$$S_i = \frac{S_f d_f}{2l_c} = \frac{(3650)(20 \times 10^{-3})}{(2)(4.5)} = 8.11 \text{ MPa.}$$

1.6 LAMINATED STRUCTURE

In many applications, FRP matrix composites are used in laminated form. Laminates are made by stacking a number of thin laminas (plies or layers) of unidirectional fibers, bidirectional fabrics, or random fibers to build the laminate thickness according to the design and performance requirements (Figure 1.20). For many aerospace laminates, each lamina is between 0.1 and 0.25 mm in thickness and contains unidirectional fibers embedded in a thin layer of polymer matrix. In a laminate, the fiber orientation with respect to the principal loading direction (see Figure 1.9) may be varied from lamina to lamina. The order in which the laminas with different fiber orientations are stacked is called the *stacking sequence* and is designed to obtain the desired stiffness and/or strength for the laminate.

As an example of laminate construction, consider a *cross-plied laminate* in which the fiber orientation angles in alternate layers are 0° and 90°. In a symmetric six-layered cross-plied laminate shown in Figure 1.21, the laminate construction is [0/90/0/0/90/0] or [0/90/0]$_S$, where the subscript S at the end of the right bracket stands for *symmetric*. This is a symmetric laminate, since for each 0 and 90° layers above its midplane, there

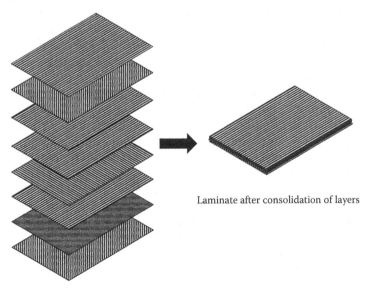

Laminate after consolidation of layers

Layers in the laminate before consolidation

FIGURE 1.20 Laminate construction.

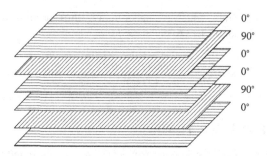

FIGURE 1.21   Symmetric cross-plied laminate of [0/90/0/0/90/0] construction.

are identical 0 and 90° layers at an equal distance below its midplane. Note that the 0 and 90° refer to the angles with respect to the direction of loading, which in this case is the x-direction. In this laminate, the fibers are aligned in the direction of loading in four layers, while in the other two layers, the fibers are oriented normal to the loading direction. The modulus of this laminate E_{xx} in the loading direction (x-direction) is $(4/6)E_L + (2/6)E_T$, which is lower than E_L given by Equation 1.10. On the other hand, the modulus of the laminate E_{yy} normal to the loading direction (y-direction) is $(2/6) E_L + (4/6)E_T$, which is greater than E_T given by Equation 1.12. Thus, while the loading direction modulus of the laminate is lower than the modulus that can be achieved if the fibers in all six layers were oriented in the loading direction, the transverse direction modulus of the laminate is improved. In many composite structures, 90° layers are purposely added to improve the transverse properties even though the longitudinal properties are slightly lowered. The 90° layers help in reducing cracking between the fibers in the 0° layers, which is often a problem if the structure contained only 0° fibers.

A 0° laminate or a symmetric 0/90 cross-plied laminate has different elastic properties, such as modulus or Poisson's ratio, and strength properties, such as tensile strength, in different directions of the laminate. In other words, these laminates are not isotropic. A laminate construction such as [0/45/−45/90] or [0/60/−60] are called *quasi-isotropic*, since they produce equal elastic properties (*note*, not strength properties) in all directions in the plane of the laminate (Figure 1.22). Their elastic properties in the thickness direction are still different from those in the plane of the laminate. Quasi-isotropic symmetric laminates are used in many applications in which their in-plane isotropic behavior is of great advantage, particularly when the loading directions are not known at the design stage or may vary during the service applications.

As an example of quasi-isotropic behavior, consider a unidirectional 0° laminate with 0° fiber orientation in all its layers and an eight-layered symmetric quasi-isotropic $[0/+45/−45/90]_S$ laminate, both containing T-300 carbon fibers in an epoxy matrix. For the 0° laminate, the elastic moduli in the longitudinal and transverse directions are 132.4 and 10.8 GPa, respectively. The modulus of this laminate depends on the direction of measurement with respect to the fiber direction. For example, if the modulus is measured at a 45° angle to the fiber direction, its value is 15.6 GPa. On the other hand, the modulus of the quasi-isotropic laminate is 52.3 GPa regardless of the direction of measurement.

Another method of achieving planar isotropy in thin laminates is by using randomly oriented fibers. While random fiber orientation tends to give equal properties

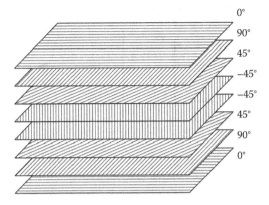

FIGURE 1.22 Construction of a symmetric quasi-isotropic laminate of [0/90/45/–45/–45/ 45/90/0] construction.

in all directions in the plane of the composite, these properties are lower than those of quasi-isotropic laminates. The random fibers can be either continuous or discontinuous. In general, elastic modulus of random continuous fiber-reinforced composites is greater than that of random discontinuous fiber-reinforced composites.

The possibility of combining different fiber orientations in different layers gives a tremendous design flexibility for laminated composite structures which is not possible with metals. Mechanical and thermal properties of the laminate can be tailored to suit the specific design requirements of the structure under consideration. In practice, however, there may be practical limitations as to how these layers should be stacked and what fiber orientations can be used. For example, unless a symmetric construction is selected, the laminate will exhibit bending and/or twisting curvatures when subjected to in-plane tensile or shear loadings and will warp under thermal loading. Another practical design recommendation is to use a *balanced* construction in which for every lamina with $+\theta$ orientation of fibers, there is an identical lamina of $-\theta$ orientation. Thus, if there is $+45°$ lamina in the laminate, there should also be $-45°$ lamina. Lamina balancing eliminates shear deformation when only normal stresses are applied on the laminate or extensional deformations when only shear stresses are applied on the laminate. The concept of combining symmetric and balanced constructions is illustrated in Figure 1.23.

One major limitation of laminated structures is the development of *interlaminar stresses*, which can cause *delamination*, a failure mode that involves the separation of layers at their interfaces as the laminate is being loaded. Interlaminar stresses are caused by the mismatch of Poisson's ratio and shear–extension coupling between the adjacent layers. Unless the fiber orientations in adjacent layers are properly selected, high interlaminar stresses can initiate delamination even at low loads, and the laminate may become structurally weak.

1.7 THERMAL PROPERTIES

PMCs are processed at elevated temperatures and then cooled down to room temperature either in the mold or outside the mold. For an epoxy matrix composite, the

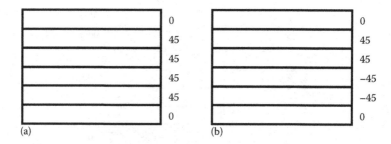

FIGURE 1.23 Balanced and unbalanced symmetric laminate constructions. (a) Unbalanced, [0/45/45/45/45/0] and (b) balanced, [0/45/45/−45/−45/0].

processing temperature is around 170°C, while for a PEEK matrix composite, the processing temperature may be as high as 350°C. Thermal properties of these composites determine not only their heating and cooling times, but also residual thermal stresses that may be generated as they are cooled down from the processing temperature to room temperature. Additionally, thermal properties may be important in designing PMCs for elevated temperature applications.

In this section, we will consider the thermal properties of PMCs, such as their coefficient of linear thermal expansion, thermal conductivity, specific heat, and thermal diffusivity. The thermal properties of selected fibers and polymers at 23°C are listed in Table 1.7. It should be noted that thermal properties of polymers may significantly vary with increasing temperature. This is shown schematically in Figure 1.24.

TABLE 1.7

Thermal Properties (at 23°C) of a Few Selected Fibers and Polymers Used in PMCs

Material	CLTE(10^{-6}/°C)	Thermal Conductivity (W/m °C)	Specific Heat (J/kg °C)
E-glass fiber	5	1.3	810
T-300 carbon fiber	−0.6 (longitudinal); 7–12 (radial)	11 (longitudinal) 2.2 (transverse)[a]	795
Kevlar 49 fiber	−2 (longitudinal); 59 (radial)	0.04 (longitudinal)	1420
Epoxy	81–117	0.19	1050
Polyester	100–180	0.17	719–920
PEEK	50	0.29	1340
PPS	50	0.29	1090
PP	80–100	0.11–0.17	1800–2400
Polyamide-6,6	80–83	0.24	1670

[a] Estimated.

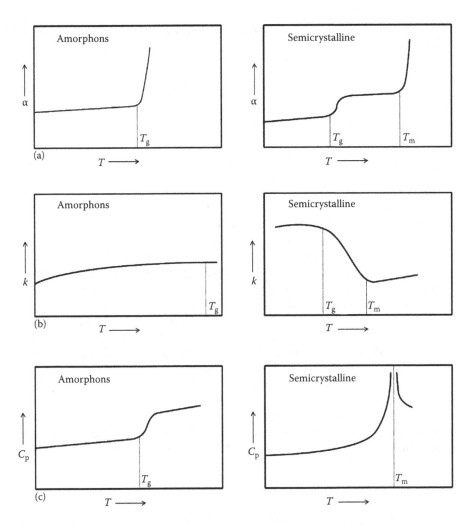

FIGURE 1.24 Thermal properties of polymers as a function of temperature. (a) Coefficient of thermal expansion (α), (b) thermal conductivity (k), and (c) specific heat (C_p).

1.7.1 COEFFICIENT OF LINEAR THERMAL EXPANSION

The coefficient of thermal expansion determines the changes in dimensions as the temperature of a material is either increased or decreased. The *coefficient of linear thermal expansion* (CLTE) α measures the change in length per unit length per unit temperature increase. The unit of the CLTE is meter per meter-degree Celsius (m/m °C). It is used in calculating both expansion due to temperature increase and contraction due to temperature decrease. The CLTE of polymers is much greater than that of fibers (Table 1.7). When unidirectional continuous fibers are added to a polymer, the CLTE in the longitudinal direction of the resulting composite is significantly reduced from that of the polymer itself, while that in the transverse direction is altered only slightly. The expressions for the longitudinal (α_L) and transverse (α_T) CTLEs are as follows:

$$\alpha_L = \frac{\alpha_{fl} E_f v_f + \alpha_m E_m v_m}{E_f v_f + E_m v_m}, \tag{1.15}$$

$$\alpha_T = (1 + \upsilon_f)\left(\frac{\alpha_{fl} + \alpha_{fr}}{2}\right) v_f + (1 + \upsilon_m)\alpha_m v_m - (\upsilon_f v_f + \upsilon_m v_m)\alpha_L, \tag{1.16}$$

where α_{fl} is the CLTE of the fiber in its length direction; α_{fr} is the CLTE of the fiber in the radial direction; α_m is the CLTE of the polymer matrix; υ_f is Poisson's ratio of the fiber; and υ_m is Poisson's ratio of the polymer matrix.

Note from Table 1.7 that for glass fibers, $\alpha_{fl} = \alpha_{fr}$; but for carbon and aramid fibers, $\alpha_{fl} \ll \alpha_{fr}$. Additionally, for carbon and aramid fibers, α_{fl} is negative. Figure 1.25 shows the variations of α_L and α_T with fiber volume fraction υ_f in a unidirectional continuous carbon fiber-reinforced epoxy composite. Note the large difference between the CLTEs in the longitudinal or fiber direction and the transverse or normal to the fiber direction. The large difference between these two CLTEs is the principal source for thermal residual stresses in laminated composite structures. For example, consider a four-layered [0/90/90/0] or [0/90]$_s$ laminate, which is being cooled from a processing temperature of 150°C to room temperature. If the layers were not joined together and were allowed to contract freely, the 0° layers will contract much less in their length direction than the adjacent 90° layers, while the reverse is true for the 90° layers. However, since they are joined together, they must contract by an equal amount. This constraint against free deformation creates

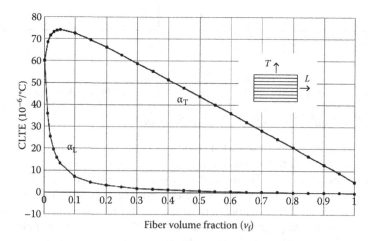

FIGURE 1.25 CLTEs in the longitudinal and transverse directions of a unidirectional continuous carbon fiber/epoxy composite as a function of fiber volume fraction (calculated using Equations 1.15 and 1.16).

compressive residual stresses in the length direction (i.e., x-direction) and tensile residual stresses in the transverse direction (i.e., y-direction) of the laminate. If the residual stresses are high, they can cause failure of the composite at lower than expected loads. In the case of the [0/90]$_s$ laminate, this early failure may be in the form of matrix cracks in the 90° layers in which the residual stresses are tensile and normal to fiber direction.

The [0/90]$_s$ laminate we considered earlier is a symmetric laminate in which the 0° layers and 90° layers are symmetrically located about the midplane of the laminate. If it were not a symmetric laminate, for example, if it had a construction such as 0/90/0/90, the thermal mismatch between the layers would cause curvature in the laminate in addition to creating thermal residual stresses as it cools down from the processing temperature.

1.7.2 THERMAL CONDUCTIVITY

Thermal conductivity is another important thermal property to consider in the processing of PMCs. *Thermal conductivity k* determines the rate of heat flow through the thickness of a material experiencing a thermal gradient across its thickness. The rhermal conductivity of polymers is much lower than that of metals, and although it can increase with the addition of fibers, it is still lower than that of metals. The thermal conductivity of semicrystalline polymers is higher than that of amorphous polymers. It should also be noted that the thermal conductivity of polymers varies with temperature; however, the variation is much smaller for amorphous polymers than for semicrystalline polymers. The large decrease in thermal conductivity occurs for semicrystalline polymers as their melting point is approached (Figure 1.24).

Among the several different types of fibers used with PMCs, carbon fibers have a relatively high thermal conductivity (Table 1.7). The thermal conductivity of glass fibers, although greater than that of most polymers, is much lower than that of carbon fibers. Thus, carbon FRPs have much higher thermal conductivity than glass FRPs. In addition to the fiber type, fiber volume fraction, fiber orientation with respect to the direction of heat flow, fiber–matrix adhesion, presence of voids, and lamination configuration are some of the other factors that control the thermal conductivity of PMCs.

The following equation, called the Lewis–Nielsen equation, is useful for predicting the thermal conductivity of a discontinuous fiber composite [14]. In this equation, it is assumed that good adhesion exists between the polymer matrix and the reinforcing fibers, and there are no voids in the material.

$$k_c = k_m \left[\frac{1 + ABv_f}{1 - Bv_f \phi} \right], \tag{1.17}$$

where,

k_c = thermal conductivity of the composite (unit: watts/meter-degree Celsius [W/m °C]).

$A = \dfrac{2l_f}{d_f}$ for uniaxially oriented fibers with heat flow in the fiber direction, 0.5 for uniaxially oriented fibers with heat flow normal to the fiber direction, and 1.58–8.38 for randomly oriented fibers with heat flow in any direction

$$B = \left(\dfrac{\dfrac{k_f}{k_m} - 1}{\dfrac{k_f}{k_m} + A} \right);$$

k_m = thermal conductivity of the matrix
k_f = thermal conductivity of the fibers
v_f = fiber volume fraction

$$\varphi = 1 + \left(\dfrac{1 - v_{fm}}{v_{fm}^2} \right) v_f$$

v_{fm} = maximum packing fraction of fibers, which depends on the type of packing. (0.907 for uniaxial hexagonal close packed, 0.785 for uniaxial simple cubic, 0.82 for uniaxial random, and 0.52 for three-dimensional random).

Note that Equation 1.17 can also be used to predict the elastic modulus of a discontinuous fiber reinforced composite by replacing k_f and k_m with E_f and E_m, respectively. Also, note that for a unidirectional continuous fiber composite, the thermal conductivity in the fiber direction becomes

$$k_L = k_m + (k_f - k_m)v_f. \tag{1.18}$$

For the transverse direction thermal conductivity of a continuous fiber composite, Springer and Tsai [15] proposed the following equation based on the assumption that fibers are cylindrical in shape and are arranged in a square array.

$$k_T = \left[1 - 2\sqrt{\dfrac{v_f}{\pi}} \right] k_m + \dfrac{k_m}{B} \left[\pi - \dfrac{4}{\sqrt{1 - (B^2 v_f/\pi)}} \left\{ \tan^{-1} \dfrac{\sqrt{1 - (B^2 v_f/\pi)}}{1 - \sqrt{(B^2 v_f/\pi)}} \right\} \right], \tag{1.19}$$

where $B \approx 2\left(\dfrac{k_m}{k_f} - 1 \right)$.

1.7.3 SPECIFIC HEAT

Specific heat of a material is defined as the amount of heat energy needed to increase the temperature of 1 kg mass of the material by 1 K. The specific heat of polymers is much lower than that of metals. It increases with increasing temperature for both amorphous and semicrystalline polymers. At the melting point of semicrystalline polymers, specific heat increases to a peak value due to the heat required to melt the crystalline phase (Figure 1.24). Above the melting point, it reduces approximately to the same level as before the melting point.

The specific heat of a composite is calculated from the following volume average equation.

$$c_c = \frac{c_f v_f + c_m v_m}{\rho_c}, \qquad (1.20)$$

where c_c, c_f, and c_m are the specific heats of the composite, fibers, and matrix, respectively. The unit of specific heat is joule per kilogram-degree Celsius (J/kg °C).

1.7.4 THERMAL DIFFUSIVITY

Thermal diffusivity is a measure of the rate of heat transport through a material and is defined by the following equation:

$$\lambda_c = \frac{k_c}{\rho_c c_c}, \qquad (1.21)$$

where λ_c is the thermal diffusivity (unit: square meters per second [m^2/s]); k_c is the thermal conductivity; ρ_c is the density; and c_c is the specific heat.

Thermal diffusivity is an important thermal parameter in the calculation of cooling time at the end of processing which is done at high temperatures for both thermoset and thermoplastic matrix composites. The higher the thermal diffusivity, the faster the heat propagation and the shorter the cooling time.

1.8 COMPOSITE MANUFACTURING PROCESSES

The manufacturing processes for making structural PMC parts are different from the traditional manufacturing processes used for making structural metal parts. For example, cold forming operations, such as bending and stretch forming, are used for manufacturing thin-walled shell-like structures of steel or aluminum alloys. Examples of these shell-like structures are fuselage sections in aircrafts and body panels in automobiles. If the matrix is a thermoset polymer, molding processes, such as vacuum bag molding, resin transfer molding (RTM), or compression molding, will be considered. If the matrix is a thermoplastic polymer, shape-forming processes, such as diaphragm forming or thermostamping, will be considered. In general, these processes require much lower pressure or force compared to cold forming for metals, and therefore, the tooling cost is lower. They are also near-net shape manufacturing processes, and in general, there is much lower material wastage in the form of scrap.

A variety of manufacturing processes exist for making PMC parts [9,16–19]. The most important of these processes to produce structural composite parts are listed in Table 1.8 and are briefly described in this section; they are discussed in

TABLE 1.8
Manufacturing Processes for Structural PMCs

Process	Common Resins	Starting Material	Major Equipment or Tool	Typical Process Conditions	Cost Issues[a]	Productivity	Part Complexity	Application Examples
Bag molding	Epoxy	Prepreg	Autoclave	120–175°C, 7 bar	H/L/H[b]	Low	Flat/curved panels	Aircraft wing sections and panels
Filament winding	Epoxy	Dry fiber tows, liquid resin	Filament winder	120–175°C, 7 bar	H/L/M	Medium	Hollow cross sections	Pressure vessels, helicopter blades
Pultrusion	Polyester	Dry fiber rovings, liquid resin	Pultrusion die, puller	150°C	M/M/L	High	Solid and hollow cross-sections	Structural sections
Compression molding	Polyester	SMC	Press, mold	150°C, 60–150 bar	H/H/M	High	Complex shapes	Automotive body panels
RTM	Polyester, epoxy	Dry fiber preform, liquid resin	Resin injection unit, clamping press, mold	25–40°C, 1–10 bar	M/L/L	Medium	Complex shapes	Aircraft components, Automotive body panels
SRIM	Polyurethane	Dry fiber preform, liquid resin components	Mixing and dispensing head, clamping press, mold	80–95°C, 20–30 bar	M/L/L	Medium	Complex shapes	Automotive body panels
Thermostamping	PEEK	Prepreg	Press, mold	380°C, 10 bar	H/H/M	Medium	Flat/curved panels	Aerospace components

[a] Equipment/mold/process.
[b] H: high, M: medium, L: low.

greater detail in Chapters 5–10. In some of these processes, fibers and matrix are directly combined in the mold or on the tool to make the composite part. Examples of these processes are filament winding, pultrusion, and RTM. In another category of processes, fibers are incorporated into the matrix to prepare a ready-to-mold sheet in a premolding operation and later used in molding the composite part. Examples of these processes are bag molding and compression molding. The selection of the manufacturing process for making a particular PMC part depends on the shape, size, and number of parts to be produced. For example, bag molding is used for making large flat or curved panels that are of relatively simple shape. This process is labor intensive and is considered if the number of panels to be produced is relatively few. RTM is much less labor intensive and can be automated and is therefore considered instead of bag molding if the number of parts to be produced is much greater. If the part is an axisymmetric tube or a pressure vessel, filament winding will be a much better choice than bag molding.

Some of the processes listed in Table 1.8 can be used for both thermoset and thermoplastic matrix composites. However, there are differences in their processing characteristics. For thermoset matrix composites, either uncured or partially cured resin is transformed into the solid state by the application of heat in the mold. The curing reaction is a thermally activated chemical reaction that generates heat. Pressure is used for the resin to flow before its viscosity becomes too high to fill the mold. Good resin flow is needed not only to fill the mold, but also to consolidate various layers in the composite and to expel air or other volatile gases from the composite. Cure time in the mold determines the production rate and can be reduced by increasing the cure temperature. For thermoplastic matrix composites, the polymer must be melted or softened by the application of heat; however, no chemical reaction takes place during processing. Transformation from the liquid state to the solid state requires cooling in the mold. Pressure is required to consolidate the layers in the composite and form the shape of the part.

1.8.1 BAG MOLDING

Bag molding is a very common composite manufacturing process in the aerospace industry. It is a slow and labor-intensive process, but it can produce parts with precise fiber orientation, low void content, and controlled fiber volume fraction.

The starting material in a bag molding process is a prepreg, which is a continuous sheet of fibers preimpregnated with a thin layer of a thermosetting polymer. The prepreg may contain either unidirectional continuous fibers or a woven fabric. The polymer in the prepreg is a partially cured (called *B-staged*) thermosetting resin. The prepreg sheet is typically 0.125–0.25 mm thick. It is made in a separate process prior to bag molding and stored at approximately −18°C. At the time of bag molding, several precut layers of prepregs are stacked on top of a mold surface, covered by a thin bag of a flexible polymer film and then cured in an autoclave or in a press at an elevated temperature. Vacuum is used in addition to pressure to help remove air or volatile matters from the bag and to discharge excess resin flowing out of the prepreg stack. Pressure, in the range of 1–10 bar, is applied to consolidate the layers as the resin is being cured.

1.8.2   COMPRESSION MOLDING

Compression molding is mostly used with randomly oriented short fiber-reinforced thermosetting polymers. The starting material for compression molding is either a *sheet molding compound* (SMC) or a *bulk molding compound* (BMC). SMC is prepared in the sheet form, whereas BMC is prepared in the form of a cylindrical mass. In both materials, the resin is in an uncured, but very viscous state. In the compression molding process, a stack of SMC layers or a length of BMC cylinder is placed in the lower half of a preheated matched mold. The upper half of the mold is quickly moved down to close the mold. As the material inside the closed mold is heated up and molding pressure is applied, it spreads outward to fill the cavity. The mold temperature is usually in the range of 130–160°C and depending on the complexity and size of the part being molded, the pressure can be as high as 100–150 bar. The mold remains closed until the material is cured.

The tooling cost in compression molding is high because of the high pressure needed for mold filling, but it is capable of producing parts with good surface finish and close dimensional tolerances. Compression molding can be used to produce complex shapes, containing ribs, bosses, holes, and other geometric variations. The production rate is high, and the process can be automated.

1.8.3   LIQUID COMPOSITE MOLDING

Liquid composite molding (LCM) is a group of processes in which a liquid thermoset resin is either injected under pressure into a dry fiber preform or pulled into it by applying vacuum and then allowed to cure to manufacture a composite part. There are several variations of the LCM processes. The most common of these processes are RTM and structural reaction injection molding (SRIM).

1.8.3.1   Resin Transfer Molding

In the RTM process, a dry fiber preform is placed in the mold cavity; the mold is closed and held in a clamping press. A liquid thermosetting prepolymer, mixed with a catalyst or a curing agent, is injected into the closed mold at a pressure ranging from 1 to 10 bar. The thermosetting prepolymer is either an epoxy, an unsaturated polyester, or a vinyl ester resin. As the liquid mix flows through the dry fibers, it wets them and displaces air from the mold. The curing reaction is completed either at room temperature or at an elevated temperature. RTM has a higher production rate than bag molding, and since the polymer is injected at a relatively low pressure, a high tonnage press is not required and the tooling cost is also not very high.

1.8.3.2   Structural Reaction Injection Molding

SRIM, like the RTM process, starts with the placing of a dry fiber preform in the mold. A resin mix is injected into the dry fiber preform after closing the mold. The difference with RTM is in the resin reactivity. The resins used for SRIM are polyurethanes or polyureas. The chemical ingredients used for making these resins are highly reactive. They are mixed using a high-speed spray impingement technique just before injecting the mix into the mold. The curing reaction for the SRIM resins is very rapid and does not require temperatures greater than 120°C. The resin injection pressure is usually in the range of 1–10 bar.

1.8.4 FILAMENT WINDING

Filament winding is a semicontinuous process of producing hollow structural parts of continuous fiber-reinforced thermoset. In this process, a band of dry continuous fibers is pulled through a liquid resin tank containing a catalyzed resin and then wound around a rotating mandrel. The resin-coated fiber band is also traversed back and forth along the length of the rotating mandrel to create a helical winding pattern. The winding angle can be varied by controlling the mandrel speed and the traversing rate of the fiber band. The part is cured in an oven, and the mandrel is removed to create a hollow shape. In some applications, such as oxygen tanks, the mandrel is not removed after curing, and it becomes a part of the structure.

1.8.5 PULTRUSION

Pultrusion is a continuous process of producing long, straight structural members, such as I-beams, hollow rectangular beams, and round tubes, containing mostly continuous fibers and a few layers of random discontinuous fibers along their lengths. In this process, collimated continuous fiber rovings are pulled first through a liquid resin tank containing a catalyzed thermosetting resin and then through a long heated die where the resin-coated fibers are gathered to form the cross-sectional shape of the structural member being produced. Curing takes place as they move along the length of the die. The cured shape is pulled out from the exit end of the die using a set a continuous belts or chains. A diamond-coated saw is used at the end of the pultrusion line to cut the pultruded member into desired lengths.

1.8.6 FORMING

Forming is the principal processing method for manufacturing continuous fiber-reinforced thermoplastic matrix composite parts. In one of the forming methods, called *thermostamping*, a stack of thermoplastic matrix composite prepreg sheets is heated to a temperature at which the thermoplastic matrix in them is transformed into the liquid state. The heated stack is placed in the mold for forming the shape under pressure. It is then cooled down to room temperature, which returns the thermoplastic matrix into its solid state. Unlike the thermosetting polymers, no chemical reaction takes place during forming.

1.9 COST ISSUES

The selection of fibers and matrix in a PMC depends on a variety of factors, such as the design requirements (maximum load, stiffness, mode of loading, etc.), operating environment (temperature, humidity, etc.), processing characteristics (such as processing time and processing temperature), cost, and availability. When two or more materials provide similar structural performance, the cost may become one of the deciding factors for material selection. The scale of production or production volume is a critical factor in cost calculation. In general, the cost per part is reduced as the production volume increases.

The major cost factors for a PMC part are the feedstock cost and the processing cost [20]. The feedstock cost has two parts: (1) material cost that includes the costs of fibers, polymer matrix, surface treatment, and additives and (2) material preparation cost. Material preparation is the preprocessing step needed for producing the form in which the fibers will be used in preparation for manufacturing the composite part; for example, prepregging for vacuum bag-molded composites and preforming a dry fiber network for liquid composite-molded composites. For some processes, such as filament winding, there is no material preparation cost, since fibers in these processes are directly used in the manufacturing of the composite part.

In the material cost category, the fiber cost is generally higher than the matrix cost. Carbon fibers provide much greater weight saving than E-glass fibers in both stiffness critical and strength critical applications. But carbon fibers are much more expensive than E-glass fibers, which is one of the reasons for selecting E-glass fibers instead of carbon fibers in current automotive composite applications. Carbon fibers are the primary reinforcing fibers in aerospace composites. The cost of high-temperature thermoplastics, such as PEEK, is also very high compared to epoxies, which is one of the reasons for fewer applications of PEEK compared to epoxies. For the material preparation cost, one must also include the cost of incorporating fibers into the matrix, which, in the case of aerospace composites, will be the cost of the prepregging. The cost of making standard unidirectional carbon fiber–epoxy prepreg is typically 1.5–2 times higher than the cost of carbon fibers. The cost of making thermoplastic prepregs is in general higher than the cost of making epoxy prepregs. If the fiber architecture is biaxial, triaxial, or random instead of unidirectional, then the cost of producing such architecture must also be considered. Preforms produced from these various fiber architectures are used in LCM. The cost of preforming is an additional cost to consider. In the case of compression molding, the material preparation cost includes the cost of making the sheet molding compound sheet.

The processing cost includes premolding preparation cost, molding cost, and postmolding operation cost. In the case of a bag molding operation with thermoset matrix composites, the premolding preparation cost includes prepreg cutting, laying up, and bag mold preparation costs. The molding cost for thermoset matrix composites is related to the curing time, which depends on the thermoset matrix selected for the composite part and can vary from several minutes to several hours. The cure time for epoxies is much longer than that of either polyesters or vinyl esters. For some epoxies, the cure time may be 5 hours or longer, whereas for polyester and vinyl ester resins, the cure time is between 5 and 10 minutes. The postmolding operations may include trimming, inspection (including nondestructive testing), and secondary operations, such as drilling of holes in the molded composite part.

Most of the aerospace composites are processed by vacuum bag molding and cured in autoclaves. It starts with laying up the prepreg layers in the mold and assembling the vacuum bag by hand. The laying up process, which involves several steps, such as cutting the prepreg layers, stacking them in proper orientations, and debulking, may take several hours and require highly skilled and trained workers. It is a labor-intensive process, and the direct labor cost is estimated to be 3–3.5 times the direct material cost. The use of microprocessor-controlled prepreg cutting and automatic tape-laying machines reduces the direct labor cost, but they are expensive equipment, which means higher capital investment. LCM processes, such as RTM, are less expensive than vacuum bag molding in terms of both equipment cost and direct labor cost. Filament winding and pultrusion are also less labor-intensive. They can be highly automated and are capable of producing a large number of parts per hour; however, they are not suitable for manufacturing three-dimensional complex shapes, such as an aircraft wing section or an automotive body panel.

An important part of the cost issue is the cost of scrap produced during the laying up operation for vacuum bag molding and making preforms for LCM. The scrap is generated when prepregs and preforms are cut and trimmed to closely fit the mold surface. There will also be some scarp generated at the end of the molding cycle as the edges of the molded part are cleaned off by trimming and machining; but, in general, this is relatively very small.

Another aspect of cost issues is the tooling cost, which includes both mold material cost and mold-making cost. If the mold includes heating and/or cooling channels and other features, such as a part ejector system, the cost becomes higher. The selection of mold material depends on the part design, processing temperature, production rate, and production volume, i.e., how many parts are to be produced per year. A list of mold materials used in the composite industry is given in Table 1.9. For high production volumes, the mold material is required to have long-term durability, especially since the processing is done at high processing temperatures, and an alloy steel mold is preferred. For low production volumes, other mold materials are considered. Among them are zinc alloys, aluminum alloys, E-glass/epoxy, carbon fiber/epoxy, and Invar. The processing temperature for thermoplastic matrix composites is higher than that for thermoset matrix composites, and therefore, an alloy steel mold is more suitable for thermoplastic matrix composites.

The other considerations for mold material selection includes the CLTE and the thermal conductivity of the mold material. The CLTE of the mold material should be close to that of the composite material so that their expansions during heating and contractions during cooling are similar. Otherwise, there will be problems in maintaining the dimensional accuracy of the molded composite part. Fiber buckling, fiber misorientation, and fiber breakage can also occur if there are differential thermal contractions. Based only on the CLTE values, composite molds made of carbon fiber/epoxy will be ideal for manufacturing carbon fiber/epoxy composite parts, but they are more prone to wear and find use only in low-volume productions. The CLTE of Invar, which is a Ni–Fe alloy, is close to that of many PMCs. For this reason, Invar molds can produce composite parts to close tolerance, but because of its high density, Invar molds are heavy. The cost of Invar molds is also quite high. Because of its high thermal mass, it also takes longer time to heat and cool an Invar mold. Invar-coated carbon fiber/epoxy is lighter in weight, and its CLTE matches that of carbon FRP. Molds made out of Invar-coated carbon fiber/epoxy are more durable than molds made out of carbon fiber/epoxy. Aluminum is another mold material with relatively low density and can be used to design

TABLE 1.9
Mold (Tool) Materials Used in Composite Manufacturing Processes[a]

Material	Density (g/cm³)	Thermal Conductivity (W/m °C)	Specific Heat (kJ/kg °C)	CLTE (10⁻⁶/°C)	Maximum Service Temperature (°C)	Tensile Modulus (GPa)	Durability	Surface Finish	Cost of Finished Mold
4340 steel	7.8	42.7	0.4	11.3	>500	200	Very high	Good	High
Cast aluminum	2.6	209	0.9	22	300	69	High	OK	Medium
Invar 36	8	10.5	0.5	1	>500	141	High	OK	Very high
Electroformed Nickel	8.9	75	0.46	13.5	>500	205	Very high		Very high
Monolithic graphite	1.9	168	0.83	5	>500	7	Low	Poor	High
Carbon fiber/epoxy	1.6	1.7	0.3	3	210	72	Low	Poor	Medium low
E-glass fiber/epoxy	2.1	0.86	0.1	14	210	21	Low	Poor	Low
Cast epoxy	2.5	1	0.7	40	180	2	Low	OK	Very low
Plaster	1.3	0.5	–	9	100	30	Low	Poor	Very low

Source: R. S. Parnas, *Liquid Composite Molding*, Hanser, Munich, 2000.

lightweight molds. It is also easy to machine, but its CLTE is much higher than that of PMCs. Several low-CLTE tool materials are available, for example, castable ceramic, bonded ceramic, and monolithic graphite. These materials are expensive, and since they are brittle materials, they tend to be easily damaged during demolding and transportation.

Next, the cost of assembly must also be considered. The use of PMCs allows significant reduction in the number of parts through parts integration and therefore reduces the parts count, number of molds, number of assemblies, and associated tooling and labor costs, as well as capital investment. But there may still be several composite parts that need to be assembled if it is a very large and/or complex structure. There may also be assembly required with metal parts. In addition, the assembly of composite parts with metal parts may be required in many applications. Mechanical assembly using rivets and bolts is the primary assembly technique used in the aerospace industry. Adhesive bonding is the other method of assembly. Although its use in assembling composite parts has increased with the development of stronger and more durable adhesives and better understanding of adhesive joint design, its main drawbacks are the need for surface preparation, long cure time, and lack of high-resolution nondestructive inspection techniques that can detect internal defects with high confidence and accuracy. Assembling with adhesives is in general more expensive than with mechanical fasteners, but even with mechanical assembly, the assembly cost is close to the part processing cost. Since close tolerance required for assembly operations are not always achieved with PMCs, it is often required to use shimming for good fits, which adds to the assembly cost.

The final cost to consider is the quality inspection cost. For high-performance parts or safety-related parts, it may be necessary to nondestructively inspect 100% of the manufactured parts for process-induced defects, such as voids, delaminations, and resin-rich areas. In the aerospace industry, the most common quality inspection technique is the ultrasonic C-scan. Depending on the part complexity and both the level and frequency of inspection needed, the cost of quality inspection may be in the range of 25–100% of the part processing cost.

In summary, the following direct cost components should be taken into consideration for manufacturing of composite parts [16,20].

1. Feedstock cost, which includes (1) costs for fiber, matrix, and all additives and (2) material preparation cost (e.g., surface treatment, prepregging, weaving, preforming)
2. Processing cost, which includes direct labor, machine operation cost, scrap cost, machine downtime for maintenance and repair
3. Tooling cost, which includes costs of the material, fabrication, and maintenance of the mold or tool
4. Assembly cost, which includes costs for preparation, material, assembly operation, and testing of the assembly
5. Quality inspection cost for both composite parts and their assembly

PROBLEMS

1. Compare the tensile modulus-to-weight ratios of E-glass fiber, T-300 carbon fiber, M-55J carbon fiber, and Kevlar 49 fiber. The fiber properties are given in Table 1.2. What is the significance of this comparison?

2. Compare the tensile strength-to-weight ratios of E-glass fiber, T-300 carbon fiber, M-55J carbon fiber, and Kevlar 49 fiber. The fiber properties are given in Table 1.2. What is the significance of this comparison?

3. A fiber of length l and diameter d is embedded in a matrix. Show that the surface area-to-volume ratio of the fiber increases as its diameter decreases. What is the significance of this result?

4. The fiber volume fraction in an E-glass reinforced PP is 40%. How many kilograms of fibers are present in the composite for every 100 kg of PP? The density of PP is 0.9 g/cm^3.

5. An SMC composite contains 25 wt.% of E-glass fibers and 30 wt.% of calcium carbonate fillers in a polyester resin. Assume that the weight percentages of other ingredients in the composite are relatively small. Calculate the density of the composite. The density of the polyester resin is 1.15 g/cm^3.

6. A resin burn-off test was conducted to determine the fiber volume fraction in a carbon fiber/epoxy composite part. A small sample weighing 0.525 g was taken from the composite part, and the resin in the sample was burned off in a muffle furnace. After the burn-off, the sample weight was 0.234 g. Calculate the fiber volume fraction and the density of the composite. Assume that the fiber and matrix densities are 1.80 and 1.3 g/cm^3, respectively.

7. In a carbon fiber/epoxy prepreg, the volume fraction of carbon fibers is determined as 60%. During processing, 10 wt.% of epoxy flows out of the prepreg. Calculate the final volume fraction of fibers in the cured composite. The fiber and resin densities are 1.9 and 1.1 g/cm^3, respectively.

8. A resin transfer-molded carbon fiber/epoxy part is required to have a fiber volume fraction of 0.5. The dry fiber preform weighs 900 g. Assuming that the fiber and resin densities are 1.9 and 1.2 g/cm^3, respectively, and there is 5 wt.% resin loss during processing, determine how much resin (by weight) needs to be injected to make each part. What will be the total weight of the part? How would you determine the resin distribution in the part after it is molded?

9. The fiber orientation tensor of a composite is given in the following. What is the nature of fiber orientation in the composite?

$$\begin{bmatrix} 0.333 & 0 & 0 \\ 0 & 0.333 & 0 \\ 0 & 0 & 0.333 \end{bmatrix}$$

10. Calculate the orientation parameters f and g for a discontinuous fiber composite in which (a) fibers are unidirectionally oriented and (b) fibers are randomly oriented.

11.   Assume that the unidirectional fibers in a PMC are arranged in a square array as shown in the following figure. Fiber cross-sections are shown at each corner of the array. If the fiber volume fraction is 60% and the fiber diameter is 7 μm, what is the distance between each fiber? What is the fiber surface area-to-fiber volume ratio in the composite?

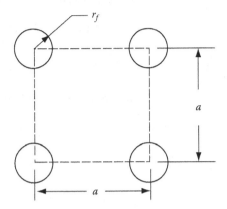

12.   Calculate the fiber volume fraction needed to produce a unidirectional continuous fiber composite that will have a transverse modulus equal to half its longitudinal modulus.

13.   The longitudinal tensile strength of a unidirectional discontinuous fiber composite needs to be at least 85% of the longitudinal tensile strength of a unidirectional continuous fiber composite. Both composites contain IM-7 carbon fibers in an epoxy matrix, and they have the same fiber volume fraction. It was determined that the interfacial shear strength is 25 MPa. What should be the minimum fiber length in the discontinuous fiber composite?

14.   A unidirectional continuous T-300 carbon fiber/epoxy is to be designed to have a zero CLTE in the longitudinal direction. Calculate the volume fraction of fibers needed for this design. What is the significance of zero CLTE? For the epoxy, $E = 3$ GPa and $\alpha = 100 \times 10^{-6}/°C$.

15.   A flat panel of a unidirectional continuous T-300 carbon fiber/epoxy is being cooled from a processing temperature of 120°C to 23°C. The panel is 1 m long, 0.5 m wide, and 2 mm thick at 120°C. The fiber volume fraction is 55%. Calculate the changes in dimensions of the part due to thermal contraction. For the epoxy, $E = 3.5$ GPa and $\alpha = 90 \times 10^{-6}/°C$. The Poisson's ratios for the fibers and matrix are 0.2 and 0.33, respectively.

16.   Assume that a unidirectional continuous fiber composite part is being cooled from a processing temperature of 170°C to room temperature. The cooling takes place by heat conduction in the thickness direction. Knowing that the cooling time is inversely proportional to the thermal diffusivity of the material being cooled, compare the cooling time of a T-300 carbon fiber/epoxy composite with that of an E-glass fiber/epoxy composite. The fiber volume fraction in both composites is 60%, and both composite parts have the same thickness. The thermal properties of the

matrix fibers are given in Table 1.7. The density of the epoxy matrix is 1.25 g/cm³.

REFERENCES

1. P. K. Mallick (ed.), *Composites Engineering Handbook*, CRC Press, Boca Raton, FL, 1997.
2. S. T. Peters (ed.), *Handbook of Composites*, Chapman & Hall, London, 1998.
3. P. Morgan, *Carbon Fibers and their Composites*, CRC Press, Boca Raton, FL, 2005.
4. G. W. Ehrenstein, *Polymeric Materials*, Hanser, Munich, 2001.
5. H. S. Katz and J. V. Milewski (eds.), *Handbook of Fillers and Reinforcements for Plastics*, Van Nostrand Reinhold, New York, 1978.
6. E. P. Plueddemann, *Silane Coupling Agents*, Plenum Press, New York, 1982.
7. L.-G. Tang and J. L. Kardos, A review of methods for improving the interfacial adhesion between carbon fiber and polymer matrix, *Polymer Composites*, Vol. 18, No. 1, pp. 100–113, 1997.
8. M. Sharma, S. Gao, E. Mäder, H. Sharma, L. Y. Wei, and J. Bijwe, Carbon fiber surfaces and composite interphases, *Composites Science and Technology*, Vol. 102, pp. 35–50, 2014.
9. S. G. Advani and E. M. Sozer, *Process Modeling in Composites Manufacturing*, CRC Press, Boca Raton, FL, 2003.
10. R. M. Jones, *Mechanics of Composite Materials*, 2nd Ed., Taylor & Francis, Philadelphia, PA, 1999.
11. P. K. Mallick, *Fiber-Reinforced Composites*, 3rd Ed., CRC Press, Boca Raton, FL, 2008.
12. R. F. Gibson, *Principles of Composite Materials Mechanics*, 2nd Ed., CRC Press, Boca Raton, FL, 2012.
13. I. M. Daniel and O. Ishai, *Engineering Mechanics of Composite Materials*, 2nd Ed., Oxford University Press, Oxford, 2005.
14. R. C. Progelhoff, J. L. Throne, and R. R. Ruetsch, Methods of predicting the thermal conductivity of composite systems: A review, *Polymer Engineering and Science*, Vol. 16, No. 9, pp. 615–625, 1976.
15. G. S. Springer and S. W. Tsai, Thermal conductivities of unidirectional materials, *Journal of Composite Materials*, Vol. 1, pp. 166–173, 1967.
16. T. G. Gutowski (ed.), *Advanced Composites Manufacturing*, John Wiley & Sons, Hoboken, NJ, 1997.
17. S. K. Mazumdar, *Composites Manufacturing: Materials, Product and Process Engineering*, CRC Press, Boca Raton, FL, 2001.
18. S. V. Hoa, *Principles of the Manufacturing of Composite Materials*, DesTech, Lancaster, PA, 2009.
19. R. S. Davè and A. C. Loos, *Processing of Composites*, Hanser, Munich, 2000.
20. M. G. Bader, Selection of composite materials and manufacturing routes for cost-effective performance, *Composites: Part A*, Vol. 33, pp. 913–934, 2002.
21. R. S. Parnas, *Liquid Composite Molding*, Hanser, Munich, 2000.

2 Fiber Architecture

Fiber architecture is the arrangement of fibers in a PMC and has a strong influence on its properties, performance, and failure behavior. It also plays a key role in fiber impregnation, matrix infiltration, formability, etc., that determine the quality of the manufactured composite part. For continuous FRPs, the fibers can be arranged in either linear, two-dimensional (2D), or three-dimensional (3D) architecture (Table 2.1). In the linear form, fibers in the composite are oriented in one direction only, i.e., fibers are parallel and unidirectional. Linear fiber architecture is often utilized for constructing laminates, which contain multiple layers of linear fiber architecture. Fiber orientation in each layer of the laminate is unidirectional, but the fiber orientation angle may vary from layer to layer to meet the design requirements of the composite structure. Laminated structures may also contain 2D fiber architecture in which fibers in each layer are oriented in two different directions. Normally, the fibers in a 2D architecture are oriented in mutually perpendicular directions as in a bidirectional fabric.

Unlike linear and 2D architectures, a 3D architecture contains fibers in the thickness direction. The presence of fibers in the thickness direction helps reduce the possibility of laminate failure by delamination (i.e., separation of layers along the interlaminar zone). High interlaminar stresses at the free edges (for example, at the edges of a hole or a cutout in a laminated composite plate) often cause the initiation of delamination in many laminates containing linear or 2D fiber architecture. Delamination can also occur due to impact and fatigue loads. It has been shown that through-the-thickness fiber reinforcement can enhance the interlaminar shear strength and impact/fatigue damage tolerance of laminated composites.

Random fiber architecture is found in the applications of PMCs in which fibers are used in discontinuous lengths. Discontinuous fibers can be either directly mixed with the polymer matrix, for example, in injection molding, or combined with a binder to form a thin planar mat. Most of the sheet molding compounds contain discontinuous fibers in random orientation. The random architecture form of fibers is utilized in compression molding and LCM processes.

Two-dimensional and 3D fiber architectures are produced using textile processes such as weaving, interlacing, intertwining, and looping continuous fibers in textile machines. The fiber architecture produced by these textile processes can be further processed to make *preforms* of complex shapes that are subsequently coated and filled with a liquid resin using a LCM process, such as RTM or SRIM. In making preforms with 2D architecture or random fiber architecture, several layers of fibers with these architectures are first stacked to build the desired thickness. The stacked layers can also be stitched together or joined together using pins (called z-pins). The stack of dry fiber layers is then shaped to the desired form in a press before transporting it to the mold where a liquid resin mix is injected into them.

TABLE 2.1

Classification of Fiber Architectures

Construction	Fiber Architecture	Fiber Angles and/or Orientations	Fiber Volume Fraction (%)	Processing Technique
Roving	Linear	0	60–80	Bundling
Yarn	Linear	Yarn surface helix angle = 5–10°	70–90	Bundling and twisting
Woven	Biaxial	0/90, crimp angle = 30–60°	~50	Weaving
	Multiaxial	0/90/±30–60, crimp angle = 30–60°	~50	
Knitted	Biaxial	Stitch yarn orientation = 30–60°	20–30	Knitting with loops
Braided	Biaxial	Braiding angle = 10–80°	50–70	Braiding
	Triaxial	Braiding angle = 10–45°	40–60	
Nonwoven	Multiaxial		20–60	Layering and stitching
Planar mat	Random	Random		Chopped strands or Continuous filaments

It should be noted that the fiber-dominated properties of composites, such as tensile modulus and tensile strength, depend on the fiber type, fiber length, fiber orientation, fiber volume fraction, as well as fiber architecture. To demonstrate their effect, consider the following first-estimate equation for the tensile modulus E of a composite:

$$E \approx \eta E_f v_f, \tag{2.1}$$

where η is the efficiency factor that depends on fiber length, fiber orientation, and fiber architecture, E_f is the fiber modulus, and v_f is the fiber volume fraction. Table 2.2 gives the typical values of the efficiency factors for several different fiber architectures. The value of η is equal to 1 for the longitudinal modulus of a unidirectional composite, but can be as low as 0.25 for some 2D fiber architectures.

TABLE 2.2

Efficiency Factors for Various Fiber Architectures

Fiber Architecture	Example	Efficiency Factor (η)
Linear	Unidirectional fibers	1.000 (in the fiber direction)
Biaxial 0/90	Woven fabric with zero crimp	0.500
Biaxial ±45	Woven fabric with zero crimp	0.250
Multiaxial 0/±45/90	Multiaxial fabric with zero crimp	0.375
Two-dimensional random	CSM	0.375

2.1 FIBER FORMS

Fibers are available from their manufacturers in different forms (Figure 2.1). Each fiber is a bundle of very thin filaments. Depending on the fiber type, the filament diameter is in the range of 5–15 μm. The number of filaments in a fiber and the form in which they are bundled together are slightly different in different fiber industries. For example, in the glass fiber industry, a bundle of continuous parallel glass filaments is called a *strand*. The number of filaments in a glass fiber strand is called the *end*, which can vary from 51 to 1632. In the carbon fiber industry, *tow* is the term used to describe a bundle of continuous parallel carbon filaments. The number of carbon filaments in a tow can vary from 1000 (1K) to higher than 50,000 (50K). The aramid fiber industry also uses the term *tow* to describe a bundle of continuous parallel aramid filaments. The selection of strand or tow size depends on the manufacturing process selected. For example, smaller tows, such as 1K or 3K, are selected for thin prepregs or woven fabrics, whereas larger tows, such as 12K, are selected for filament winding. Larger tow size increases the production rate in filament winding, but it becomes more difficult for the polymer matrix to wet out all the filaments in the tow if the tow size is too large.

Another term used in the glass fiber industry is *roving*, which is a bundle of parallel continuous glass filaments or strands with little or no twist (Figure 2.1b). The term *yarn* is used in all three fiber industries to describe bundles of filaments with twists (Figure 2.1b). There are three types of yarns: discontinuous filament yarns, staple yarns, and continuous filament yarns (Figure 2.1c), the latter being used in the majority of PMC applications. In many yarn constructions, two or more single yarns are twisted together to form plied yarns. Several plied yarns can be combined together to form even thicker yarns.

The fibers are designated by their linear density, which represents their mass per unit length and is expressed in the unit of tex, where 1 tex is equivalent to 10^{-6} kg/m [1]. Another designation of fibers is called the *yield*, which represents the length per unit mass, and therefore, it is the inverse of tex. For complete fiber designation, it is also necessary to indicate the fiber form. As an example, consider the designation used for describing glass fiber yarns shown in Figure 2.2. In this figure, the glass fiber yarn designation is ECD-900-2/2. This is an E-glass yarn containing 5 μm diameter filaments. The strand

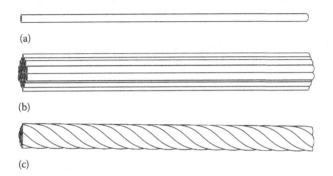

(a)

(b)

(c)

FIGURE 2.1 Fiber forms: (a) filament, (b) roving or tow, and (c) yarn.

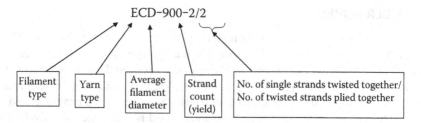

ECD-900-2/2

Filament type	Yarn type	Average filament diameter	Strand count (yield)	No. of single strands twisted together/ No. of twisted strands plied together

First letter: filament type (e.g., E for electrical glass or E-glass)
Second letter: type of yarn (e.g., C for continuous filament)
Third letter: average filament diameter (e.g., letter D for a filament diameter of 0.00023 in.)
(Note: in SI units, the diameter will be given in number, which in this case is 5, standing for 5 m)
First number: strand count (e.g., 900 is for 900 × 100 = 90,000 yards per lb) (Note: in SI units, the strand count will be given in Tex, which in this case is 5.5, standing for 5.5 g per 1000 m)
Second numbers: number of single strands twisted together/number of twisted strands plied together (e.g., 2/2 stand for 2 single strands twisted together and 2 twisted strands plied together)

FIGURE 2.2   E-glass fiber yarn nomenclature.

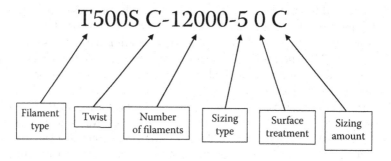

T500S C-12000-5 0 C

Filament type	Twist	Number of filaments	Sizing type	Surface treatment	Sizing amount

FIGURE 2.3   Nomenclature used by a carbon fiber manufacturer.

count is 900, which is multiplied by 100 to give the strand weight as 900,000 yards/lb, which is equivalent to 5.5×10^{-6} kg/m or 5.5 tex. The approximate number of filaments in this strand is 102, which can be calculated from the strand weight. Finally, the two numbers separated by the diagonal line, which in this example is 2/2, indicate the total number of strands and how they are plied together. The first number gives the number of single strands twisted together to form a thread and the second number gives the number of threads in the yarn. Thus, in this particular example, there are two single strands per thread and two threads per yarn, and therefore, the total number of strands in this yarn is 2 × 2 or 4. The designation used for describing carbon fiber yarns is given in Figure 2.3.

Example 2.1

Calculate the tex and yield for a T-300 carbon fiber tow with 6000 filaments. Assume that the average filament diameter is 7 μm and the density of the material is 1.76 g/cm³.

Solution:

Filament radius = 3.5 μm = 3.5 × 10⁻⁶ m
Density = 1.76 g/cm³ = 1760 kg/m³
Mass of each filament per unit length = π (3.5 × 10⁻⁶ m)²(1760 kg/m³) = 67.70 × 10⁻⁹ kg/m
Therefore, tex = 6000 filaments × (67.70 × 10⁻⁹ kg/m per filament) = 406.2 × 10⁻⁶ kg/m or 406 and yield = $\dfrac{1}{406 \times 10^{-6}\,\text{kg/m}}$ = 2460 m/kg = 2.46 m/g

An important parameter for determining the properties of a composite is the fiber volume fraction in the composite. For the composite to produce high strength, modulus, and other properties, it is important that the liquid polymers infiltrate into the fiber bundle, coat each filament in the bundle, and occupy the open spaces between the filaments in the bundle. Liquid polymer infiltration into the interior of the fiber bundle becomes more difficult as the number of filaments in the bundle increases.

In the case of a roving, the number of fibers in the roving and their packing determine the fiber volume fraction in the roving, and therefore, the open space in the roving that needs to be filled by the liquid polymer. If the fibers in a roving are packed in concentric rings as shown in Figure 2.4 [2] and if the filament shape is assumed to be circular, the fiber volume fraction in the roving can be calculated from the following equation [3]:

$$v_{fR} = \frac{3N_r(N_r - 1) + 1}{(2N_r - 1)^2},\tag{2.2}$$

where v_{fR} is the fiber volume fraction in the roving and N_r is the total number of concentric rings containing N_f fibers. Equation 2.2 assumes that each ring in the roving

FIGURE 2.4 Filament packing in a roving or tow. (With kind permission from Springer Science+Business Media: *Handbook of Composites*, Textile preforming, 1998, 397–424, F. K. Ko and G. W. Du (S. T. Peters, ed.).)

is completely filled with the fibers. The relationship between N_f and N_r is given by Equation 2.3:

$$N_r = \frac{1}{2} + \sqrt{\frac{1}{4} + \frac{(N_f - 1)}{3}}. \qquad (2.3)$$

For a large number of fibers, the fiber volume fraction approaches 0.75.

The fiber volume fraction in a yarn is different from that in a roving, since the fibers in a yarn are not all aligned in one direction; instead, they are helically wound as illustrated in Figure 2.5. The helix angle depends on the radial position of the filament in the yarn. The maximum helix angle φ occurs at the outer surface of the yarn and is given by

$$\varphi = \tan^{-1}\left[\pi(D - d)T_o\right], \qquad (2.4)$$

where D is the yarn diameter, d is the fiber diameter, and T_o represents the number of twists per unit length of the yarn, which for yarns used in the composite industry is less than 100 per meter. The fiber volume fraction v_{fY} in the yarn can be defined as

$$v_{fY} = \frac{nd^2}{D^2}, \qquad (2.5)$$

where n is the number of fibers in the yarn. Combining Equations 2.4 and 2.5, the fiber volume fraction in the yarn can be written in terms of the fiber diameter, twist level, and surface helix angle. This is shown in the following equation:

$$v_{fY} = n\left[1 + \frac{\tan\varphi}{\pi d T_o}\right]^{-2}. \qquad (2.6)$$

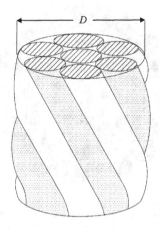

FIGURE 2.5 Yarn construction.

Since the fibers in a yarn are in twisted form, the tensile strength and modulus of a yarn are lower than those of strands, rovings, or tows, which contain straight parallel fibers. Fibers in all fiber forms are subject to further degradation of mechanical properties when they are woven, knitted, or braided to make fabrics. As a result, the strength and modulus of fibers in a fabric can be as low as 50% of the original values.

Fibers are also available and used in discontinuous lengths. Discontinuous fibers are produced by cutting or chopping continuous fibers in small lengths. Depending on the application and manufacturing process used, the fiber length can vary from a few millimeters to several hundred millimeters. For example, for conventional injection molding processes, the fiber length is between 1 and 3 mm. For compression molding, the fiber length can range up to 25 mm. For liquid injection molding using random fiber mats, the fiber length can be 50 mm or greater. Longer fiber lengths produce higher tensile strengths, but the limit on the maximum fiber length is controlled by the limitations of the selected manufacturing process and the equipment or machineries being used.

2.2 LINEAR FIBER ARCHITECTURE

Linear fiber architecture is created when fibers are oriented all in the same direction. The unidirectional fibers are either embedded in a matrix, such as in a prepreg, or are stitched together to keep them in place. The linear fiber architecture is utilized in making laminates of multiple laminas or layers of unidirectional continuous fibers. The laminate is manufactured by stacking multiple prepreg layers to build the required laminate thickness and consolidating them by the application of heat and pressure. Depending on the design requirement, the fiber orientation angle may be varied from layer to layer. A few examples of laminate constructions with each layer containing linear fiber architecture are given in the following.

1. *Unidirectional laminate* in which fiber orientation angle is the same in all the layers: If there are n layers in the laminate with fiber orientation angle θ, then the laminate construction is $[\theta/\theta/\theta/\theta/, ..., /\theta/\theta/\theta/\theta\ \theta]$ or simply $[\theta]_n$.
2. *Angle ply laminate* in which fiber orientation angle alternates between θ and $-\theta$: The laminate construction in this case will take the form $[\theta/-\theta/\theta/-\theta/\theta/-\theta/, ...]$ or simply $[\pm\theta]_n$, where n is the number of $\pm\theta$ layers.
3. *Cross ply laminate* in which fiber orientation angle alternates between 0 and 90°: The laminate construction in this case will take the form $[0/90/0/90/0/90/, ...]$ or simply $[0/90]_n$, where n is the number of 0/90 layers.
4. *Symmetric laminate* in which for each layer above the midplane of the laminate, there is an identical layer at an equal distance below the midplane (Figure 2.6): The identical layers must have the same fiber orientation angle, thickness, and material.

Some examples of symmetric laminates and their abbreviated notations are given in the following. The letter S used as a subscript outside the

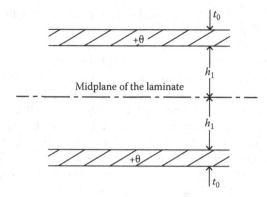

FIGURE 2.6 Symmetric laminate construction.

brackets in these examples indicates that the laminate is symmetric about its midplane.

1. [0/45/90/90/45/0] or [0/45/90]$_S$
2. [0/45/−45/90/90/−45/45/0] or [0/±45/90]$_S$
3. [0/45/−45/45/−45/−45/45/−45/45/0] or [0/(±45)$_2$]$_S$
4. [45/−45/45/−45/−45/45/−45/45] or [±45]$_{2S}$
5. [0/90/0/90/0/90/90/0/90/0/90/0] or [0/90]$_{3S}$
6. [45/−45/90/−45/45] or [±45/$\overline{90}$]$_s$

   Symmetric laminates are used in many applications to avoid bending/twisting deformations when in-plane normal and shear loads are applied and to avoid in-plane normal/shear deformations when bending and twisting moments are applied.

5. *Quasi-isotropic laminate* with basic laminate construction of either [0/±45/90] or [0/±60]: Examples of symmetric quasi-isotropic laminate constructions are shown in the following using their abbreviated notations.

   a. [0/±45/90]$_{16S}$
   b. [±45/0/90]$_{32S}$
   c. [0/±60]$_{10S}$

   Note that the in-plane elastic properties, such as elastic modulus, shear modulus, and Poisson's ratio of a quasi-isotropic laminate are direction independent; however, their strength properties can be direction dependent.

2.3 TWO-DIMENSIONAL FIBER ARCHITECTURE

Two-dimensional fiber architecture is available in three different forms: woven fabrics, knitted fabrics, and braided fabrics [3]. The fibers in these forms are oriented in two or more directions on a 2D plane. The advantage of a laminate containing layers with 2D architecture is that it can improve the transverse properties; however, its longitudinal properties will be lower compared to the longitudinal properties of a laminate containing only unidirectional 0° layers.

2.3.1 BIAXIAL WOVEN FABRICS

Biaxial woven fabrics consist of two sets of yarns interlaced at right angles to each other in a single plane to create a single 2D or biaxial layer. The set of yarns running parallel to the fabric length direction is called *warp* (lengthwise), and the set of yarns running 90° to the fabric length direction is called *fill* or *weft* (crosswise) (Figure 2.7). These fabrics are manufactured by weaving or interlacing the warp and fill yarns in a textile loom. In this machine, the warp yarns are oriented parallel to the machine direction, but they are separated from each other by a small gap or opening called a shed. The filling yarns are inserted through the shed, one yarn at a time, using a shuttle to form the weave pattern (Figure 2.8). Then the filling yarns are pushed toward the fabric edge for uniform packing. The weaving speed for glass and carbon fiber fabrics can be as high as 500 insertions per minute.

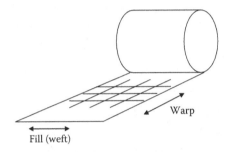

FIGURE 2.7   Warp and fill (weft) directions of a biaxial woven fabric.

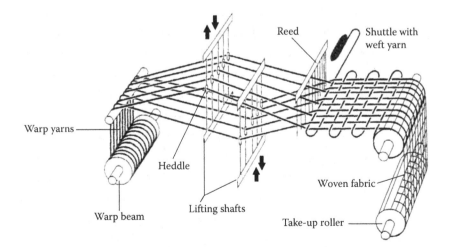

FIGURE 2.8   Fabric weaving machine.

The following nomenclature is used in industry to characterize biaxial or 0/90 woven fabrics:

1. Yarn type in both warp and fill directions.
2. Weave style, which specifies the repetitive manner in which the warp and fill yarns are interlaced in the fabric.
3. Count, which refers to the number of yarns per centimeter or inch width of the fabric in the warp and fill directions. For example, a fabric count of 60 × 52 means 60 yarns per inch in the warp direction and 52 yarns per inch in the fill direction. Note that a greater yarn count in a direction produces higher modulus and strength in that direction.
4. Nominal aerial weight in grams per square meter or ounces per square yard.
5. Weight distribution (in percentage) in the warp and fill directions.
6. Fabric thickness, measured in millimeters or thousandths of an inch (mil).
7. Fabric width, in millimeters or inches.
8. Finish.

An example of woven carbon fabric specifications is shown in Table 2.3.

Common weave styles used in woven fabrics are shown in Figure 2.9 and are described in the following:

1. *Plain weave*, in which warp and fill yarns are interlaced over and under each other in an alternating fashion
2. *Satin weave*, in which each warp yarn is interlaced over several fill yarns and under one fill yarn
3. *Basket weave*, in which two or more warp yarns are interlaced over and under two or more fill yarns in an alternating fashion
4. *Twill weave*, in which one or more warp yarns are interlaced over and under two or more fill yarns.

TABLE 2.3
Example of Woven Fabric Specification

Woven Fabric Specification	Example
1. Yarn type	Warp: AS4-3K
	Fill: AS4-3K
2. Weave style	Plain
3. Count	Warp: 2.33 yarns/cm (6 yarns/in.)
	Fill: 2.33 yarns/cm (6 yarns/in.)
4. Nominal aerial weight	104 g/m^2 (3.07 oz./yd^2)
5. Weight distribution	Warp: 50%
	Fill: 50%
6. Thickness	0.09 mm (4 mil or 0.004 in.)
7. Width	1000 mm (39.4 in.)
8. Finish	Compatible with epoxy

Note: The fiber type in the example is AS4 carbon fiber.

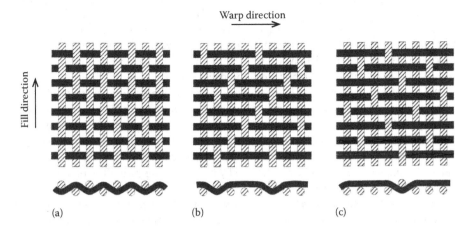

FIGURE 2.9 Common weave styles: (a) plain weave, (b) 3 × 1 twill weave, and (c) 5-H satin weave.

The important considerations in selecting the fabric style for an application are as follows.

1. Draping characteristics, which determine how well the fabric may be draped or fitted on the mold surface, especially at the corners
2. Stability, which determines the amount of fiber movement or displacement in the plane of the fabric during resin flow and slippage during draping
3. Porosity, which depends on the amount and openness of gaps between the interlaces, and therefore, it influences the amount of resin flow through the fabric
4. Crimp, which is the up and down undulation of fibers in the fabric.

It can be seen in Figure 2.9 that among the four weave styles, the plain weave has the greatest number of interlaced yarns, and the satin weave has the least number of interlaced yarns. In other words, plain weave has a higher number of crimps than satin weave. High interlacing density in plain weave fabrics prevents yarn slippage during handling and processing (for example, under high resin injection pressure in RTM), and because of this, the amount of fiber movement will be relatively small during resin flow through them. In other words, the plain weave fabrics are very stable. For this reason, they are preferred in many wet layup applications and commonly used for flat panel laminates, such as circuit board laminates. On the other hand, satin weave fabrics are more pliable than plain weave fabrics and more easily conform to contoured or complex surfaces for which fabric *drapeability* is an important fabric selection consideration. The lower amount of yarn interlacing in a satin weave allows higher fiber packing density and creates fewer crimps than in a plain weave. Their loose weave structure allows easier resin penetration. However, they are not considered suitable for applications in which a high level of preform stability may be needed.

There are several types of satin weaves. The most common types are the four-harness (4-H) satin (also called crowfoot), five-harness satin (5-H) (Figure 2.9), and eight-harness (8-H) satin. The harness number refers to the total number of yarns crossed over and passed under before repeating the pattern. In a 4-H satin, one warp yarn crosses over three fill yarns and under one fill yarn. In a 5-H satin weave, one warp yarn crosses over four fill yarns and under one fill yarn. In an 8-H satin weave, one warp yarn crosses over seven fill yarns and under one fill yarn. The 8-H satin is more pliable than the other two and, because of its higher fabric count, produces a higher strength for the composite.

A layer of woven fabrics can be regarded similar to, but not the same as, two layers of 0/90 layup in a traditional cross-plied laminate containing linear fiber architecture in each layer. However, it is important to note that the mechanical properties of woven fabric laminates in the warp and fill directions are, in general, lower than the corresponding properties of a 0/90 cross-plied laminate. The principal reason for this is the fiber undulation or crimps in the woven fabrics created by the interlacing yarns. The properties of woven fabric composites depend on the weave construction and lamination pattern, as well as the stacking sequence. Thus, for example, a laminate containing 49 end × 30 end woven crowfoot style fabrics has tensile moduli of 16.5 GPa in the warp direction and 6.9 GPa in the fill direction when the weaves are parallel. When the weaves are crossed, the tensile modulus is equal in both warp and fill directions, and its value is 23.4 GPa.

Woven fabric layups are used in dry form in LCM, such as RTM. Woven fabrics are also used in making prepregs. In both cases, the resin flow required for fiber wetting strongly depends on the weave pattern. The property that determines resin flow is called the *permeability*, which increases with increasing fiber undulation. Thus, for example, for the same fiber volume fraction, an 8-H satin weave has a 50% higher permeability and, therefore, resin flow, than a crowfoot weave. Thus, the choice of weave pattern not only affects the mechanical properties, but also the processing characteristics of the composite.

Fiber volume fraction in woven fabrics can be experimentally determined using the resin burn-off method. For theoretical calculation of fiber volume fraction in a plain biaxial weave, Dow and Ramnath [4] assumed the unit cell geometry shown in Figure 2.10. They assumed that the yarns have a circular cross-section, and their diameter and the pitch length are the same in both fill and warp directions. On the basis of these assumptions, the expression for fiber volume fraction is

$$v_f = \frac{\pi}{4} \kappa \left[\frac{\frac{2l}{d} + 4\alpha}{(L/d)^2 (T/d)} \right], \tag{2.7}$$

where d is the yarn diameter; L is the pitch length; T is the fabric thickness; l is the dimension shown in Figure 2.10; κ is the fiber packing fraction; and α is the yarn inclination angle with the fabric plane = $\tan^{-1} \left[\frac{2}{\sqrt{(L/d)^2 - 3}} \right] - \tan^{-1}(d/L)$.

Since the fabric thickness is very close to twice the yarn diameter, i.e., $T \approx 2d$ and $\tan \alpha \cong (d/l)$, Equation 2.7 can be simplified to write

$$v_f = \frac{\pi}{4} \kappa \frac{1 + 2\alpha \tan \alpha}{(L/d)^2 \tan \alpha}. \tag{2.8}$$

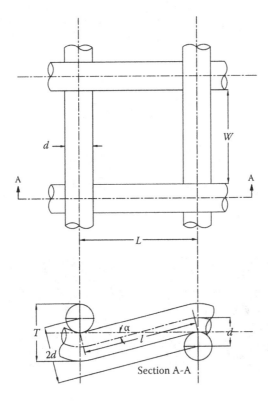

FIGURE 2.10 Unit cell geometry of a woven fabric. (With kind permission from Springer Science+Business Media: *Handbook of Composites*, Textile preforming, 1998, 397–424, F. K. Ko and G. W. Du (S. T. Peters, ed.).)

2.3.2 KNITTED FABRICS

Knitted fabrics are produced by interlooping one or more yarns on a knitting machine, similar to that used for making garments. The interloops are formed by a row of closely spaced needles which move back and forth to pull yarns through previously formed loops (Figure 2.11). There are two basic types of knits: weft knits and warp knits (Figure 2.12). In weft knits, a single yarn is fed in the width or cross-machine direction of the knitting machine, which forms a row of knit loops across the width of the fabric. In warp knits, multiple yarns are fed in the length or machine direction and each yarn forms a line of knit loops in the length direction of the fabric. By controlling the stitch density, a wide range of pore geometry can be generated. Because of the interlooped structure, the maximum fiber packing density in knitted fabrics is lower than that in woven fabrics.

Knitted fabrics are more flexible and conformable than woven fabrics and are therefore more suitable for making parts with complex shapes. Because of their looped structure, they can be stretched by a considerable amount to cover curved mold surfaces, particularly around the corners without the need for cutting small pieces and overlapping them. This also reduces material wastage and the manufacturing

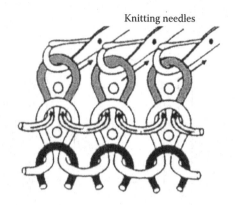

FIGURE 2.11   Fiber knitting process.

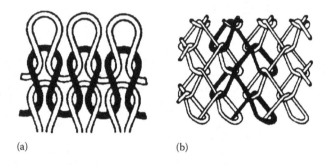

(a)   (b)

FIGURE 2.12   Basic types of knits: (a) weft knit and (b) warp knit.

cost of complex-shaped parts. The knitting parameters that are controlled to vary the drapeability of knitted fabrics are the loop density (number of loops per unit length of the fabric), the length of the loops, their areal density, and the tightness of the loops. However, because of the high degree of yarn curvature, knitted fabrics produce lower strengths for the composite compared to the woven fabrics.

2.3.3   MULTIAXIAL WOVEN FABRICS

Multiaxial woven fabrics are either triaxial or quadriaxial. In *triaxial woven fabrics*, three yarn systems are interwoven, typically at 0, +60°, and −60° angles to one another in a single plane (Figure 2.13). In another variation of multiaxial woven fabrics, 0, 45°, 90°, and −45° yarns are interlaced to form *quadriaxial* woven fabrics. Both triaxial and quadriaxial fabrics produce planar isotropic laminates, whereas the properties of a biaxial fabric laminate are highly direction dependent.

2.3.4   BRAIDED FABRICS

Biaxial braided fabrics, simply called *braids*, contain two sets of continuous yarns symmetrically interwoven about the braiding axis, one in the +θ direction and the

FIGURE 2.13 Triaxial woven fabric. (With kind permission from Springer Science+ Business Media: *Handbook of Composites*, Textile preforming, 1998, 397–424, F. K. Ko and G. W. Du (S. T. Peters, ed.).)

other in the $-\theta$ direction (Figure 2.14a). The angle θ, called the braiding angle, is less than 90°. Triaxial braids contain a third set of yarns that is oriented along the braiding axis, i.e., in the 0° direction (Figure 2.14b). Braided fabrics are highly conformable and shear resistant.

The braiding process is illustrated in Figure 2.15. The braider contains a number of fiber carriers (called *cops*), usually between 72 and 144, which rotate around the braider in a maypole fashion, interlocking yarns under each other. Depending on the relative rotating speed of the braider and the mandrel, the braiding angle θ can be controlled from ±10° to ±80°. A lower braiding angle produces higher stiffness along the braid axis, whereas a higher braiding angle produces a higher stiffness in the hoop direction. The braiding process can produce a variety of shapes including hollow tubes, flat bars, as well as many other complicated sections. Multilayer braided fabric can also be produced by braiding back and forth. The fiber type as well as

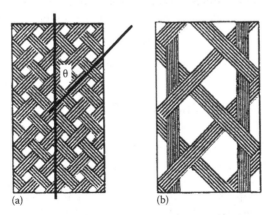

FIGURE 2.14 Braided fabric: (a) biaxial braid and (b) triaxial braid.

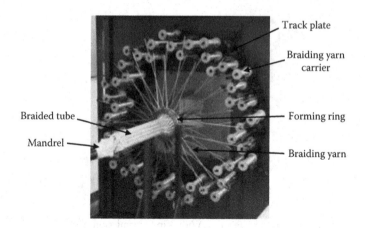

FIGURE 2.15 Braiding process.

braiding angle in each layer can be varied as needed. The triaxial braid is formed by feeding axial yarns on to the mandrel.

Tubular braids are formed on a removable mandrel. If the mandrel has the shape of a cone, the parameters that control the braiding angle are the number of carriers, braiding yarn width, mandrel radius, and half-cone angle of the mandrel. For a cylindrical mandrel, the half-cone angle is zero. For this case, the fiber volume fraction of a biaxial braid can be calculated from the following equation:

$$v_f = \frac{\kappa . w_y . N_c}{4\pi . R_m . \cos\theta}, \tag{2.9}$$

where κ is the fiber packing fraction; w_y is the yarn width; N_c is the number of fiber carriers; R_m is the mandrel radius; and θ is the braiding angle.

The ratio of the total width of either the $+\theta$ or $-\theta$ yarns and the mandrel perimeter is defined as the braid tightness factor β. To avoid yarn jamming, $0 < \eta < 1$. Since

$$\beta = \frac{w_y . N_c}{4\pi R_m},$$

Equation 2.9 can now be rewritten as

$$v_f = \frac{\kappa . \beta}{\cos\theta}. \tag{2.10}$$

Tubular braids with 15–35° braiding angle are highly suited for axisymmetric components loaded in torsion. Braids can be easily deformed in the radial direction; this feature of braids can be used to conform to other shapes.

2.3.5 NONCRIMP FABRICS

Noncrimp fabrics (NCFs) [5] have multiple layers of straight fibers, but the angular orientations of the fibers can vary from layer to layer. The layers are fixed in place, either by using up-and-down stitches (Figure 2.16) or by using a binder. Polymer fibers, such as polyester, nylon, and polyethylene, are used as the thread for stitching. A number of different stitching patterns are available. The thread density and location can also be varied. The binder can be an adhesive tape or a thin thermoplastic polymer, such as a thermoplastic polyester, that is melted and fused using a hot press.

NCFs can be unidirectional, biaxial (0/90 and 45/–45), triaxial (0/45/–45 and –45/90/45), or quadriaxial (0/–45/90/+45). They can also be combined with random fiber layers, such as a chopped strand mat (CSM). Figure 2.17 shows a quadriaxial NCF produced by up-and-down stitching of 0, 90°, 45°, and –45° layers. Unlike the woven fabrics, the fibers in NCFs are not undulated and crimped, which gives them an advantage over woven fabrics in terms of both strength and modulus. Because of straight fibers in NCFs, the flow of liquid resin through the layers is much better than through the layers of woven fabrics.

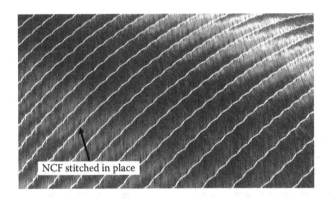

FIGURE 2.16 Photograph of a NCF. (Courtesy of Chomarat North America., Williamston, SC.)

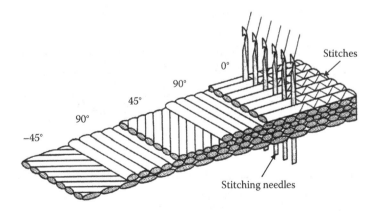

FIGURE 2.17 Quadriaxial NCF produced by the LIBA process.

2.4 THREE-DIMENSIONAL ARCHITECTURE

Three-dimensional fiber architecture [6–8] is obtained by weaving or stitching yarns in the thickness direction of the fabric. Thus, a composite with 3D fiber architecture contains both in-plane and out-of-plane fiber orientations. The out-of-plane fibers in the thickness direction provide higher strength and stiffness in that direction. They also provide high resistance to delamination and, thus, improve the damage tolerance of the composite in low-velocity impact and fatigue loadings. However, the in-plane modulus and strength of the composite are generally lower when a 3D architecture is used in comparison to a composite with a 2D architecture. A few examples of 3D fiber architecture are shown in Figure 2.18 [9].

There are five different methods of producing a 3D architecture: weaving, knitting, braiding, stitching, and z-pinning. The first three processes are similar to the 2D weaving, knitting, and braiding processes with some modifications to the machines on which they are performed. In this section, stitching and z-pinning processes are described.

2.4.1 STITCHING

Stitching is a relatively simple process of producing a 3D architecture. In this process, a needle with a stitch thread at its end is inserted up and down through a stack of dry woven fabrics or uncured prepreg layers in an industrial sewing machine. The stitch density is usually in the range of 3–10 stitches per cm². The material in the stitch thread can be glass, carbon, aramid (Kevlar), or other fiber yarns. Aramid

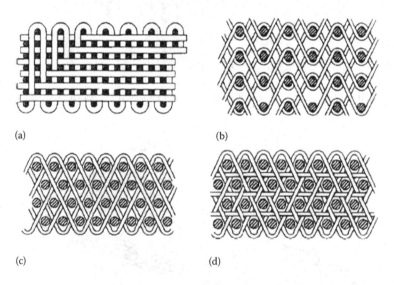

(a)   (b)

(c)   (d)

FIGURE 2.18 Examples of 3D fiber architecture. (a) 3D, (b) angle interlock, (c) 3x, and (d) 3x with warp stuffer yarns. (From S. Wilson, W. Wenger, D. Simpson, and S. Addis, *Resin Transfer Molding*, SAMPE Monograph No. 3 (W. P. Benjamin and S. W. Beckwith, eds.), Society for the Advancement of Material and Process Engineering, Covina, CA, 1999.)

yarns are relatively easy to use in stitching operations and do not easily break during stitching; however, they have a tendency to absorb moisture from the application environment, which can reduce their strength and may affect the mechanical properties of the composite. Glass and carbon yarns do not have the moisture absorption problem, but because they are more brittle, they break more easily during stitching. They also tend to fray and form stitch knots more easily.

Stitching has several advantages over weaving or knitting. It can be used with both dry fabrics or resin-impregnated fabrics. It can also be preferentially used in local areas where stitching can prevent or reduce delamination, for example, at the free edges of a composite and around a hole or a cutout. The stitch density, stitch pattern, and thread diameter can be varied to fit the design need. It can also be used to combine several separate components to construct a complex shape.

2.4.2 Z-PINNING

Another process for making a 3D architecture is called *z-pinning*. In this process, metal or pultruded composite pins with a diameter between 0.2 and 1 mm are inserted through the thickness of a stack of uncured prepreg layers or dry woven fabrics to provide reinforcement in the thickness direction (z-direction) (Figure 2.19). Ultrasonically assisted z-pinning or UAZ is the most common process for inserting the pins [10]. In this process, the pins are first arranged on a foam carrier, which is placed on top of the prepreg stack. The pins are driven from the foam carrier into the stack using an ultrasonically actuated tool. Under the compressive waves created by the tool, the foam carrier collapses as the pins enter the stack. Heat generated at the interface of the tool and the prepreg softens the resin, which helps in the insertion process. The collapsed foam carrier is then removed using a blade. The prepreg stack with the z-pins is then cured using vacuum bagging and autoclaving.

While z-pinning has several advantages, it also has a few disadvantages. It is a relatively inexpensive process, and even at pin volume fractions of 4%, it can produce

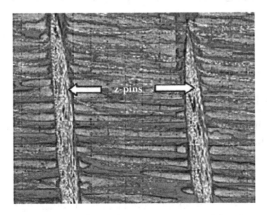

FIGURE 2.19 Z-pinned composite laminate. (Reprinted from *Composites: Part A*, 38, A. P. Mouritz, Review of z-pinned composite laminates, 2383–2397, Copyright (2007), with permission from Elsevier.)

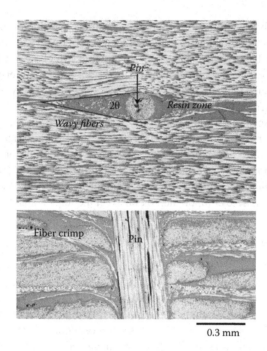

0.3 mm

FIGURE 2.20 Fiber crimp, fiber waviness, and resin-rich area caused by z-pinned composite laminate. (Reprinted from *Composites: Part A*, 38, A. P. Mouritz, Review of z-pinned composite laminates, 2383–2397, Copyright (2007), with permission from Elsevier.)

significant improvement in delamination toughness, low-velocity impact resistance, damage tolerance, and pin bearing strength. There is also a significant improvement in through-thickness modulus and strength, but the in-plane modulus and strength decrease and so does the fatigue performance. The reasons for lower in-plane properties are the fiber waviness (Figure 2.20), crimping, and fiber breakage caused by the insertion of z-pins, presence of resin-rich areas surrounding the z-pins, and residual thermal stresses set up on curing due to the difference in thermal contractions of the pin material and the composite laminate.

2.5   RANDOM FIBER ARCHITECTURE

Random fiber architecture includes nonwoven fiber mats containing in-plane randomly oriented fibers and is used in applications requiring planar isotropic properties, i.e., equal properties in all directions. Depending on the manufacturing process used for making the random fiber architecture, the fiber length can vary from 1 to 200 mm or longer. The cost of manufacturing random fiber architecture is lower, since it can be produced at a faster rate compared to the other types of fiber architecture. It also provides better drapeability and conformation to the shape of the mold compared to woven fabrics. However, it is important to keep in mind that because of random fiber orientation and discontinuous fiber lengths, the tensile strength and

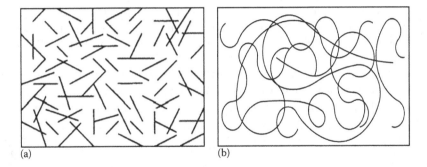

FIGURE 2.21   Random fiber architectures: (a) CSM and (b) CFM.

modulus of random fiber composites are lower than those of composites containing either linear or 2D fiber architecture.

The planar random fiber architecture is available in the form of a random fiber mat, which can be made with either short fiber strands or continuous fiber strands. The former is called the chopped stand mat (CSM) and the latter is called the *continuous filament mat* (CFM). They are shown in Figure 2.21. Both mats are used in liquid injection molding processes. CSM is also used in the pultrusion process in which it is combined with continuous fiber rovings.

CSMs are produced by chopping continuous fiber strands into small lengths and depositing the chopped fibers on a moving horizontal belt. The fiber length is usually between 25 and 50 mm. Longer fibers can be used if higher strengths are needed for the composite part; however, longer fibers reduce their drapeability in complex molds with small corners. The randomly oriented fibers are mostly evenly distributed in the mat and are held together in place by a binder, which can be either a thermoplastic polyester or a partially cured thermosetting resin. The glass fiber CSMs are available in a variety of widths, and their weights range between 2.6 and 12 g/m^2.

CFMs are produced by depositing continuous strands in a swirl pattern onto a moving horizontal belt. A binder is then applied to keep the fibers in place. In general, a CFM-reinforced composite has higher strengths than a CSM-reinforced composite because of longer fiber lengths. It also has a higher resistance to fiber movement during flow of liquid resin under pressure in LCM processes. CFMs are also the primary reinforcement in many glass mat thermoplastic sheets.

2.6   SELECTION OF FIBER ARCHITECTURE

The selection of fiber architecture starts with the design and performance requirements of the composite part, since the properties of the molded material depends on the fiber architecture. Its role on manufacturability, quality, and cost of the composite part should also be considered at the part design stage.

Three aspects of fiber architecture are important in manufacturing high-quality composite parts: conformability, fiber movement, and permeability. *Conformability* refers to how well the fiber architecture conforms to the mold surface during draping without tearing or significant movement and displacement of the fibers from their intended positions. Since resin is injected into the mold under pressure or pulled into the mold by vacuum, fiber movement can occur during resin flow through the fiber network. The fiber architecture must also be robust enough to maintain its geometry during handling, storage, and transfer to the mold surface. If the fiber movement significantly alters the fiber arrangement, there will be variations in fiber orientation and fiber volume fraction in the composite part. *Permeability* refers to the ease with which liquid resin can flow through the gaps and interstices between the fibers in the fiber architecture. The resin flow is important for good fiber wetting, proper resin distribution in the composite part, and removal of air or other volatiles between the fibers as well as from the mold.

Intra-ply shear is the primary mechanism of fiber movement in a 2D fiber architecture, such as a fabric, during draping, and/or resin flow [11]. Intra-ply shear occurs due to the sliding and rotation of parallel tows at the weft–warp crossovers (Figure 2.22) as the fabric is draped over a curved mold surface. Since the fabric has a very low resistance to bending, intra-ply shear allows the fabric to conform to the mold surface without folding or wrinkling.

The intra-ply shear behavior of fabrics is determined using the picture frame test (Figure 2.23). In this test, a square fabric specimen is clamped to four clamping bars that are connected by hinges at their ends. Initially, the fibers in the fabric specimen are directed parallel and perpendicular to the clamping bars. The clamping bars form a square picture frame which is loaded in tension in a tensile testing machine. One corner of the frame is connected to the fixed crosshead, and its opposite corner is connected to the moving crosshead of the testing machine. When a tensile load is applied, the distance between these two vertical corners increases and the distance between the two horizontal corners decreases. The change of shape of the picture frame from a square to a rhombus imparts shear deformation to the fabric.

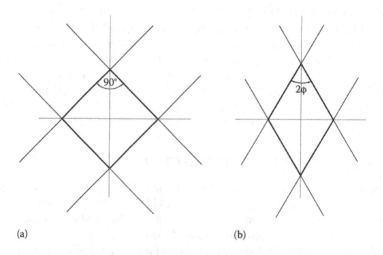

(a)   (b)

FIGURE 2.22   Intra-ply shear deformation: (a) shape before deformation and (b) shape after intra-ply shear deformation.

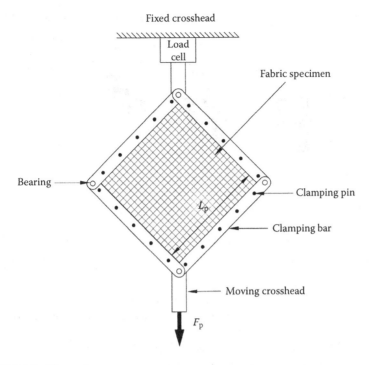

FIGURE 2.23   Picture frame test.

During the picture frame test, the tensile force F_p required to deform the fabric and the crosshead displacement d_p are recorded. The shear force and shear angle are calculated using these two recorded data and the following equations:

$$F_s = \frac{F_p}{2\cos\phi},$$

$$\theta = \frac{\pi}{2} - 2\phi, \tag{2.11}$$

where F_s is the shear force; θ is the shear angle $\phi = \cos^{-1}\left(\frac{1}{\sqrt{2}} + \frac{d_p}{2L_p}\right)$; F_p is the recorded tensile force on the picture frame; d_p is the recorded crosshead displacement; and L_p is the picture frame length.

Figure 2.24 schematically shows a shear load–shear angle plot obtained in a picture frame test. Initially, the yarns in the fabric are orthogonal to each other. As the intra-ply shear is initiated, the yarns begin to rotate and slide over each other. Friction between the yarns at the crossovers and viscous drag if a liquid resin is present contributes to the resistance to shear deformation, which is still relatively low at this level of loading. As the load is increased, the adjacent yarns start to come in contact and press against each other, which results in yarn compaction and causes shear stiffness

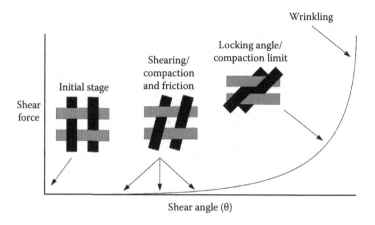

FIGURE 2.24 Shear force–shear angle plot describing shear characteristic of a fabric in a picture frame test.

to increase. If the load is increased further, the yarns become locked. Loading beyond locking causes out-of-plane buckling of the fabric, and then the deformation is not due to shear only. The load now increases very rapidly to a high value.

Figure 2.25 shows the shear load vs. shear angle plots of a plain fabric, a 4-H satin fabric, and a twill fabric. The difference in shear behavior of these three fabrics can be clearly observed in this figure. The plain fabric has a higher resistance to shear deformation than the 4-H and twill fabrics. Resistance to shear also increases with increasing shear rate, but it decreases with increasing temperature.

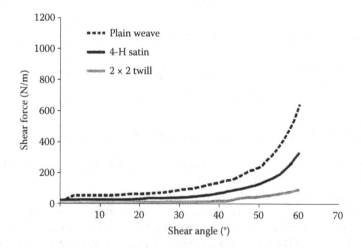

FIGURE 2.25 Shear force vs. shear angle plots of three different fabrics. (From A. C. Long and M. J. Clifford, *Composites Forming Technologies* (A. C. Long, ed.), Woodhead, Cambridge, UK, 2007.)

In addition to intra-ply shear, several other mechanisms of deformation are observed in woven fabrics during draping (Figure 2.26). Major among these deformations are (1) intra-ply rotation, (2) ply-to-ply sliding, (3) interfiber slip, (4) fiber buckling, (5) fiber extension, (6) ply compaction, and (7) ply bending. In-plane tensile deformation is relatively small compared to intra-ply shear, and its role becomes important only after the tows in the fabric straighten. As the fabric is pressed against the mold surface, some of the yarns in the fiber tows may experience high tensile stresses and break, which can then produce weak areas in the molded part. Ply-to-ply

Deformation mode	Schematic representation	Principal characteristics
In-plane shear		Rotation of crisscrossing tows
Intra-ply rotation		Rotation of plies relative to each other
Intra-ply slip	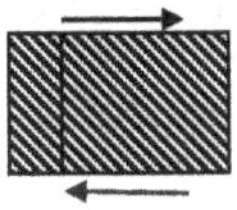	Relative linear movement between individual layers
Interfiber slip	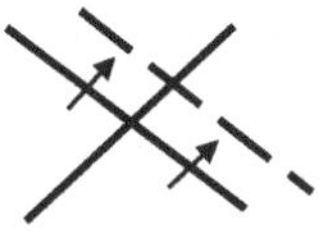	Individual tow movement due to shear
Fiber buckling		Individual tow buckling due to axial compression
Fiber extension		Individual tow extension due to axial tension
Ply compaction		Thickness reduction due to compaction resulting in increase in fiber volume fraction
Ply bending		Bending of individual plies

FIGURE 2.26 Deformation mechanisms observed in woven fabrics during draping.

sliding is controlled by the coefficient of friction at the interface between the plies. There can also be sliding at the interface between the mold surface and the ply in contact with the mold surface. If the friction is high, low resistance to sliding can cause in-plane buckling of the fiber tows and out-of-plane wrinkling of the fabric. The parameters that control the friction coefficients are mold surface roughness, fiber tow surface roughness, presence of binders, presence of liquid resin, and processing temperature. Ply bending is critical when a fabric is draped on a curved surface, especially with a small radius of curvature. Resistance to ply bending is relatively low for a dry fabric, but can be significantly high for a prepreg containing the fabric. Ply compaction occurs as the pressure is applied on the fabric or a stack of fabrics during draping and during forming. Variation in compaction can cause variation in fiber volume fraction and permeability to resin flow between the fabric layers.

PROBLEMS

1. Calculate the yield (which is the length per unit mass) of glass fiber rovings that contain 20 strands per roving. Each strand in the roving has 204 filaments, and the average filament diameter is 10 μm.

2. A composite panel of thickness t is made of a stack of N plies of equal thickness. The number of fiber yarns per unit width of each ply is n and the yarn yield is my. Show that the fiber volume fraction in the panel can be written as

$$v_f = \frac{nNm_y}{t\rho_f},$$

where ρ_f is the fiber density.

3. Assume that the filaments in a circular carbon fiber tow is arranged as shown in Figure 2.4. If there are 10,000 filaments in the tow, what are the fiber volume fraction and void volume fraction in the tow?

4. Calculate the volume fraction of fibers in a 12 K carbon yarn with 7 μm filament diameter and 80 twists per meter. The helix angle in the yarn is 15°.

5. How many layers are present in each of the following symmetric laminates?
   a. $[0/90/30_2/-30_2]_{40S}$
   b. $[(0_4/(\pm60)_2]_{20S}$
   c. $[\pm15]_{30S}$

6. A symmetric quasi-isotropic laminate with 80 layers is to be constructed using 0°, 45°, −45°, and 90° fiber orientations. Write four different orders in which the laminate can be constructed and their abbreviated notations. If additional 0° fiber layers are added to the laminates mentioned earlier to increase its longitudinal stiffness, will they be considered quasi-isotropic?

7. The yarn count in a plain-woven fabric of 3 K carbon fibers is 9.45 per centimeter (warp) × 9.45 cm (fill). The filament diameter is 7 μm and the fiber density is 1.76 g/cm². Estimate the aerial weight of the fabric in grams per square meter.

8. Compare the constructions of 4-H, 5-H, and seven-harness satin weaves. Comment on their draping and resin flow characteristics.
9. What are the advantages and disadvantages of using a CSM instead of a plain woven fabric in the construction of the hood of a car?
10. A woven fabric is being used to make a preform of the shape of a helmet. What are the different deformation modes that may be observed in making the preform? Which of the three fabrics may produce the best result: (1) plain fabric, (2) 4-H stain, or (3) 2 × 2 twill?

REFERENCES

1. B. Wulfhorst, T. Gries, and D. Viet, *Textile Technology*, Hanser, Munich, 2006.
2. F. K. Ko and G. W. Du, Textile preforming, in *Handbook of Composites*, 2nd Ed. (S. T. Peters, ed.), pp. 397–424, Springer, Dordrecht, Netherlands, 1998.
3. T. W. Chou and F. K. Ko, *Textile Structural Composites*, Elsevier Science, Amsterdam, 1989.
4. N. F. Dow and V. Ramnath, *Analysis of Woven Fabrics for Reinforced Composite Materials*, NASA Contract Report No. 178275, National Aeronautics and Space Administration, Washington, DC, 1978.
5. S. V. Lomov (ed.), *Non-Crimp Fabric Composites*, Woodhead, Oxford, 2011.
6. L. Tong, A. P. Mouritz, and M. K. Bannister, *3D Fibre Reinforced Composites*, Elsevier Science, Oxford, 2002.
7. A. P. Mouritz, M. K. Bannister, P. J. Falzon, and K. H. Leong, Review of applications for advanced three-dimensional fibre textile composites, *Composites: Part A*, Vol. 30, pp. 1445–1461, 1999.
8. R. Kamiya, B. A. Cheeseman, P. Popper, and T.-W. Chou, Some recent advances in the fabrication and design of three-dimensional textile preforms: A review, *Composites Science and Technology*, Vol. 60, pp. 33–47, 2000.
9. S. Wilson, W. Wenger, D. Simpson, and S. Addis, "SPARC" 5 axis, 3D woven, low crimp preforms, in *Resin Transfer Molding*, SAMPE Monograph No. 3 (W. P. Benjamin and S. W. Beckwith, eds.), Society for the Advancement of Material and Process Engineering, Covina, CA, pp. 101–113, 1999.
10. A. P. Mouritz, Review of z-pinned composite laminates, *Composites: Part A*, Vol. 38, pp. 2383–2397, 2007.
11. A. C. Long and M. J. Clifford, Composite forming mechanisms and materials characterization, in *Composites Forming Technologies* (A. C. Long, ed.), Woodhead, Cambridge, UK, pp. 1–21, 2007.

3 Matrix Materials

The matrix used in a PMC is either a thermosetting or a thermoplastic polymer. In the composite industry, both thermosetting and thermoplastic polymers are often called *resins*. The majority of continuous FRPs in aerospace, sporting goods, automotive, and many industrial applications (for example, wind turbine blades, gasoline storage tanks, and electrical housings) use a thermoset resin, such as an epoxy, a vinyl ester, a polyester, or a polyurethane resin. On the other hand, the majority of discontinuous or short FRPs used today contain a thermoplastic polymer, such as a polypropylene and a polyamide. The difficulty of incorporating continuous fibers in thermoplastic polymers is due to their high viscosity, which is typically 100–1000 times greater than the viscosity of thermosetting polymers at the time of fiber incorporation (see Table 3.1). Short fibers, up to 3 mm in length, can be relatively easily incorporated in thermoplastic polymers using melt processing methods, such as extrusion compounding and injection molding. Although continuous fiber-reinforced thermoplastic polymers are used in a few aerospace and automotive applications, their overall use is still relatively small compared to that of continuous fiber-reinforced thermosetting polymers.

In this chapter, we discuss the general processing characteristics of thermosets and thermoplastics commonly used in PMCs. Among the thermosetting polymers discussed are epoxies, polyesters, vinyl esters, bismaleimides (BMIs), and polyurethanes. Among the thermoplastic polymers are PEEK, PPS, polyetherimide (PEI), polyamide imide (PAI), PP, and polyamide-6 (PA-6). Table 3.2 lists the processing methods used with these polymers, and Table 3.3 provides information on their processing temperatures. There are many other thermosets and thermoplastics available for composite applications [1,2]. Their processing characteristics can be found in their product literature or by contacting their manufacturers.

3.1 THERMOSETTING POLYMERS

There are a variety of thermosetting polymers that are used for making composite parts [1]. Starting materials used for making a thermosetting polymer are low-molecular weight organic prepolymers that are, in most cases, low-viscosity liquids. The molecules in the prepolymers are relatively small in length and are not chemically linked. The transformation from the liquid state to the solid state involves a chemical reaction that takes place mostly during the composite part manufacturing stage. Catalysts or other reactive molecules may be needed to initiate and/or accelerate the chemical reaction. As curing proceeds, pre-polymer molecules become chemically linked with each other to form a 3D molecular mass with viscosity approaching infinity (Figure 3.1).

The chemical reaction transforming a low-viscosity prepolymer liquid to a solid thermosetting polymer is called *curing* or *cross-linking*, which is a time- and

TABLE 3.1

Representative Viscosities of Thermosetting and Thermoplastic Polymers at the Fiber Incorporation Stage

Polymer Type	Polymer	Melting Temperature T_m (°C)	Viscosity (Pa s)
Thermoset	Epoxy resin at 25°C	–	2–20
Thermoset	Unsaturated polyester resin mixed with styrene at 25°C	–	0.4–1
Thermoset	Vinyl ester resin mixed with styrene at 25°C	–	0.4–1
Thermoplastic	PEEK at 390°C and 1000 s^{-1} shear rate	343	100
Thermoplastic	PP at 190°C and 1000 s^{-1} shear rate	167	90
Thermoplastic	PA-6 at 240°C and 1000 s^{-1} shear rate	220	400

Note: For comparison: At 20°C, the viscosity of water is 1 mPa s, and the viscosities of motor oil (SAE 50) and pancake syrup are 1540 and 2500 mPa s, respectively.

TABLE 3.2

Processing Methods for Thermosetting and Thermoplastic PMCs

Polymer	Manufacturing Processes
Epoxies	Vacuum bag molding, filament winding, pultrusion, RTM
Polyesters	Compression molding, filament winding, pultrusion, RTM
Vinyl esters	Compression molding, filament winding, pultrusion, RTM
BMIS	Vacuum bag molding, filament winding
Polyurethanes	SRIM
PEEK	Injection molding, diaphragm forming, thermostamping
PPS	Injection molding, diaphragm forming, thermostamping
PP	Injection molding, compression molding, thermostamping
PA-6	Injection molding, thermostamping

temperature-dependent phenomenon. The properties of the cured polymer depend on the amount of chemical conversion, called the *degree of cure* and denoted by α_c. The degree of cure can be controlled by varying the cure time, temperature, or both. The length of time required to cure a composite part to the desired degree of cure is called the *cure cycle*. Since the cure cycle determines the production rate and therefore the production cost, it is important to achieve the desirable degree of cure in the shortest possible time. It should be noted that both cure temperature and cure cycle depend on the composition of the resin mix, which includes the resin chemistry, curing agent and catalyst reactivity, and presence of accelerator or inhibitor. Curing agents used with epoxies and several other thermoset prepolymers are called hardeners. They not only take part in the cuing reaction, but also form cross-links in the cured polymer. Catalysts used with polyester, vinyl ester, and other thermoset

TABLE 3.3
Typical Processing Temperatures of Thermosetting and Thermoplastic Polymers

Polymer	Type	Molecular Morphology	Melting Temperature T_m (°C)	Typical Processing Temperature Range (°C)
Epoxy	Thermoset	Amorphous	–	120–190
BMI	Thermoset	Amorphous	–	
Polyester	Thermoset	Amorphous	–	120–150
Vinyl ester	Thermoset	Amorphous	–	120–150
Polyurethane	Thermoset	Amorphous	–	60–120
PEEK	Thermoplastic	Semicrystalline	343	360–400
PPS	Thermoplastic	Semicrystalline	282	300–345
PEI	Thermoplastic	Amorphous	–	327–415
PP	Thermoplastic	Semicrystalline	170	260
PA-6	Thermoplastic	Semicrystalline	220	260–280

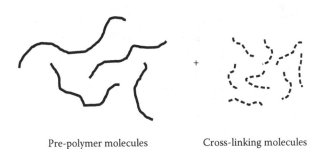

Pre-polymer molecules   Cross-linking molecules

Cross-link

Cross-linked polymer molecules

FIGURE 3.1 Schematic of a cross-linking reaction.

prepolymers initiate the curing reaction; but they do not form cross-links in the cured polymer. An accelerator is sometimes added to speed up the curing reaction, while an inhibitor is used to slow it down so that processing does not become difficult at the early stages of composite part manufacturing. In addition to these ingredients, the liquid resin may be mixed with a solvent, primarily to reduce the resin viscosity, which may be needed for proper resin flow and impregnation of fiber bundles with the resin. For some thermosetting resins, such as polyesters and vinyl esters, reactive solvents are used not only to reduce their viscosity, but also to participate in the curing reaction. In this case, the solvent molecules form the chemical links or cross-links between the polyester or vinyl ester molecules.

The initiation of a curing reaction needs a supply of heat energy, which is called the *activation energy*. The curing reaction itself is exothermic, meaning that heat is generated as the cross-links are formed. The generated heat accelerates the curing reaction, which in turn increases the rate of heat generation. Since the polymer being formed has low thermal conductivity, the generated heat is not efficiently conducted away through the thickness of the thermoset part. As a result, the temperature inside the part increases above the cure temperature. The increase in temperature, which can be measured by inserting a thermocouple in the reacting mass, is called an *exotherm*. Figure 3.2 shows two exotherm profiles that were observed during the curing reaction of an epoxy resin with a curing agent. The peak in the exotherm profile occurs due to the fact that as the reaction proceeds with increasing time, the number of epoxy prepolymer molecules is reduced and the reaction becomes increasingly difficult as it changes from a spontaneous to a diffusion-controlled mode. Thus, after the degree of cure reaches a certain level, the rate of reaction decreases. In Figure 3.2, the effect of cure temperature on the peak exotherm can be observed. At a lower cure temperature, the time to reach peak exotherm is longer and the temperature at peak exotherm is also lower.

One characteristic of a cross-linking thermosetting polymer is its *gel time* or time to gelation, which is defined as the time at which an abrupt and irreversible

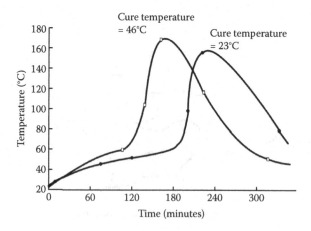

FIGURE 3.2   Exotherm plots of an epoxy resin at two different cure temperatures.

transformation from the liquid state to a gel-like state occurs. At this instant, the network formation incipiently begins and viscosity tends to increase very rapidly (Figure 3.3). The transformed material, with a degree of cure α_{gel}, starts to develop elastic properties that are not present in the pregel mix. The reaction continues beyond the gel time, but the material loses its ability to flow, and the processing of the polymer becomes increasingly difficult. For this reason, gel time is considered as the upper limit of the working life of the resin mix, and flow of the resin mix in the mold or through dry fiber network must be complete before the gel time is reached. Factors that control the gel time of a resin mix are its formulation (for example, selection of hardener or initiator type, reactive diluents, inhibitor, and accelerator), cure temperature, and rate of temperature increase.

There are several other characteristics that define the degree of cure of thermosetting polymers. They are listed in Table 3.4. $T_{g,gel}$ is the glass transition temperature of the material at the gel point, and the corresponding conversion is α_{gel}. As curing continues, the glass transition temperature T_g of the material continues to increase. *Vitrification* is the point at which T_g of the curing material becomes equal to the cure temperature and the material transforms from a rubbery gel to a glassy gel. For a fully cured system, $\alpha_c = 1$, and the glass transition temperature is $T_{g,\infty}$. The cure temperature is selected to fall between $T_{g,gel}$ and $T_{g,\infty}$. The degree of cure at any time after the curing reaction is initiated depends on the cure temperature

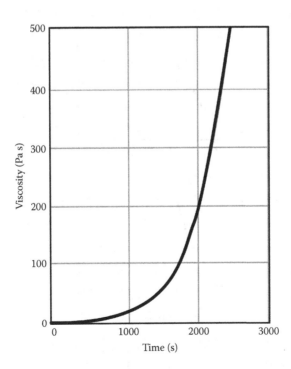

FIGURE 3.3 Viscosity of a polyester resin as a function of time during curing reaction.

TABLE 3.4
Cure Parameters of Thermosetting Polymers

Cure Parameter	Symbol
Degree of cure	α_c
Degree of cure at gel point	α_{gel}
Gel time	t_{gel}
Time to vitrification	t_{vit}
T_g of uncured thermoset (at $\alpha = 0$)	$T_{g,0}$
T_g of gel time (at $\alpha = \alpha_{gel}$)	$T_{g,gel}$
T_g of fully cured thermoset (at $\alpha = 1$)	$T_{g,\infty}$

(Figure 3.4) [3]. It is possible to remove a thermosetting matrix composite part from the mold before the thermosetting polymer is 80–90% cured. The full degree of cure can be achieved by *post-curing* it in an oven outside the mold where it is heated for a few hours at an elevated temperature.

Another term used in thermosetting polymer specification is called *pot life*. It is defined as the time it takes for the initial viscosity of the resin mix to double

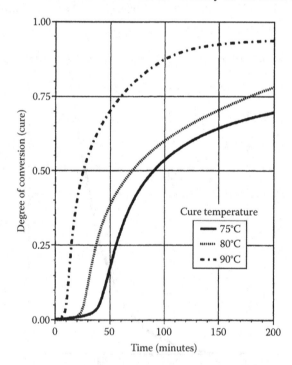

FIGURE 3.4 Degree of conversion (cure) vs. time for a polyester resin at three different cure temperatures. (S. V. Muzumdar and L. J. Lee: Chemorheological analysis of unsaturated polyester-styrene copolymerization. *Polymer Engineering and Science.* 1996. 36. 943–952. Copyright Wiley-VCH Verlag GmbH & Co. KGaA. Reproduced with permission.)

TABLE 3.5
Effects of Temperature on the Pot Life of an Epoxy Resin Mix

Parameter	Resin-Hardener 1 Combination[a] (100:24 mix ratio by weight)	Resin-Hardener 2 Combination[b] (100:25 mix ratio by weight)
Viscosity (at 22°C)	1750 mPa s	1325 mPa s
Pot Life at		
18°C	36 minutes	86 minutes
22°C	24 minutes	62 minutes
29°C	14 minutes	24 minutes

Note: The epoxy resin is the same for both combinations and the resin viscosity at 22°C is 3000 mPa s.
[a] Viscosity at 22°C is 165 mPa s.
[b] Viscosity at 22°C is 50 mPa s.

(or quadruple for lower viscosity resin mix) after the resin is mixed with either the hardener or the catalyst. The pot life of a resin mix depends on the temperature of the mix, chemistry of the ingredients, mixing ratio of the ingredients, and mass of the resin mix. The effect of temperature on the pot life of an epoxy resin at approximately the same mixing ratio of different hardeners is demonstrated in Table 3.5. Usually, the pot life is measured at 23°C with a 100 g mass of the resin mix, and the time is measured from the moment the resin mix is prepared. The working life of the resin mix is longer than the pot life, since even when its viscosity has doubled or quadrupled, it may be still fluid enough to be applied to fibers or other substrates without great difficulty.

3.1.1 EPOXIES

Epoxies are the most common matrix material in high-performance composites in aerospace, sporting goods, and many other applications. Among the various thermosetting polymers used in PMCs, they are known for their high strength, durability, excellent resistance to chemicals and solvents, and, in general, high performance at elevated temperatures. They also offer excellent adhesion to a wide variety of fibers, fillers, and other materials, which is the reason epoxies are being used as adhesives in many structural applications. They exhibit low curing shrinkage (in the range of 1–5%) and do not produce any volatiles during the curing reaction. The principal disadvantages are their relatively high cost and long cure time.

Epoxy matrix is produced by the chemical reaction of a low-molecular weight epoxy prepolymer and a curing agent, commonly called a hardener [4]. The epoxy prepolymer can be in either liquid or solid form. The curing agent does not act like a catalyst; instead, its function is to coreact and join the epoxy molecules. The amounts of the epoxy prepolymer and the curing agent are accurately weighed and then mixed thoroughly to create a uniform resin mix. The amount of curing agent does not influence the cure rate, but it has an effect on the properties of the cured epoxy polymer. Other ingredients that may be mixed with the starting liquid are diluents to reduce its viscosity, flexibilizers to improve the impact strength of the cured epoxy, fillers to reduce cost and curing shrinkage, and pigments for coloration.

The epoxy resin contains a number of reactive epoxide rings in its molecule. Each epoxide ring contains one oxygen atom and two carbon atoms (Figure 3.5). The functionality of an epoxy resin is represented by the number of epoxide rings in its molecule. The curing agent breaks the epoxide rings and initiates the curing process. In most common epoxy resins, the curing reaction is initiated by the application of heat to the resin mix. As the chemical reaction continues, an increasing number of epoxy molecules are cross-linked, forming a 3D network structure and transforming the resin mix from a low-viscosity liquid to a solid polymer matrix. As this transformation takes place, the viscosity increases, which ultimately reaches so high that the transformed material behaves like a solid.

At the beginning of the curing reaction, the epoxy is in the *A-stage*. The curing reaction at this stage can be slowed down by external means (e.g., by lowering the temperature markedly below the room temperature) when only a few cross-links have formed. The curing reaction will now either cease or progress at a very slow rate. At this stage of curing, the resin viscosity is higher than the initial viscosity and the resin is in a soft, tacky form, called the *B-stage*. The B-staged form can be maintained for several weeks or months by storing the mixed resin in a freezer. The curing reaction can be completed (*C-staged*) later by heating the B-staged resin to an appropriate cure temperature. This two-stage curing process is used in bag molding which uses prepregs containing B-staged epoxy resin. The prepreg can be stored at −18°C for several months. The epoxy resin in the prepreg is later cured at an elevated temperature in an autoclave to manufacture epoxy-based composite parts. Prepregging and the bag molding process are discussed in Chapter 5.

One very common epoxy resin is diglycidyl ether of bisphenol-A (DGEBA), which is a bifunctional molecule having two epoxide groups, one at each end of the molecule, that can react (Figure 3.6). As the number of repeating units, represented by n in Figure 3.6, increases, the molecular weight of the DGEBA prepolymer increases and its viscosity also increases. When n is close to 2, the DGEBA prepolymer is a solid, but it can be melted by heating it to temperatures higher than 40°C. With higher n, the number of cross-links formed during the curing reaction is low and the resulting cured epoxy becomes less useful as a matrix material. A polyfunctional curing agent, such as a diethylene triamine (DETA), shown in Figure 3.6, is added to DGEBA in stoichiometric or near-stoichiometric quantity to obtain the maximum degree of cross-linking. DETA is a very low-viscosity organic liquid at room temperature. It is thoroughly mixed with DGEBA using a high-speed stirrer

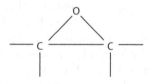

FIGURE 3.5 Epoxide ring.

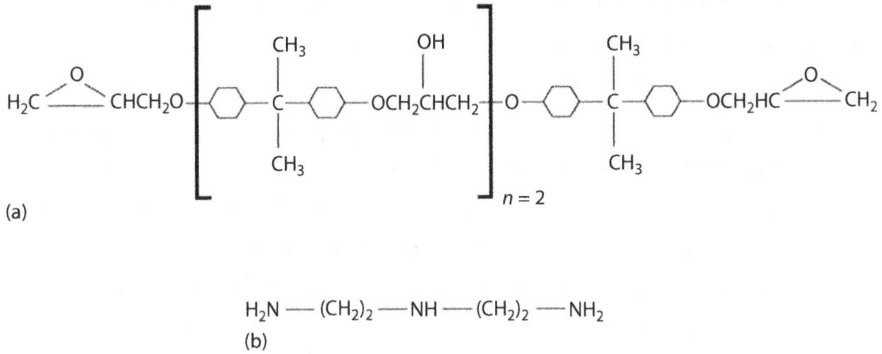

(a)

(b)

FIGURE 3.6 Molecules of (a) diglycidyl ether of bisphenol-A (DGEBA) epoxy prepolymer and (b) a curing agent (diethylene triamine [DETA]).

before incorporating fibers in the liquid resin mix. The curing reaction of the DGEBA epoxy and the DETA curing agent is illustrated in Figure 3.7. The active hydrogen (H) atoms in the amine (NH) groups of a DETA molecule react with the epoxide groups in the DGEBA molecules to open up the epoxide rings and initiate the curing reaction. As the reaction continues, DGEBA molecules are cross-linked with each other via the DETA molecules and a 3D networked molecule is formed.

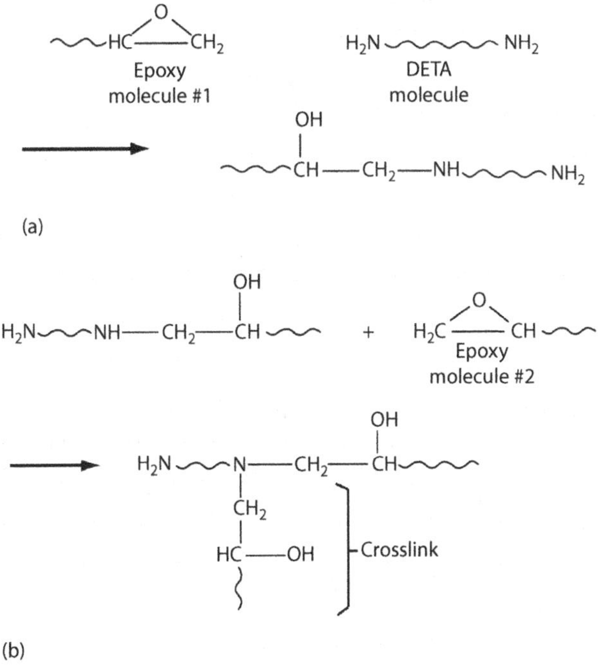

(a)

(b)

FIGURE 3.7 Curing reaction of DGEBA epoxy and DETA curing agent. (a) Initiation of curing reaction: opening of the epoxide ring. (b) Formation of crosslinks.

For polyamine curing agents, such as DETA, the stoichiometric quantity is calculated using the following three steps.

Step 1: Calculate the amine hydrogen equivalent weight (AHEW), which is the molecular weight of the amine divided by the number of active hydrogen atoms in each amine molecule:

$$AHEW = \frac{\text{molecular weight of the amine}}{\text{number of active hydrogen atoms in the amine}}.$$

Step 2: Calculate the epoxide equivalent weight (EEW) of the epoxy prepolymer, which is the molecular weight of the epoxy molecule divided by the number of epoxide groups in the epoxy molecule:

$$EEW = \frac{\text{molecular weight of the epoxy molecule}}{\text{number of epoxide groups in the epoxy molecule}}.$$

Step 3: Calculate the stoichiometric amount of amine to be added to the epoxy prepolymer. This is given as parts by weight of amine curing agent per 100 parts by weight of epoxy or simply parts per hundred (phr). The following equation is used to calculate the *phr*:

$$phr = \frac{AHEW}{EEW}.$$

Example 3.1

Calculate the stoichiometric amount of DETA to be added to a DGEBA for which $n = 1$.

Solution:

Step 1: Calculate the molecular weight of the DETA molecule.
   Referring to Figure 3.6, we can observe that each DETA molecule has 4 carbon atoms, 3 nitrogen atoms, and 13 hydrogen atoms. Since the atomic weights of carbon, nitrogen, and hydrogen atoms are 12, 14, and 1 g/mol, respectively, the molecular weight of DETA is $(4 \times 12) + (3 \times 14) + (13 \times 1) = 103$ g/mol.
   Step 2: Calculate AHEW.
   There are five active hydrogen atoms in each DETA molecule. Therefore,

$$AHEW = \frac{103 \text{ g/mol}}{5} = 20.6 \text{ g/mol}.$$

Step 3: Calculate EEW.
First, calculate the molecular weight of the DGEBA molecule with $n = 1$.
Each DGEBA molecule with $n = 1$ contains 39 carbon atoms, 7 oxygen atoms, and 44 hydrogen atoms. Therefore, its molecular weight is $(39 \times 12) + (7 \times 16) + (44 \times 1) = 624$ g/mol.
There are two epoxide groups in each DGEBA molecule. Therefore,

$$EEW = \frac{624 \text{ g/mol}}{2} = 312 \text{ g/mol}.$$

Step 4: Calculate phr.

$$phr = \frac{312 \text{ g/mol}}{20.6 \text{ g/mol}} = 15.1.$$

Therefore, the stoichiometric amount of DETA is 15.1 g for each 100 g of the DGEBA considered in this example. When the two are mixed, the total weight is 115.1 g of which 13% is DETA and 87% is DGEBA.

The curing reaction of DGEBA epoxy with DETA can take place over several days at 23°C or can be completed in 1–2 hours at 100°C. There are many different curing agents available for reacting with epoxy resins, a brief list of which is given in Table 3.6. Many of them require elevated temperatures for the initiation of the curing reaction.

TABLE 3.6
Curing Agents Commonly Used with Epoxy Resins

Curing Agent	Acronym	Physical State at 23°C	Parts per Hundred of Resin[a] (phr)	Suggested Cure Cycle
Diethylene triamine	DETA	Liquid	10.9	Gel at 23°C plus several days at 23°C or 1–2 hours at 100°C
Triethylene triamine	TETA	Liquid	12.9	Gel at 23°C plus several days at 23°C or 1–2 hours at 100°C
Metaphenylene diamine	MPDA	Solid	14.3	Gel at 55°C plus 2 hours at 125°C and 2 hours at 175°C
Diaminao diphenyl sulfone	DDS	Solid	30	1 hour at 150°C plus 3 hours at 220°C
Hexahydrophthalic anhydride	HHPA	Solid	60–75[b]	2 hours at 100°C plus 2–6 hours at 150°C
Dodecenyl succinic anhydride	DDSA	Liquid	95–130[b]	2 hours at 100°C plus 4–6 hours at 150°C

[a] When used with a DGEBA epoxy resin.
[b] Requires a suitable accelerator, usually a tertiary amine (0.5–3 wt.%).

The chemistry of the curing agent influences the reaction rate, viscosity increase, types of chemical bonds formed, the degree of cross-linking achieved, and, ultimately, the properties of the cured epoxy. The selection of the proper curing agent is important, since it influences the pot life, gel time and cure cycle, and, therefore, speed of the manufacturing process. It is important to note that the nonstoichiometric amount of the curing agent in the resin mix and inadequate mixing are often the root causes of improper curing of the epoxy resin, and they should be given proper attention.

Two other ingredients used in the epoxy resin mix for some composite applications are an accelerator and a diluent. Accelerators are added to increase the reaction rate and reduce gel time. Diluents are primarily added to reduce viscosity, which may be required for good resin flow and fiber wetting. One method of reducing viscosity is to increase the resin mix temperature, but it may also reduce the pot life. Because of this, instead of increasing the temperature, a low-viscosity diluent is selected. The diluents can be of either a reactive type or nonreactive type. Reactive diluents are typically monofunctional low-molecular weight epoxies. However, the addition of diluent can reduce the mechanical properties and glass transition temperature of the cured resin, and therefore, the amount to be added must be properly determined based on the application being considered.

A variety of epoxy resins are available for use in composite material applications (Table 3.7). DGEBA is the most commonly used epoxy resin in most composite and adhesive applications. Specifications of a typical DGEBA epoxy resin are given in Table 3.8. In high-performance aerospace and military applications, epoxies with higher functionality, such as tetraglycidyl 4,4′-diaminodiphenylmethane (TGDDM), are selected. The molecular structures of these epoxy resins vary in a number of different ways, such as the number and location of the epoxide groups in the molecule, the functionality of the epoxide groups, and the presence of aromatic groups in the molecule. Among these, the number and location of the epoxide groups in the molecule is one of the factors that determine the cross-link density, i.e., the number of cross-links per unit of the molecular weight of the cured epoxy. For example, the TGDDM molecule shown in Table 3.7 has a functionality of 4, which is higher than that of a DGEBA molecule, and because of this, it has a higher cross-link density. The other factors are the curing agent reactivity

TABLE 3.7

Epoxy Resins Used in Composite Material Applications

Epoxy Type	Chemical Name	Typical Epoxy Equivalent[a] (g/eq)	Functionality[b] (eq./mol)	Viscosity at 23°C (Pa s)
DGEBA	Diglycidyl ether of bisphenol-A	190	2	3.5
TGAP	Triglycidyl *p*-amino phenol	95	3	0.6
TGDDM	Tetraglycidyl 4,4′-diaminodiphenylmethane	100	4	94.5

[a] Epoxy equivalent is the amount of resin that contains 1 mol of epoxy.

[b] Functionality is the number of epoxide groups (which is also the number of reactive sites) per molecule.

TABLE 3.8

Specifications of a Typical DGEBA Epoxy Resin and its Properties after Curing

Material	Specification	Example
Resin type: DGEBA epoxy	Visual	Clear, pale yellow liquid[a]
	Epoxy content	5.30–5.45 eq/kg
	Density at 25°C	1.15–1.20 g/cm^3
	Viscosity at 25°C	10,000–12,000 mPa s
	Flash point	>200°C
Hardener type: Anhydride	Visual	Clear liquid[a]
	Density at 25°C	1.20–1.25 g/cm^3
	Viscosity at 25°C	50–100 mPa s
	Flash point	195°C
Accelerator type: Imidazole	Visual	Clear liquid[a]
	Density at 25°C	0.95–1.05 g/cm^3
	Viscosity at 25°C	≤50 mPa s
	Flash point	92°C

Properties of the Cured Epoxy: Cured for 4 Hours at 80°C and Postcured for 8 Hours at 140°C	
Glass transition temperature	148–153°C
Tensile modulus	3.1–3.3 GPa
Tensile strength	83–93 MPa
Ultimate elongation	5–7%
Flexural strength	125–135 MPa
Bending deflection at maximum load	10–18 mm
CLTE	$(55–57) \times 10^{-6}/°C$

Note: Mix ratio: resin/hardener/accelerator: 100/90/1 (parts by weight). Initial mix viscosity: 600–900 mPa s at 25°C. Viscosity: (1) 1500 mPa s after 3.5–4.5 hours and 3000 MPa after 7–8 hours at 25°C. (2) 1500 mPa s after 52–57 minutes and 3000 mPas after 60–65 minutes at 80°C. Pot life: 95–105 hours at 23°C and 4–5 hours at 40°C. Gel time: 140–160 minutes at 80°C and 3–5 minutes at 140°C. Typical cure cycle: 2–4 hours at 80°C and postcure for 2–8 hours at 140°C.

[a] Storage temperature for all three liquids is 2–40°C.

(which depends on the number of active hydrogen atoms in the curing agent molecule) and the reaction conditions, which include cure temperature and cure time. The higher the cross-link density, the higher the heat deflection temperature, glass transition temperature, hardness, modulus, and brittleness of the epoxy matrix. Thermal stability and chemical resistance also increase with increasing cross-link density, but they are highly improved if the epoxy molecules themselves contain aromatic groups.

Brittleness or low strain at failure is an inherent problem with highly cross-linked epoxy polymers. For making damage-tolerant epoxy-based composites for which high strain at failure is desirable, epoxy resins are blended with either elastomer particles or a ductile thermoplastic polymer. One of the elastomers is carboxyl-terminated butadiene-acrylonitrile (CTBN), which is dissolved in epoxy resin at the mixing stage. During curing, CTBN particles are developed, which form a flexible dispersed phase in the epoxy matrix and provides resistance to crack propagation. Although fracture toughness is improved by the addition of CTBN, both glass transition temperature and modulus are decreased. The addition of ductile thermoplastic (such as PEI and polysulfone) particles to the epoxy resin improves fracture toughness, but does not reduce the other properties.

3.1.2 POLYESTERS

Polyesters are the most commonly used thermosetting polymers in automotive, electrical, commercial, and consumer product applications because of their relatively low cost and a good balance of mechanical, electrical, and chemical properties. For many of these applications, short glass fibers are used as the reinforcement, and compression molding, as the principal processing method. Compared to epoxies, they have higher curing shrinkage, but lower viscosity and faster cure time. High curing shrinkage (which is in the range of 5–12%) is helpful in releasing the composite part from the mold surface after the molding is complete, but must be accounted for in the design of the mold if a polyester is selected as the matrix. Even though polyesters have lower mechanical properties than epoxies, they are selected in mass-production applications because of their faster cure time.

The starting prepolymer for thermosetting polyesters is an unsaturated polyester which contains a number of unsaturated carbon–carbon double bonds in its molecule (Figure 3.8). These double bonds are the cross-linking sites during the curing reaction. The unsaturated polyester molecules are produced by the reaction of an unsaturated organic acid, such as maleic anhydride or phthalic anhydride, with a glycol, such as ethylene glycol, diethylene glycol, and propylene glycol.

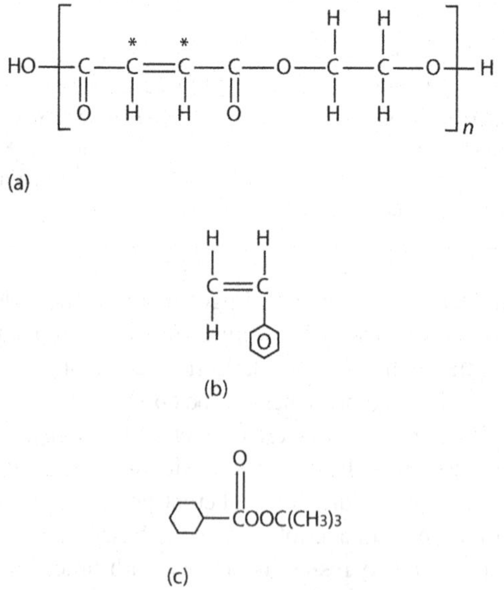

(a)

(b)

(c)

FIGURE 3.8 Principal ingredients in the preparation of a thermosetting polyester resin: (a) unsaturated polyester molecule, (b) styrene molecule, and (c) *tert*-butyl peroxybenzoate (TBPB) molecule, which acts as a catalyst.

Figure 3.9 shows an unsaturated polyester molecule obtained by the reaction of a maleic anhydride and ethylene glycol. A saturated organic acid, such as an orthophthalic acid or an isophthalic acid is added to modify the chemical structure of the unsaturated polymer molecules between the cross-linking sites. The name of the polyester resin is derived from the name of the saturated acid used, for example, orthophthalic acid for orthophthalic polyester and isophthalic acid for isophthalic polyester. Saturated acids do not contain any carbon–carbon double bonds and, therefore, do not provide any cross-linking sites. However, their presence increases the distance between the double bonds in the unsaturated polyester molecules and thus reduces the density of cross-linking, which in turn affects the properties of the cured polymer.

Even though unsaturated polyester is a low-viscosity liquid, its viscosity is further reduced by mixing it with a reactive monomer, which acts as both a diluent and a cross-linking agent. Styrene is the most common monomer used, since it is relatively inexpensive and it has a very low viscosity. It serves as an excellent diluent to achieve proper processing viscosity for polyester resins. Each styrene molecule contains one carbon–carbon double bond in its main molecular chain (Figure 3.8), which acts as the cross-linking site. The disadvantages of styrene are that it is a highly flammable volatile liquid, and the emission of styrene vapor into the environment during mixing and composite manufacturing is considered to be a health hazard. Proper precautions, such as the use of active ventilation and wearing of protective face masks, must be taken while working with styrene.

The curing reaction of the mix of a polyester resin and styrene monomer is initiated by adding a small quantity of a catalyst, such as benzoyl peroxide (BP) for elevated-temperature curing and methyl ethyl ketone peroxide (MEKP) for room-temperature curing. If BP is used as the catalyst, the application of heat rapidly decomposes the catalyst into free radicals that react with the C=C bonds in styrene as well as polyester molecules. The styrene free radicals combine with each other and join with the polyester molecules at their unsaturation points, eventually forming cross-links between the polyester molecules (Figure 3.10). As the number of cross-links increases with time, the material changes from a liquid state to a solid state.

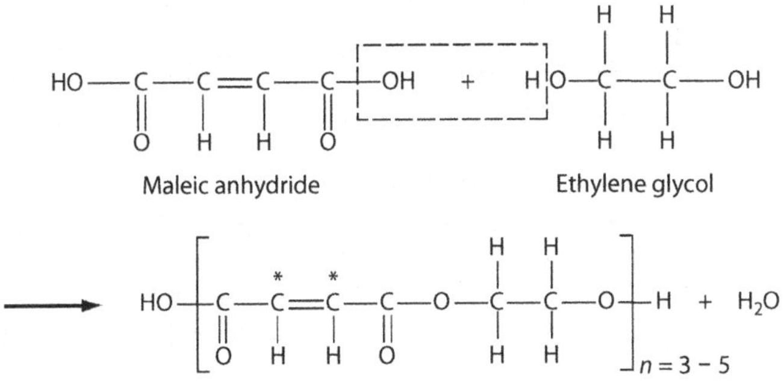

FIGURE 3.9 Unsaturated polyester molecule formed by the reaction of maleic anhydride and ethylene glycol (note the unsaturation points on the molecule shown by asterisks).

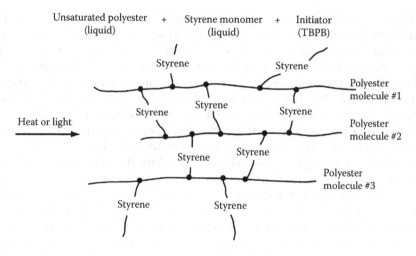

FIGURE 3.10 Schematic representation of a cross-linked polyester resin.

The cure time for polyester resins depends on the decomposition rate of the catalyst, which can be increased by increasing the cure temperature. However, for a given resin-catalyst system, there is an optimum cure temperature at which all the free radicals generated from the catalyst are utilized in the curing reaction. Above this temperature, free radicals are formed so rapidly that wasteful side reactions occur and deterioration of the curing reaction is observed. Below this temperature, the curing reaction is very slow. The decomposition rate of a catalyst can be increased by adding small quantities of an accelerator, such as cobalt naphthanate, which essentially acts as a catalyst for the primary catalyst. A list of ingredients mixed with unsaturated polyester resins and their functions are given in Table 3.9.

TABLE 3.9

Ingredients Mixed with Unsaturated Polyester Resins and Their Functions

Ingredient	Primary Functions	Examples
Reactive diluent	Reduces viscosity of the prepolymer and participates in the curing reaction	Styrene, vinyl toluene, divinyl benzene
Catalyst or initiator	Acts as the reaction initiator	MEKP, cumene hydroperoxide, BP, TBPB
Promoter	Accelerates the curing reaction and/or promotes lower-temperature cure	Cobalt naphthenate, cobalt octoate, demethylaniline, copper naphthenate
Inhibitor	Increases gel time and reduces curing reaction rate	Butyl catechol, hydroquinone, toluhydroquinone, benzoquinone
Filler	Reduces cost and controls mold shrinkage	Calcium carbonate, talc, mica
Low-profile additive	Reduces mold shrinkage and improves surface quality	Polyvinyl acetate, polyethylene, polystyrene

The curing reaction of unsaturated polyester with styrene monomer is basically a free radical chain-growth polymerization. The initiation of the curing reaction starts with the decomposition of the catalyst into free radicals; however, in the beginning stages, most of the free radicals are consumed by the inhibitor if it is present in the resin formulation. Polymerization starts only after the inhibitors are all reacted. After the inhibition stage, free radicals generated by the decomposition of the remaining catalyst react with styrene monomer and polyester molecules, and the cross-linking reaction starts to propagate. Depending on the catalyst and accelerator types and levels as well as the cure temperature, gel times ranging from a few minutes to over an hour can be obtained. However, if an excessive amount of accelerator is used, gelation can start early, which can create problems in resin flow in the mold.

As with epoxy resins, the properties of cured polyester resins depend on the cross-link density. The modulus, glass transition temperature, and thermal stability increase with increasing cross-link density, while strain-at-failure and impact resistance decrease. The major factor controlling the cross-link density is the number of cross-linking sites in the polyester resin. The simplest way to control the number of cross-linking sites is to vary the weight ratio of the unsaturated polyester and the saturated acid. In general, the tensile strength and tensile elongation increase, but the heat deflection temperature decreases with the increasing weight ratio of saturated acid to unsaturated polyester molecules (Figure 3.11). The type of saturated acid used also has an influence on the properties and processing of polyester resins. For example, if adipic acid is used instead of orthophthalic or isophthalic acid, the stiffness of polyester molecules is reduced, since it contains no aromatic rings in its molecule. The type of glycol used in the polymerization reaction also controls the stiffness of polyester molecules. For example, diethylene glycol lowers the stiffness, whereas propylene glycol, by virtue of the pendant methyl groups in its structure, increases the stiffness of polyester molecules.

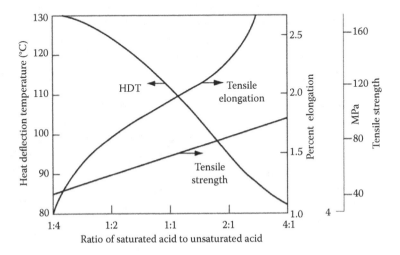

FIGURE 3.11   Effect of saturated acid to unsaturated acid weight ratio on the properties of a cured polyester resin.

Depending on the application, the amount of styrene monomer mixed with the unsaturated polyester resin varies between 30 and 50 phr. The higher the amount of styrene, the lower the viscosity of the polyester–styrene mix and the better the fiber wet out. Styrene also controls the curing characteristics. Since each styrene molecule contributes one unsaturation point (which is equivalent to 9.6 moles per 1000 g of styrene), the total number of unsaturation points in the mix becomes higher as the styrene content is increased. As a result, the higher the amount of styrene, the higher the amount of heat generation. Both peak exotherm temperature and curing time are also increased.

Specifications of a typical polyester resin are given in Table 3.10. It should be noted that the properties of a cured polyester polymer depends on the amount of styrene mixed with the unsaturated polyester resin. The variation in the properties of cured polyester with the increasing molar ratio of styrene to polyester unsaturation is shown in Figure 3.12. If the molar ratio is too high, there may not be enough unsaturation

TABLE 3.10
Specifications of a Typical Polyester Resin[a]

Appearance	Pink
Volatile content	40–43%
Viscosity at 30°C	400–600 mPa s
Density	1.13 g/cm³
Gel time at 30°C	8–10 minutes
Cure time at 30°C	14–25 minutes
Peak temperature	226°C
Flash point	32°C
Shelf life	3 months (min.)

[a] Orthophthalic polyester resin premixed with styrene.

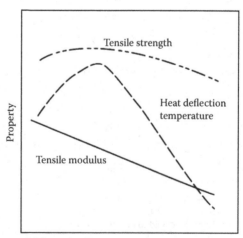

Molar ratio of styrene to polyester unsaturation

FIGURE 3.12   Effect of increasing styrene content on the properties of a cured polyester resin.

points in the polyester molecules where polystyrene molecules can attach. This may promote self-polymerization of the polystyrene molecules, and the resulting material may behave more like polystyrene instead of a properly cured polyester.

3.1.3 Vinyl Esters

Vinyl esters possess some of the good characteristics of epoxies, such as tensile strength and chemical resistance, and some of the good characteristics of polyesters, such as low viscosity and short cure times. However, they have curing shrinkage in the range of 5–10%, which is higher than that of epoxies. Their adhesion characteristic is also not as good as that of epoxies. They are more expensive than polyesters and are usually selected in mass-production applications if performance higher than that of polyesters is needed. Because of their high corrosion resistance, vinyl esters are the preferred matrix in composites used for chemical tanks, pipes, scrubbers, etc.

Vinyl ester molecules are produced by the chemical reaction of an unsaturated carboxylic acid, such as acrylic acid and methacrylic acid, and an epoxy resin (Figure 3.13). The C=C double bonds (unsaturation points) occur only at the ends of a vinyl ester molecule, and therefore, cross-linking can take place only at the ends, as shown schematically in Figure 3.13. Because of fewer cross-links, a cured vinyl ester polymer is more flexible and has a higher fracture toughness than a cured polyester resin which has a larger number of cross-links in its molecule. The presence of ester groups in vinyl ester and polyester molecules makes them susceptible to hydrolysis in the presence of water. However, since vinyl ester molecules contain fewer ester groups than polyester molecules, they have a higher resistance to degradation from water exposure. Another unique characteristic of a vinyl ester molecule is that it contains a number of OH (hydroxyl) groups along its length (Figure 3.14). These OH groups can form hydrogen bonds with similar groups on the glass fiber surface resulting in excellent wet-out and good adhesion with glass fibers.

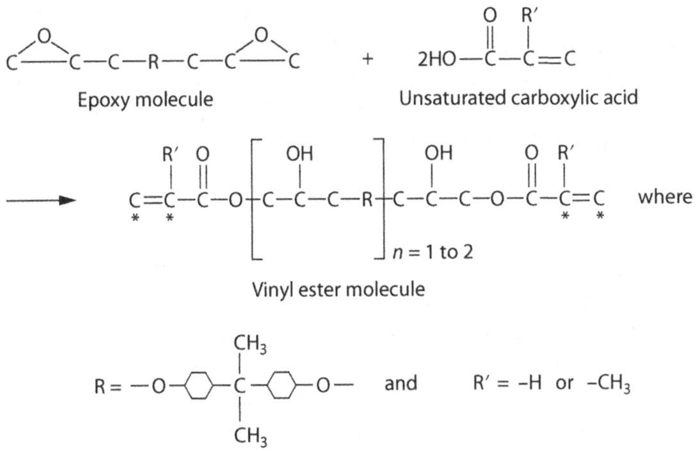

FIGURE 3.13 Vinyl ester molecule (the asterisks denote the cross-linking sites).

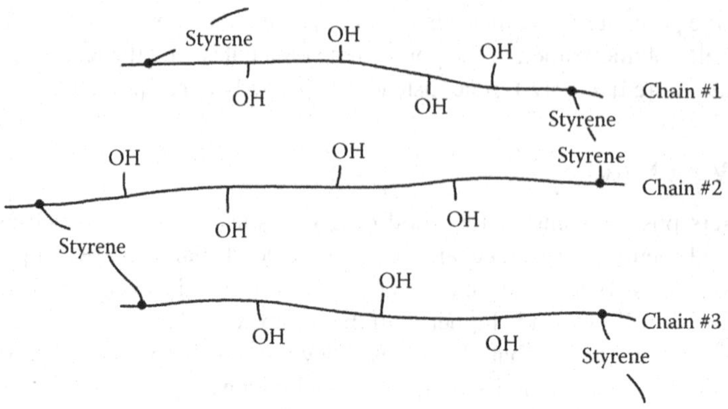

FIGURE 3.14 Schematic representation of a cross-linked vinyl ester resin.

The processing of vinyl ester resins for manufacturing of composite parts is very similar to that of polyester resins. Like polyester resins, vinyl ester resins are also mixed with a reactive diluent, such as styrene monomer, and various other ingredients (see Table 3.8). Curing of a vinyl ester resin also requires the addition of a catalyst, such as *ter*-butyl peroxybenzoate (TBPB). The curing reaction of a vinyl ester resin is very similar to that of an unsaturated polyester resin. During polymerization, styrene molecules coreact with the vinyl ester molecules to form cross-links between the unsaturation points in vinyl ester molecules. The variation of properties of a cured vinyl ester with increasing amounts of styrene is similar to that of a cured polyester, which is shown in Figure 3.12.

3.1.4 BISMALEIMIDE

BMI is selected for applications requiring a service temperature of 130–230°C. BMI prepolymers (monomers) are prepared by the reaction of maleic anhydride with a diamine. One commercially available BMI prepolymer has the chemical formula shown in Figure 3.15. It contains two maleimide end groups. The cross-linking reaction takes place with or without the presence of a catalyst at the carbon–carbon (C=C) double bonds in these end groups.

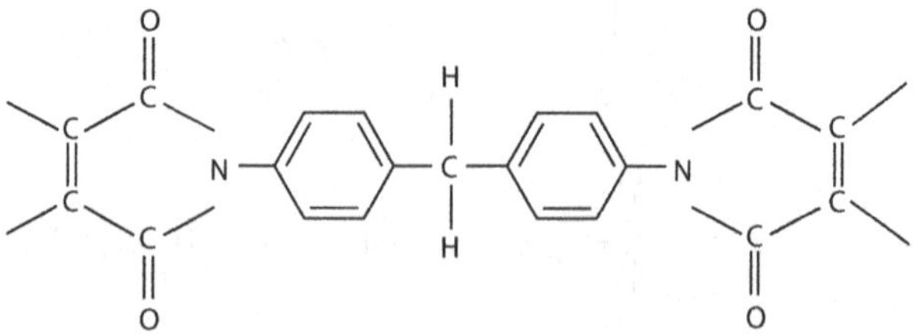

FIGURE 3.15 BMI molecule.

BMI monomers are mixed with reactive diluents to reduce their viscosity and other coreacting monomers, such as vinyl, acrylic, and epoxy, to improve the toughness of cured BMI. The handling and processing techniques for BMI resins are similar to those for epoxy resins. The curing of BMI occurs through addition-type homo-polymerization or copolymerization that can be thermally induced at 170–190°C. In some cases, a catalyst is added to increase the rate of cure. As with epoxies, the cure time depends on cure temperature, and no volatiles are produced during curing. To increase the properties of the BMI polymer, a postcure of 5–10 hours at 210–250°C is recommended.

On curing, BMI polymers not only offer high temperature resistance, but also high chemical and solvent resistance. However, these materials are inherently very brittle due to their densely cross-linked molecular structure. As a result, their com-posites are prone to excessive microcracking. One useful method of reducing their brittleness without affecting their heat resistance is to combine them with one or more tough thermoplastic polyimides. The combination produces a semiinterpen-etrating network polymer, which retains the easy processing characteristics of a ther-moset and exhibits the good toughness of a thermoplastic. Although the reaction time is increased, this helps in broadening the processing window and causes less problems in manufacturing large or complex composite parts.

3.1.5 POLYURETHANES

Polyurethanes are used in a composite manufacturing process called SRIM. In the basic *reaction injection molding* (RIM) process, a mixture of highly reactive chemi-cals is injected into a closed mold cavity at high speeds. The chemicals are mixed just before injecting the liquid mix into the cavity. The chemical reaction starts immedi-ately after mixing, but most of it takes place as the liquid mix flows in the cavity and may continue even after the cavity is filled. The reaction product is a polyurethane.

The two chemicals used in the production of polyurethanes are a diisocyanate and a polyol (Figure 3.16). A glycol or an aromatic diamine is added to serve as a chain

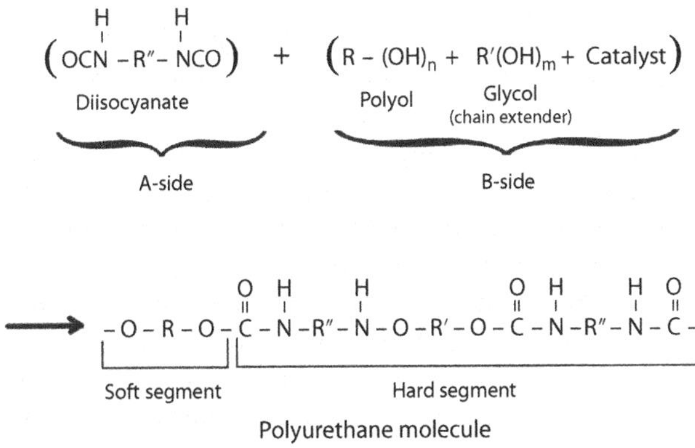

FIGURE 3.16 Chemical reaction of polyol, glycol and diisocyanate molecules to produce a polyurethane molecule.

extender. The reaction that takes place in the mold produces a polymer molecule that consists of alternating hard rigid segments and soft flexible segments, as shown in Figure 3.16. The hard segments have a glass transition temperature higher than 100°C, and the soft segments have a glass transition temperature lower than −30°C. The ratio of soft and hard segments in the polyurethane molecules determines the characteristic properties of polyurethane polymers. As the number of flexible segments in the molecules increases, the elongation at break, impact strength, and resistance to low temperature increase, but the hardness, elastic modulus, scratch resistance, and temperature resistance decrease. Compared to the glycol chain-extended systems, the diamine chain-extended systems have higher modulus and they are more thermally stable due to the presence of aromatic rings in their molecules.

Polyurethanes are available as both thermoplastic and thermosetting polymers. When stoichiometric amounts of diisocyanate, polyol, and bifunctional extender are used, the reaction produces a thermoplastic polyurethane. If excess amounts of diisocyanate or a trifunctional extender is used, the resulting material is a thermosetting polyurethane.

In a typical RIM process, the diisocyanate is called the A-side chemical and a polyol blended with the chain extender and a catalyst is called the B-side chemical. An internal mold release agent is also added to the B-side chemical. The A-side and B-side chemicals are rapidly and thoroughly mixed just before the mixed liquid is injected into the mold cavity.

The SRIM process uses the basic RIM technology to inject the liquid mix of A-side and B-side chemicals into a dry fiber preform. In most SRIM applications, either chopped strand mat or continuous strand mat glass fiber preforms are used. Even though polyurethane is a low modulus, soft material, glass fiber-reinforced polyurethane produced by the SRIM process, it has high enough stiffness for many automotive applications, such as body panels, lift truck beds, and hoods.

3.2   THERMOPLASTIC POLYMERS

Thermoplastic polymers selected for aerospace composite applications are high-temperature polymers, such as PEEK, PPS, PEI, and PAI. These polymers are usually reinforced with continuous carbon fibers or fabrics and are considered alternatives for epoxies in some aerospace applications, since they have higher fracture toughness and crack growth resistance. They also provide excellent heat and chemical resistance and their processing time is much shorter than that of epoxy matrix composites. Thermoplastic polymers selected for automotive composite applications are lower-cost polymers, such as PP and PA-6. These polymers are reinforced with randomly oriented short glass fibers for applications such as instrument panels, inner door panels, and under-the-hood components. Composites containing randomly oriented continuous glass fibers in the PP matrix are used in bumper beams and seat structures. Because of their higher chemical resistance, semicrystalline polymers are preferred over amorphous polymers in many of these applications in both aerospace and automotive industries.

Thermoplastics are supplied by polymer manufacturers in the polymerized form. They are available as pellets (or granules), which are the starting materials for injection molding and extrusion processes. Short fibers, typically 1–3 mm in length, are first mechanically blended with the thermoplastic pellets and then compounded in a single screw extruder or a twin screw extruder. During the compounding process, the pellets melt into a viscous liquid which mixes with the fibers and coats their surfaces. The melt-compounded mix is then extruded through a strand die to form continuous strands or through a sheet die to form a continuous thin sheet. The cooling arrangement outside the extruder transforms the liquid polymer in the extruded material into its solid state. The short fiber-reinforced composite part is produced by either injection molding or thermoforming. For injection molding, the extruded strands are first pelletized to make small pellets, which are then fed into an injection molding machine to produce injection-molded parts. During the injection molding process, the thermoplastic in the pellets goes through a cycle of melting and solidification. For thermoforming, the extruded sheet is cut into required lengths and heated to above the glass transition temperature of the thermoplastic so that it is in a highly softened state. The heat-softened sheet is then placed in a thermoforming machine where vacuum and/or pressure is applied to make the thermoformed part.

The viscosity of liquid thermoplastic polymers during melt processing, such as extrusion and injection molding, is the most important parameter for controlling fiber wet-out and polymer flow in the mold. Viscosity is reduced as the temperature and shear rate are increased. However, too high a temperature may cause thermal degradation of the polymer, and very high shear rates may not be achieved within the practical range of the processing equipment being used. In a melt processing operation, a combination of processing temperature and shear rate is selected to produce the optimum viscosity required for good fiber wet-out at the fiber incorporation stage and good flow in the mold cavity during the mold filling stage. As the liquid polymer is cooled from the processing temperature to room temperature, it starts to transform into a solid polymer. The microstructure of the polymer after solidification depends on the cooling rate. For example, the amount of crystallinity in semicrystalline polymers, such as PEEK and PP, decreases with increasing cooling rate. For these polymers, the tensile strength and modulus increase and the failure strain decreases with increasing amount of crystallinity. Several other properties, such as impact resistance and heat resistance, are also affected by the amount of crystallinity.

Continuous fiber-reinforced thermoplastics are more difficult to produce than short fiber-reinforced thermoplastics, mainly due to the difficulty of wetting the fiber surface with highly viscous thermoplastic liquid. However, several different processes have been developed to produce thin sheets of continuous fiber-reinforced thermoplastics. These processes are discussed in Chapter 10. A few thermoplastic polymer composite sheets with continuous fiber reinforcement are now available in which these processes are utilized.

General characteristics of a few key thermoplastics used in the aerospace and automotive industries are briefly described here. More details can be found in References [2] and [5].

3.2.1 POLYETHER ETHER KETONE (PEEK)

PEEK is a linear aromatic thermoplastic polymer based on the repeating unit in its molecules shown in Figure 3.17. Continuous carbon fiber-reinforced PEEK composites are known in the industry as aromatic polymer composite or APC.

PEEK is a semicrystalline polymer with a maximum achievable crystallinity of 48% when it is slowly cooled from its melt. Amorphous PEEK is produced if the melt is quenched. At normal cooling rates, the crystallinity is between 30% and 35%. The presence of fibers in PEEK composites tends to increase the amount of crystallinity to a higher level, since the fibers act as nucleation sites for crystal formation. Increasing the crystallinity increases both modulus and yield strength of PEEK, but reduces its strain-at-failure (Figure 3.18).

PEEK has a glass transition temperature of 143°C and a crystalline melting temperature between 332°C and 343°C. Melt processing of PEEK requires a temperature range of 360–400°C. The maximum continuous use temperature is 260°C. The outstanding property of PEEK is its high fracture toughness, which is 50–100 times higher than that of epoxies. Another important advantage of PEEK is its low water absorption, which is less than 0.5% at 23°C compared to 4–5% for conventional aerospace epoxies.

PEEK has an exceptional resistance to both heat and chemicals. It retains good mechanical properties even at 200°C for prolonged periods. It does not dissolve

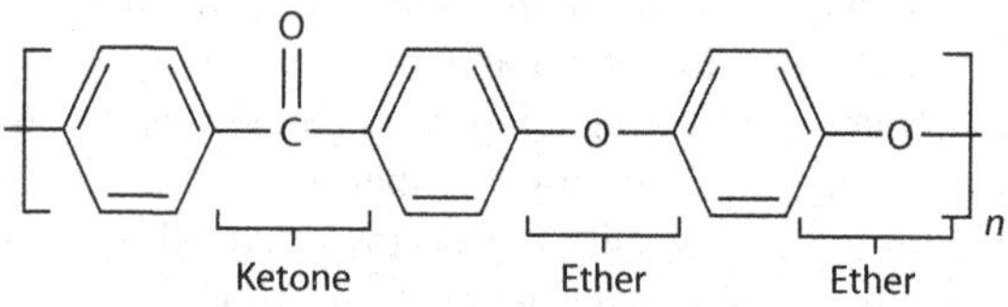

FIGURE 3.17   Repeating unit of a PEEK molecule.

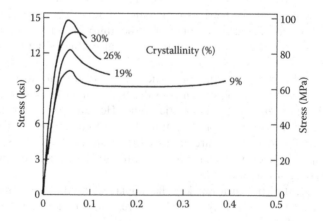

FIGURE 3.18   Effect of crystallinity on the stress–strain diagram of PEEK.

in common solvents. However, it may absorb some of these solvents, most notably methylene chloride. The amount of solvent absorption decreases with increasing crystallinity.

3.2.2 POLYPHENYLENE SULFIDE (PPS)

PPS is a semicrystalline polymer with repeating units containing an aromatic ring and a sulfur atom in its molecules (Figure 3.19). PPS is normally 65% crystalline. It has a glass transition temperature of 85°C and a crystalline melting point of 285°C. Its relatively high crystallinity is attributed to the chain flexibility and structural regularity of its molecules. The melt processing of PPS requires heating the polymer in the temperature range of 300–345°C.

3.2.3 POLYETHER IMIDE (PEI)

PEI is a high-temperature amorphous polymer containing repeating units of ethers and imides (Figure 3.20). It has a glass transition temperature of 217°C. For melt processing, PEI must be heated between 327°C and 413°C. Its tensile strength is among the highest of thermoplastic polymers, and its properties are not much affected over a wide temperature range. Although it is an amorphous polymer, its resistance to chemicals, including most corrosive automotive fluids, is comparable to that of many semicrystalline polymers. It has an exceptionally high flame resistance and oxygen index with very low smoke generation.

3.2.4 POLYAMIDE IMIDE (PAI)

PAI is an amorphous thermoplastic with the repeating unit shown in Figure 3.21. When fully polymerized, its glass transition temperature is 275°C, but its melt processing becomes very difficult. Because of this, PAI is supplied in less than fully

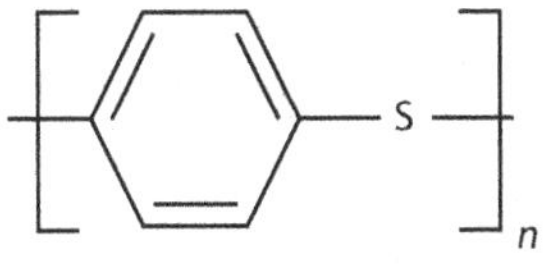

FIGURE 3.19 Repeating unit of a PPS molecule.

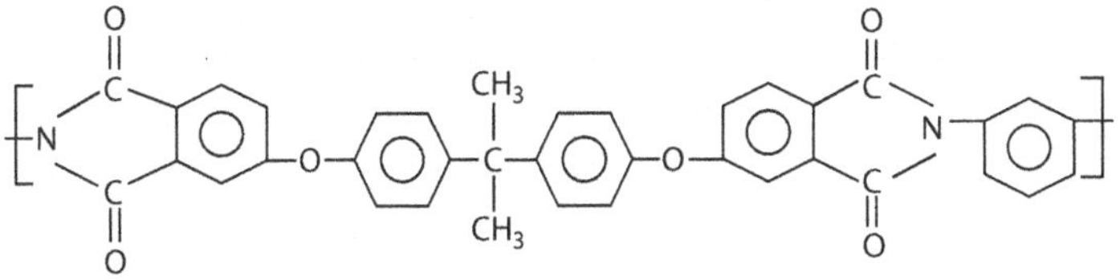

FIGURE 3.20 Repeating unit of a PEI molecule.

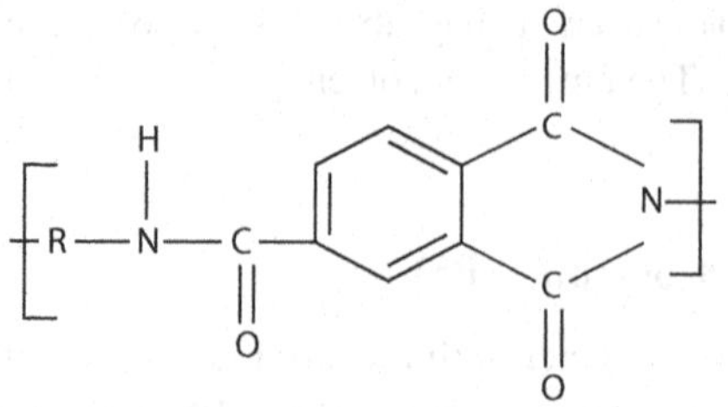

FIGURE 3.21   Repeating unit of a PAI molecule.

polymerized condition. After melt processing at or above 350°C, the material is heated in an oven outside the mold to achieve close to full polymerization.

3.2.5   POLYPROPYLENE (PP)

PP is a semicrystalline thermoplastic with a very simple chemical repeating unit in its molecules, which is shown in Figure 3.22. One of its attributes is its very low density, which is 0.9 g/cm^3. Its glass transition temperature is around −10°C, and melting temperature is between 168°C and 176°C. The degree of crystallinity in PP is in the range of 70–80%. Melt processing requires heating the material to about 260°C. PP becomes brittle and loses its impact strength at temperatures close to −10°C. Ethylene–propylene copolymers are used for applications requiring low-temperature impact performance; however, their modulus and elevated temperature performance are lower than those of PP.

Because of its low cost and excellent chemical resistance, PP and its copolymers are found in many household (such as bottles and food containers), commercial (such as toys), industrial (such as storage boxes, appliance parts, films, and fibers), and automotive (such as fan blades, bumpers, and dashboards) applications. In many of these applications, mineral fillers, such as talc and mica, and short glass fibers are used as reinforcement for the PP matrix.

PP is used as the matrix material for glass mat thermoplastics (GMTs), which is available in sheet form, typically 3.7 mm in thickness. GMT usually contains either randomly oriented chopped E-glass fibers (25–100 mm in length) or randomly oriented continuous E-glass fibers in a swirl pattern. GMT is also available

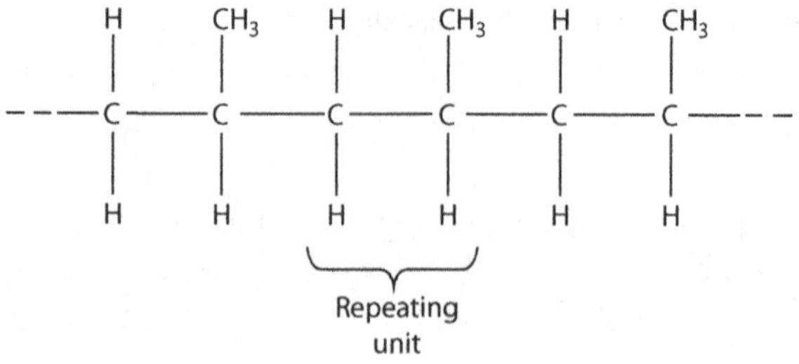

FIGURE 3.22   Repeating unit in a PP molecule.

with unidirectional continuous E-glass fibers as well as bidirectional E-glass fabric. PP is the most commonly used thermoplastic in GMT; however, other thermoplastics are also used. Compression molding is the common manufacturing process for making GMT parts.

In a commercially available fabric called Twintex, PP filaments are combined with E-glass filaments to produce commingled fiber rovings that are then woven into a bidirectional fabric. Both compression molding and thermostamping are used to make composite parts using this fabric. At the temperature and pressure used in these processes, the PP filaments are melted and spread to wet the glass fibers. PP thus becomes the matrix in the composite made from the commingled fiber rovings.

Another application of PP is found in *self-reinforced* PP. In this material, PP filaments are used to reinforce the PP matrix. The propylene molecules are highly oriented in PP filaments, which gives them the high modulus and strength needed to serve as the reinforcement for the matrix. In one of the processes for making self-reinforced PP, a woven fabric of closely spaced PP filaments is heated to a high enough temperature to melt the skin of each filament. On cooling, the melted skin recrystallizes to form the matrix. The density of self-reinforced PP is roughly 30% lower than that of PP GMT. One outstanding property of self-reinforced PP is its notched impact strength, which is nearly three times higher than PP GMT.

3.2.6 POLYAMIDE-6 (PA-6)

PA-6, also called nylon-6, is a semicrystalline thermoplastic polymer used in many industrial, commercial, and automotive applications, such as gears, bearings, and electrical switches. It is one of the many aliphatic polyamides that are characterized by the presence of amide groups in their molecules. The repeating unit of PA-6 is shown in Figure 3.23. Its glass transition temperature is between 50°C and 60°C, and the melting temperature is between 215°C and 230°C. The degree of crystallinity in PA-6 is in the range of 35–45 wt.%. Melt processing of PA-6 requires heating it to a temperature between 250°C and 290°C. The viscosity of liquid PA-6 is relatively low compared to that of many thermoplastics. Like most polyamides, it will absorb moisture from the environment, which will affect its properties and dimensional stability. In terms of mechanical properties, both strength and modulus of PA-6 are reduced, while impact strength and strain-at-failure are increased with increasing moisture absorption. It is recommended that PA-6 be dried before melt processing, such as injection molding; otherwise, the moisture present in the material may cause voids in the molded part.

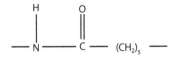

FIGURE 3.23   Repeating unit in a polyamide-6 (PA-6) molecule.

PROBLEMS

1. Calculate the molecular weight of a DGEBA epoxy resin with $n = 2$.
2. Calculate the stoichiometric amount of triethylene tetramine (TETA) to be mixed with a DGEBA epoxy for which $n = 0$. The chemical formula of TETA is given in the following:

$$H_2N\text{--}C_2H_4\text{--}NH\text{--}C_2H_4\text{--}NH\text{--}C_2H_4\text{--}NH_2.$$

3. The pot life of a 100 g mass of an epoxy/hardener combination at 23°C is recorded as 15 minutes. Will the pot life of a 500 g mass of the same epoxy/hardener combination at 23°C be higher, lower, or the same? Explain.
4. TGDDM epoxy has a higher temperature resistance than DGEBA epoxy. Why?
5. What are the factors to be considered when selecting the epoxy resin/hardener combination for a carbon fiber-reinforced epoxy application in the race car industry?
6. A resin mixture contains 200 g of DGEBA epoxy, 80 g of DETA, and 300 g of a solvent.
   a. Calculate the percentage of each ingredient in the mixture.
   b. Calculate the phr of each ingredient in the resin.
   c. Determine if this is a stoichiometric mixture.
7. The repeating unit of PEEK is shown in Figure 3.17. If the average molecular weight of a batch of PEEK polymer is 20,000 g/mol, how many repeating units are present in the molecule?
8. The viscosity η of liquid thermoplastic polymers increases with average molecular weight according to the following equation:

$$\eta = CM_w^{3.4},$$

where C is a constant (which depends on the polymer type) and M_w is the average molecular weight. Compare the viscosities of two PPs, one containing 1000 repeating units and the other containing 5000 repeating units. The repeating unit of PP is shown in Figure 3.22.

REFERENCES

1. D. Ratna, *Handbook of Thermoset Resins*, Smithers, Shawbury, UK, 2009.
2. I. Rubin (ed.), *Handbook of Plastic Materials and Technology*, John Wiley & Sons, New York, 1990.
3. S. V. Muzumdar and L. J. Lee, Chemorheological analysis of unsaturated polyester-styrene copolymerization, *Polymer Engineering and Science*, Vol. 36, No. 7, pp. 943–952, 1996.
4. H. Lee and K. Neville, *Handbook of Epoxy Resins*, McGraw-Hill, New York, 1967.
5. J.-M. Charrier, *Polymeric Materials and Processing*, Hanser, Munich, 1990.

4 Processing Fundamentals

Processing of PMCs involves a combination of complex fluid flow and thermome-chanical phenomena that determines the quality of the composite part and, ulti-mately, its mechanical properties and structural performance. Fluid flow takes place under pressure and determines if the liquid polymer has flowed through each layer of the dry fiber stack and wetted the fibers if LCM is used or has filled the mold cav-ity if compression molding is used. Good fluid flow is required for consolidation of prepreg layers, and removal of air between them, if a bag molding process is used. Process-induced defects, such as voids, uneven distribution of fibers, resin-starved areas, and interlayer cracks, are influenced by fluid flow. Thermal condition deter-mines if the part is sufficiently cured in the case of a thermoset matrix composite or has melted during heating and developed the desired molecular structure during cooling in the case of a thermoplastic matrix composite. Mechanical properties, such as tensile modulus and strength, and thermal properties, such as glass transition temperature and thermal conductivity, depend on the degree of cure in a thermoset matrix composite and the degree of crystallinity in a semicrystalline thermoplastic matrix composite. Several process-induced effects, such as shrinkage, warpage, and residual stresses, also depend on the thermomechanical condition during processing.

The quality of PMC parts can be improved by the proper selection of materials and processing parameters that influence fluid flow and thermomechanical phenomena during their processing. The objective of this chapter is to consider these parameters and examine their influence on the processing, process-induced defects, and post-processing characteristics of both thermoset and thermoplastic matrix composites. It should be noted however that both part design and tool design have major influence on fluid flow and thermomechanical conditions, and their roles should not be overlooked.

4.1 CURE CYCLE FOR THERMOSETTING POLYMERS

The transformation of an uncured or a partially cured thermoset resin into a cured thermoset matrix in a composite part involves a chemical reaction at an elevated temperature for a specific length of time. The chemical reaction is called the curing reaction, during which the polymer molecules are joined by cross-links between them. A high cure temperature is required to initiate and sustain the chemical reac-tion during the transformation process. Pressure is applied to provide the force needed for the flow of the resin or the fiber–resin mixture in the mold and for the consolidation of individual layers into a laminate. The magnitude and duration of these two important process parameters have significant effects on the quality and performance of the cured composite part. The length of time required to properly cure the part is called the cure cycle. It should be noted that the cure cycle depends on a number of factors, including resin chemistry, catalyst reactivity, cure tempera-ture, and presence of inhibitors or accelerators in the resin mix. Since the cure cycle determines the production rate for a part, it is desirable to achieve the proper degree of cure in the shortest amount of time. The proper degree of cure is often selected by balancing the properties of the cured part vs. the production rate and cost rather than just considering the properties that are achievable if there was a full degree of cure.

4.1.1 DEGREE OF CURE

The degree of cure is a measure of the conversion from uncured to cured state. It is defined as the ratio of cross-links formed at any given time to the maximum number of cross-links that can form at the selected cure temperature. It influences physical, thermal, and mechanical properties of a thermoset polymer. Among the physical and thermal properties, density, glass transition temperature, and thermal conductivity increase with increasing degree of cure, whereas heat capacity decreases with increasing degree of cure [1]. Among the mechanical properties, modulus, strength and elongation at failure, and fracture toughness of epoxy resins are all affected by the degree of cure. Not all these properties increase with increasing degree of cure; some have shown a decreasing trend [2,3]. Therefore, it is suggested that the degree of cure should be selected to provide the optimum values of the properties that are needed for a given application.

4.1.1.1 Measurement of Degree of Cure

Several different experimental methods are available for measuring the degree of cure of thermosetting polymers. They include differential scanning calorimetry, Fourier transform infrared spectroscopy, and dielectrometry [4]. The most common among these methods is the differential scanning calorimetry, which has been used by numerous investigators [5–9] to determine the heat evolved in the curing reactions of epoxy, vinyl ester, and polyester resins and to relate them to the degrees of cure during their curing process. Experiments are performed in a differential scanning calorimeter (DSC) in which a small uncured resin sample, weighing only a few milligrams and thoroughly mixed with the curing agent or initiator, is heated either isothermally (i.e., at a constant temperature) or dynamically (i.e., nonisothermally with temperature increasing at a constant rate). The instrumentation in DSC monitors the rate of heat generation as a function of time and records it. Figure 4.1 schematically illustrates the rate of heat generation curves for isothermal and dynamic heating. Both isothermal and dynamic measurements are used to determine the cure characteristics of thermosetting polymers. Isothermal tests are more time consuming than dynamic tests. For isothermal tests, the sample is heated up very quickly to the test temperature, and then the measurements are started. Because of very rapid heating, the initial 10% of the isothermal cure data may not be sufficiently accurate [5].

The total heat generation to complete a curing reaction (i.e., 100% degree of cure) is equal to the area under the rate of heat generation–time curve obtained in a dynamic heating experiment. It is expressed as

$$H_R = \int_0^{t_f} \left(\frac{dQ}{dt} \right) dt, \qquad (4.1)$$

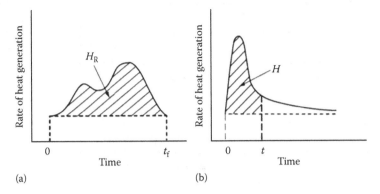

FIGURE 4.1 Schematic representation of the rate of heat generation in (a) dynamic and (b) isothermal heating of a thermoset resin in a DSC.

where H_R is the heat of reaction (unit: J/g); $(dQ/dt)_d$ is the rate of heat generation in a dynamic experiment; and t_f is the time required to complete the reaction.

The amount of heat released in time t at a constant cure temperature T is determined from isothermal experiments. The area under the rate of heat generation–time curve obtained in an isothermal experiment is expressed as

$$H = \int_0^t \left(\frac{dQ}{dt} \right) dt, \tag{4.2}$$

where H is the amount of heat released in time t and $(dQ/dt)_i$ is the rate of heat generation in an isothermal experiment conducted at a constant temperature T.

The degree of cure α_c at any time t is defined as

$$\alpha_c = \frac{H}{H_R}. \tag{4.3}$$

Figure 4.2 shows a number of curves relating the degree of cure α_c to cure time for a vinyl ester resin at various cure temperatures. From this figure, it can be seen that α_c curves have a sigmoidal shape, indicating that α_c slowly increases in the beginning as well as near the completion of the curing reaction. This figure also shows that α_c depends not only on time, but also on cure temperature. As the cure temperature is increased, α_c increases, but if the cure temperature is too low, α_c may not reach a 100% level for any reasonable length of time. The rate of cure $d\alpha_c/dt$, obtained from the slope of α_c vs. t curve, is plotted in Figure 4.3. The rate of cure curve is bell-shaped, and depending on the cure temperature, it exhibits a maximum value at 10–40% of the total cure achieved. Higher cure temperatures increase the rate of cure and produce the maximum degree of cure in shorter periods.

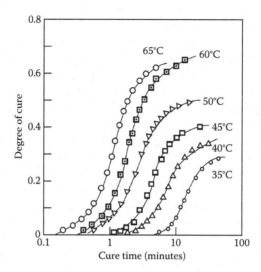

FIGURE 4.2 Degree of cure (α_c) for a vinyl ester resin at various cure temperatures. (C. D. Han and K.-W. Lem: Chemorheology of thermosetting resins: IV. The chemorheology and curing kinetics of vinyl ester resin. *Journal of Applied Polymer Science*, 1984, 29, 1879–1902. Copyright Wiley-VCH Verlag GmbH & Co. KGaA. Reproduced with permission.)

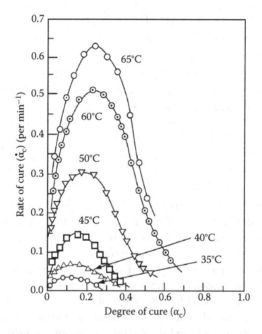

FIGURE 4.3 Rate of cure for a vinyl ester resin at various cure temperatures. (C. D. Han and K.-W. Lem: Chemorheology of thermosetting resins: IV. The chemorheology and curing kinetics of vinyl ester resin. *Journal of Applied Polymer Science*, 1984, 29, 1879–1902. Copyright Wiley-VCH Verlag GmbH & Co. KGaA. Reproduced with permission.)

4.1.1.2 Cure Models

Kamal and Sourour [10] proposed the following empirical equation for the isothermal cure rate of a thermosetting resin.

$$\frac{d\alpha_c}{dt} = \left(k_1 + k_2 \alpha_c^m\right)(1 - \alpha_c)^n, \tag{4.4}$$

where k_1 and k_2 are reaction rate constants and m and n are constants describing the order of reaction. Equation 4.4 assumes an autocatalytic reaction, meaning that the reaction products act as a catalyst in the subsequent stages of reaction. The parameters m and n do not significantly vary with the cure temperature, but k_1 and k_2 strongly depend on the cure temperature. With the assumption of a second-order reaction (i.e., $m + n = 2$), Equation 4.4 has been used to describe the isothermal cure kinetics of epoxy, unsaturated polyester, and vinyl ester resins. The values of k_1, k_2, m, and n are determined by the nonlinear least-squares curve fit to the $d\alpha_c/dt$ vs. α_c data. Typical values of these constants for a number of thermosetting resins are listed in Table 4.1.

The reaction rate constants k_1 and k_2 in Equation 4.4 follow the Arrhenius equation:

$$\begin{aligned}
k_1 &= A_1 \exp\left(-\frac{E_1}{RT}\right), \\
k_2 &= A_2 \exp\left(-\frac{E_2}{RT}\right),
\end{aligned} \tag{4.5}$$

where A_1 and A_2 are constants; E_1 and E_2 are the activation energies (unit: J/mol); T is the absolute value of the cure temperature (unit: K); and R is the universal gas constant, which has a value of 8.314 J/mol K.

TABLE 4.1
Kinetic Parameters for Various Resin Systems

Resin	Temperature (°C)	Kinetic Parameters in Equation 4.4			
		k_1 (min^{-1})	k_2 (min^{-1})	m	n
Polyester	45	0.0131	0.351	0.23	1.77
	60	0.0924	1.57	0.40	1.60
Low-profile polyester (with	45	0.0084	0.144	0.27	1.73
20% polyvinyl acetate)	60	0.0264	0.282	0.27	1.73
Vinyl ester	45	0.0073	0.219	0.33	1.76
	60	0.0624	1.59	0.49	1.51

Source: K.-W. Lem and C. D. Han: Thermokinetics of unsaturated polyester and vinyl ester resins. *Polymer Engineering and Science*, 1984, 24, 175–184. Copyright Wiley-VCH Verlag GmbH & Co. KGaA. Reproduced with permission.

Equation 4.4 indicates that when $\alpha_c = 0$, the cure rate is equal to k_1. It can also be derived from Equation 4.4 that if $k_1 = 0$, the maximum cure rate occurs when the degree of cure becomes equal to $m/(m + n)$. In general, the curing reaction of many thermosetting resins is autocatalytic in nature. However, the curing reaction becomes more diffusion controlled when vitrification begins at the later stages of the reaction. As a result, the experimental values of the degree of cure and the cure rate near the completion of the curing reaction do not match with the values predicted from Equation 4.4. For diffusion-controlled reactions, Chern and Poehlein [11] proposed the following modification to Equation 4.4:

$$\frac{d\alpha_c}{dt} = f(\alpha_c)\left(k_1 + k_2\alpha_c^m\right)(1-\alpha_c)^n, \tag{4.6}$$

where $f(\alpha_c)$ is called the diffusion factor and is given by

$$f(\alpha_c) = \frac{1}{1+\exp\left[C(\alpha_c - \alpha_{cr})\right]}. \tag{4.7}$$

In Equation 4.7, C and α_{cr} are temperature-dependent constants. α_{cr} is called the critical reaction rate at which diffusion becomes the controlling factor. Note that for $\alpha_c \ll \alpha_{cr}$, $f(\alpha_c)$ approaches unity and the effect of diffusion is negligible. As α_c approaches α_{cr}, $f(\alpha_c)$ decreases and approaches zero, which indicates that the reaction has effectively ended.

Example 4.1

The degree of cure of a vinyl ester resin used in a resin transfer-molded part is given by the following:

$$\alpha_c = \frac{kt}{1+kt}$$

where $k = Ae^{-\frac{E}{RT}}$, t is the cure time in minutes, and T is cure temperature in Kelvins. A and E have been determined as $A = 1.25 \times 10^6$ min^{-1} and $E = 41{,}570$ J/mol.

Assuming that the presence of fibers does not influence the cure kinetics, determine the temperature which should be used to achieve 80% cure in 2 minute cure time.

Solution:

Step 1: Since $\alpha_c = 0.80$ and $t = 2$ minutes, calculate k from the degree of cure equation.

$$\alpha_c = 0.80 = \frac{k.(2)}{1+k.(2)}$$

which gives k as 2 min^{-1}.

Step 2: Use $k = Ae^{-\frac{E}{RT}}$ to calculate T.

$$k = 2 = 1.25 \times 10^6 e^{-\frac{41,570}{(8.314)T}}$$

This gives $T = 374.7$ K $= 101.7°C$.

4.1.1.3 Factors Affecting Cure Cycle

In addition to time and temperature, there are several other factors that affect the degree of cure and, therefore, the cure cycle [4]. For example, in the case of epoxies, the type of hardener (but not the amount) influences the cure rate. Promoters or accelerators are often added to many commercially available polyester and vinyl ester resins. Although the final degree of cure is not affected by their presence, the cure rate is increased.

Fillers are used with thermosetting resins to reduce their curing shrinkage and increase their modulus and hardness after curing. It is observed that the filler concentration or type does not affect the reaction exponents m and n, but influences the reaction rate through the Arrhenius rate constants k_1 and k_2 in Equation 4.4.

The reaction rate is also affected by the chemical composition of the reactants. For example, for an unsaturated polyester/styrene system, the rate as well as the composition of the polymer network are highly affected by the molar ratio of styrene and polyester C=C bonds.

Many composite manufacturing processes use moderate to high pressure for compaction and consolidation. Increasing the pressure tends to increase the cure rate as well as the total heat of reaction in both epoxy and polyester resins [12,13], but above a certain pressure level, the curing reaction slows down with increasing pressure.

4.1.2 Gel Time

Gelation corresponds to the incipient formation of an infinite molecular network. As the gel time approaches, there is a rapid increase in the formation of cross-links, which shows up as a rapid increase in the viscosity of the curing resin. The resin flow becomes increasingly difficult as the resin starts to transform from a liquid state to a gelatin state. The resin continues to cure and harden after it has started to gel until it becomes fully cured. It is generally recommended that the liquid resin flow needed for producing good-quality parts be complete before the gel time is reached.

The gel time of a resin–catalyst combination is determined by the gel time test. In this test, a measured amount (typically 10 g) of a thoroughly mixed resin–catalyst combination is poured into a standard test tube. The temperature rise in the material is monitored as a function of time by means of a thermocouple while the test tube is suspended in a water bath maintained at 82°C.

A typical temperature–time curve (also known as exotherm curve) obtained in a gel time test is illustrated in Figure 4.4. On this curve, point A indicates the time required for the resin–catalyst mixture to attain the water bath temperature. The beginning of temperature rise indicates the initiation of the curing reaction. As the curing reaction begins, the liquid mix begins to transform into a gel-like mass. Heat generated by the exothermic curing reaction increases the mix temperature, which in turn causes the catalyst to decompose at a faster rate and the reaction to proceed

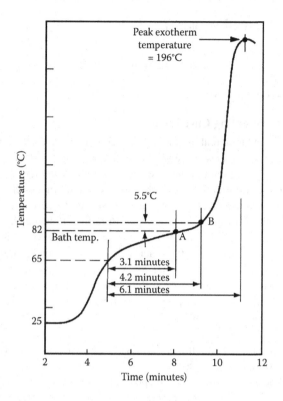

FIGURE 4.4   Typical temperature–time curve obtained in a gel time test.

at a progressively increasing speed. Since the rate of heat generation is higher than the rate of heat loss to the surrounding medium, the temperature rapidly rises to a high value. As the curing reaction nears completion, the rate of heat generation is reduced and a decrease in temperature follows. The exothermic peak temperature observed in a gel time test is a function of the resin chemistry and the resin–hardener or resin–catalyst ratio. The slope of the exotherm curve is a measure of cure rate, which primarily depends on the hardener or catalyst reactivity.

Shortly after the curing reaction begins at point A, the resin viscosity increases very rapidly owing to the increasing number of cross-links formed by the curing reaction. The time at which a rapid increase in viscosity ensues is called the gel time and is indicated by point B in Figure 4.4. According to one standard, the time at which the exotherm temperature increases by 5.5°C above the bath temperature is considered the gel time. It is sometimes measured by probing the surface of the reacting mass with a clean wooden applicator stick every 15 s until the reacting material no longer adheres to the end of a clean stick.

4.1.3   CURE CYCLE MONITORING

While the DSC measurement for the degree of cure can be used at the resin development or characterization stage and for resin quality control, real-time cure cycle monitoring during the actual molding operation can be done using a dielectric monitoring technique. In dielectric analysis or simply dielectrometry, the loss factor of the curing resin is measured using two electrodes placed on the top and bottom surfaces of the mold or on the top and bottom surfaces of the part being molded. The probes are connected to an alternating electric field, and the resulting sinusoidal current is measured. Since the resin is a dielectric material, the combination of the probes and fiber–resin system between them forms a parallel-plate capacitor. The charge accumulated in this capacitor depends on the ability of the dipoles and ions in the resin molecules to follow the electric field at different stages of curing. The loss factor of the resin represents the energy expended in aligning the dipoles with the electric field and moving the ions toward the electrode of opposite polarity. It increases with increasing degree of dipole and ion mobility.

Figure 4.5 shows a typical loss factor diagram as a function of time during the molding of an epoxy prepreg. In the beginning of the cure cycle, the resin viscosity in the prepreg is relatively high so that the dipole alignment and ion movement are restricted. This results in a low loss factor. As the temperature of the prepreg increases with time, the resin viscosity decreases and the loss factor increases due to greater dipole and ion mobility. When the gel point is reached and vitrification begins, the resin viscosity starts to rapidly increase and the loss factor starts to decrease. A peak in loss factor is observed at the gel point. At the full degree of cure, the loss factor levels off to a nearly constant value. The loss factor depends on the cure temperature. Both peak loss factor and time to reach the peak decrease with increasing cure temperature [14].

The loss factor depends on the frequency of the electric field application. The higher the frequency, the lower the loss factor. At any given frequency, it increases with cure time, reaches a peak, and then reduces to a constant value. A frequency-independent parameter, called ion viscosity, can be derived from the loss factor data and used to monitor the cure phenomenon. It initially decreases with time, shows a minimum, and then increases (Figure 4.5). At the end of cure, the slope of the ion viscosity curve reaches a plateau indicating a significant decrease in molecular

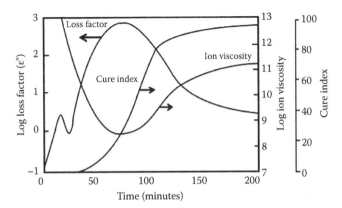

FIGURE 4.5   Dielectric loss factor, ion viscosity, and cure index as a function of time.

mobility and the end of the curing reaction. In general, it correlates well with the viscosity development in the curing polymer. For optimum sensitivity, a frequency of 1 kHz is recommended for dielectric measurements [15].

In dielectric monitoring used for automatic process control, the electrodes are placed on the mold surfaces in the thickest parallel section of the part, and ionic viscosity is recorded as a function of time. In-mold dielectric monitoring of parts as they are being molded can not only provide information on cure advancement, but also be used to automatically open the mold at the end of the cure cycle or adjust the process parameters to detect part-to-part variation in curing.

4.2 VISCOSITY

The viscosity of a fluid is a measure of its resistance to flow under shear stresses. Low-molecular weight fluids, such as water and motor oil, have low viscosities and readily flow. High-molecular weight fluids, such as liquid polymers, have high viscosities and require high shear stresses for flow.

The two most important factors determining the viscosity of a fluid are the temperature and shear rate. For all fluids, the viscosity decreases with increasing temperature. In general, the shear rate does not have any influence on the viscosity of very low-molecular weight fluids, whereas it tends to either increase or decrease the viscosity of high-molecular weight fluids. The viscosity as a function of shear rate for these three types of fluids is shown in Figure 4.6. The fluid with a constant viscosity regardless of the shear rate is called a Newtonian fluid. The other two types are non-Newtonian fluids—the one with decreasing viscosity with increasing shear rate is shear thinning, and the one with increasing viscosity with increasing shear rate is shear thickening. The range of shear rates used for polymer processing is $10\text{–}10{,}000\ s^{-1}$, and in this range, liquid polymers behave as shear-thinning fluids as their viscosity decreases with increasing shear rate.

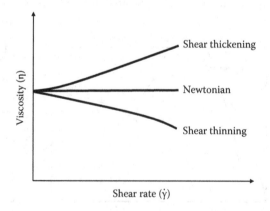

FIGURE 4.6 Viscosity vs. shear rate curves for Newtonian, shear thinning, and shear thickening fluids.

4.2.1 Thermoplastic Polymers

Figure 4.7 shows the variation in viscosity η of a liquid thermoplastic polymer with increasing shear rate $\dot{\gamma}$ at several different temperatures. It can be seen in this figure that viscosity at very low shear rates does not vary much with shear rate, and the polymer behaves like a Newtonian fluid. The near-constant viscosity at very low shear rates is called the zero-shear viscosity and is denoted by η_o. Viscosity starts to decrease as the shear rate is increased, which is the typical shear thinning behavior. At very high shear rates, the polymer starts to behave again as a Newtonian fluid, as its viscosity becomes nearly a constant. The effect of increasing temperature on the viscosity is also evident in Figure 4.7.

The viscosity η of a liquid thermoplastic polymer can be represented by the following relationship.

$$\eta = \eta(\dot{\gamma}, T), \tag{4.8}$$

where $\dot{\gamma}$ and T are the shear rate and temperature, respectively, during processing. The simplest model to represent the viscosity–shear rate relationship of liquid thermoplastic polymers in the shear thinning range is the power law equation [16]:

$$\eta = K\dot{\gamma}^{(n-1)}, \tag{4.9}$$

where η is the viscosity (unit: Pa s); $\dot{\gamma}$ is the shear rate (unit: s^{-1}); K is the constant (independent of shear rate, but is a function of temperature); and n is the power law index.

In the power law equation, K is a function of temperature, but n does not depend on temperature. Note that $n = 1$ for a Newtonian fluid and $0 < n < 1$ for a shear-thinning fluid.

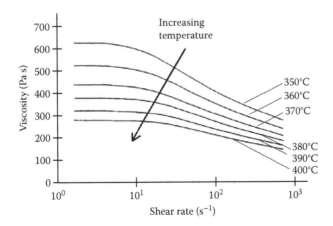

FIGURE 4.7 Viscosity of a liquid thermoplastic polymer (PEEK in this figure) as a function of shear rate at several different temperatures.

There are several other viscosity models that are used to correlate the viscosity of liquid thermoplastic polymers with shear rate [16]. One of these models, known as the Carreu model, is as follows:

$$\eta = \eta_o \left[1 + (\lambda \dot{\gamma})^2 \right]^{(n-1)/2}, \tag{4.10}$$

where η_o is the zero-shear viscosity (see Figure 4.8) and λ is the material constant (unit: seconds).

The effect of temperature on the viscosity of a thermoplastic is often represented by the Arrhenius equation:

$$\eta = A \exp\left(\frac{E}{RT}\right), \tag{4.11}$$

where η is the viscosity at temperature T, E is the activation energy, and R is universal gas constant.

Note that T in Equation 4.11 is the absolute temperature in kelvins, and the activation energy E has the unit of joules per mole. Another equation for the effect of temperature on viscosity is known as the WLF equation, named after Williams, Landel, and Ferry, who first proposed this equation for time-dependent viscoelastic characterization of polymers. The WLF equation is as follows:

$$\log_{10} \frac{\eta}{\eta_r} = \frac{-C_1(T - T_r)}{C_2 + (T - T_r)}, \tag{4.12}$$

where η is the viscosity at temperature T, η_r is the viscosity at a reference temperature T_r, and C_1 and C_2 are constants that depend on the reference temperature. If T_g is used

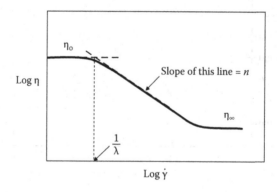

FIGURE 4.8 Variation of viscosity of a liquid thermoplastic polymer as a function of shear rate.

TABLE 4.2

Factors Affecting the Viscosity of Liquid Thermoplastic Polymers

Factor	Effect on Viscosity
Temperature	Viscosity decreases with increasing temperature.
Shear rate	Viscosity decreases with increasing shear rate.
Pressure	Viscosity increases with increasing pressure.
Average molecular weight	Viscosity increases with increasing average molecular weight.
Molecular weight distribution	Viscosity increases with decreasing molecular weight distribution.
Addition of fillers and short fibers	Viscosity increases with increasing fillers and short fibers.
Addition of flow improvers	Viscosity decreases with increasing flow improvers.

as the reference temperature, then $C_1 = 17.44$ and $C_2 = 51.6$ for a temperature range of $T_g < T \leq T_g + 100°C$.

It is important to note that the viscosity of liquid thermoplastic polymers depends on a number of other factors. Among them are (1) molecular weight, (2) molecular weight distribution, (3) presence of fillers and short fibers, and (4) presence of flow improvers. The effects of these parameters are listed in Table 4.2.

4.2.2 THERMOSETTING POLYMERS

The viscosity of a thermosetting polymer is influenced not only by temperature and shear rate, but also by the degree of cross-linking, which is a function of both time and temperature. The starting material for a thermosetting polymer is a low-viscosity fluid. However, its viscosity increases with curing and approaches infinity as it transforms into a solid mass. The variation in viscosity during isothermal curing of an epoxy resin is shown in Figure 4.9 [17]. Similar viscosity–time curves are also observed for polyester [18] and vinyl ester resins [9]. In all cases, viscosity increases with increasing cure time and temperature. The rate of viscosity increase is low at the early stages of curing. After a threshold degree of cure is achieved, the resin viscosity increases at a very rapid rate. The time at which this occurs is called the *gel time*. The gel time is an important molding parameter, since the flow of resin in the mold becomes increasingly difficult at the end of this period.

The viscosity η of a thermosetting polymer during the curing process is a function of cure temperature T, shear rate $\dot{\gamma}$, and the degree of cure α_c.

$$\eta = \eta (T, \dot{\gamma}, \alpha_c). \tag{4.13}$$

It is important to note that the viscosity function for thermosets is significantly different from that for thermoplastics. Since no chemical reaction occurs during the processing of a thermoplastic polymer, its viscosity depends only on temperature and shear rate. There are no satisfactory viscosity models for thermosetting polymers that correlate all three parameters that appear in Equation 4.13.

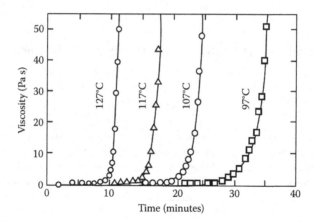

FIGURE 4.9 Variation of viscosity of an epoxy resin as a function of time at different isothermal cure temperatures. (M. R. Kamal: Thermoset characterization for moldability analysis. *Polymer Engineering and Science*, 1974, 14, 231–239. Copyright Wiley-VCH Verlag GmbH & Co. KGaA. Reproduced with permission.)

A number of important observations can be made from the viscosity of thermosetting polymers reported in the literature:

1. Studies on the shear rate effect on the viscosity of thermosetting polymers show that polyesters and polyurethanes essentially exhibit a Newtonian behavior, i.e., their viscosity does not vary with shear rate. Epoxies, on the other hand, are highly shear thinning, i.e., their viscosity decreases with increasing shear rate.
2. A B-staged epoxy or a thickened polyester resin has a much higher viscosity than the neat resin at all stages of curing.
3. The addition of fillers, such as $CaCO_3$, to the neat resin increases its viscosity as well as the rate of viscosity increase during curing. On the other hand, the addition of thermoplastic additives (such as thermoplastic powder as low-profile additives to polyester and vinyl ester resins) tends to reduce the rate of viscosity increase during curing.
4. The increase in viscosity with cure time is less if the shear rate is increased. This phenomenon is more pronounced in B-staged or thickened resins than in neat resins. Fillers and thermoplastic additives also tend to increase the shear rate effect.
5. At a constant shear rate and for the same degree of cure, η vs. $1/T$ plot is linear (Figure 4.10). This suggests that the viscous flow of a thermoset polymer is an energy-activated process. Thus, its viscosity as a function of temperature can be written as

$$\eta = \eta_\infty \exp\left(\frac{E}{RT}\right), \tag{4.14}$$

where η is the viscosity (Pa s), E is the flow activation energy (cal/g mol), R is the universal gas constant, T is the cure temperature (K), and η_∞ is a constant.

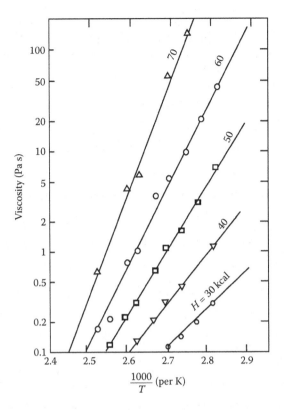

FIGURE 4.10 Viscosity–temperature relationships for an epoxy resin at different levels of cure (represented by the heat of reaction H_R). (M. R. Kamal: Thermoset characterization for moldability analysis. *Polymer Engineering and Science*, 1974, 14, 231–239. Copyright Wiley-VCH Verlag GmbH & Co. KGaA. Reproduced with permission.)

It should be noted that the activation energy for viscous flow increases with increasing degree of cure and approaches a very high value near the gel point.

A modified form of viscosity equation takes into account the degree of cure [6] and is given as

$$\eta = \eta_o \exp\left(\frac{E}{RT} + K_o \alpha_c\right), \tag{4.15}$$

where α_c is the degree of cure and K_o is a constant which is independent of temperature. Equation 4.15 can be written in the following form:

$$\ln \eta = A + K_o \alpha_c, \tag{4.16}$$

where $A = \ln \eta_0 + \dfrac{E}{RT}$.

To determine η_0, E, and K_0, experiments are run in which the viscosity η and degree of cure α_c are measured at different cure temperatures [6]. Linear least-square fits are first made between $\ln \eta$ and α data to determine A and K_0, and then between A and $1/T$ to determine η_0 and E.

Roller [19] developed the following expression for the viscosity of a curing B-staged epoxy resin under isothermal conditions.

$$\ln \eta = \ln \eta_\infty + \frac{E_1}{RT} + tk_\infty \exp\left(-\frac{E_2}{RT}\right), \quad (4.17)$$

where t is the time; E_1 and E_2 is the activation energies; k_∞ is a kinetic parameter that depends on the resin type, curing agent concentration, etc.; and η_∞ is the constant.

Equation 4.17 applies when curing takes place under isothermal conditions. For curing under nonisothermal conditions, Equation 4.17 is modified to write

$$\ln \eta = \ln \eta_\infty + \frac{E_1}{RT} + \int_0^t k_\infty \exp\left(-\frac{E_2}{RT}\right) dt. \quad (4.18)$$

4.2.3 VISCOSITY MEASUREMENT

The test methods available for measuring the viscosity of both thermoplastic and thermosetting polymers fall into two categories [16]: (1) steady shear in which viscosity is measured as a function of steady shear rate over a range of temperatures and (2) dynamic shear in which complex viscosity is measured as a function of frequency at a constant temperature. Steady shear tests use either a capillary rheometer, parallel plate rheometer, or cone-and-plate rheometer. Dynamic shear tests, also known as isothermal dynamic frequency sweeps, use either a dynamic mechanical analyzer or a torsion pendulum. The measurements in dynamic shear tests are made in the linear viscoelastic region in which the viscosity response is strain independent.

4.3 FIBER SURFACE WETTING

In order for the fibers and the matrix to interactively work and produce high mechanical and thermal properties for a composite, each filament in the fiber bundle (roving, tow, or yarn) must be wetted by the matrix, and a good bond must be formed at the interface between the two. Typically, strong bonds are needed for improved transverse tensile strength and interlaminar shear strength; on the other hand, very strong bonds tend to reduce impact strength and fracture toughness.

One of the requirements for good filament surface wetting is low viscosity of the liquid polymer so that it can flow inside the fiber bundle and uniformly spread on the filament surfaces. Good resin flow is also required to displace air that is entrapped in the interstices between the filaments. Another important requirement for both good wetting and bonding is that the filament surfaces have higher surface energy than the matrix. This can be explained by considering the spreading of a liquid on a solid surface in the presence of a vapor, which, in our case, will be the entrapped air.

Consider, for example, a liquid drop placed on a solid surface as shown in Figure 4.11. At each corner of the liquid drop, three phases are present—a liquid, a solid, and a vapor—each with its own surface energy per unit area, γ_{LS}, γ_{SV}, and γ_{LV}. For equilibrium among the three phases, the following equation must be satisfied:

$$\gamma_{LS} \cdot dA + \gamma_{LV} \cos\varphi \cdot dA = \gamma_{SV} \cdot dA,$$

which gives

$$\cos\varphi = \frac{\gamma_{SV} - \gamma_{LS}}{\gamma_{LV}}. \tag{4.19}$$

In Equation 4.19, φ is called the contact angle. The unit of surface energies is joules per square meter (J/m^2).

The work of adhesion, defined as the energy required to reversibly separate the liquid from the solid, is given by the following equation:

$$W = \gamma_{SV} + \gamma_{LV} - \gamma_{LS} = \gamma_{LV}(1 + \cos\varphi). \tag{4.20}$$

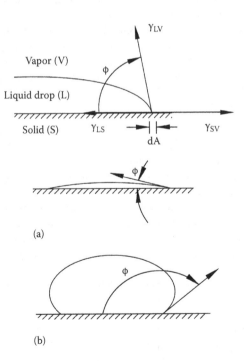

FIGURE 4.11 Wetting of a solid surface by a liquid drop (top): (a) good wetting and (b) poor wetting.

TABLE 4.3

Surface Energies of Fibers and Polymers

Material	Surface Energy (mJ/m²)
Glass fiber (fresh)	500–600
Carbon fiber	40–70
Kevlar fiber	44
Epoxy	43
Polyester	35
PP	30

For complete spreading on the surface of the solid, $\varphi = 0$, and as per Equation 4.19, $\gamma_{SV} = \gamma_{LV} + \gamma_{LS}$. Although complete spreading may not achieved in all cases, it is important to note from Figure 4.11 and Equation 4.19 that for spontaneous spreading and good wetting, φ should be as small as possible, which requires that $\gamma_{SV} \gg \gamma_{LV}$. In other words, the surface energy of the solid must be higher than that of the liquid.

In general, polymers have a lower surface energy than fibers (Table 4.3), and therefore, it is expected that polymers will spontaneously spread on the fiber surface, and there will be excellent wetting of the fibers by the liquid polymer. However, if the fiber surface is contaminated or contains moisture, its surface energy may be reduced and the wettability may be affected.

Although the surface energy is important for wetting the fiber surface, it may not be sufficient for good bonding between the fibers and the matrix. To improve fiber–matrix bonding, the fiber surface is modified to form chemical bonds with the matrix. For glass fibers, sizing is applied on the glass filament surfaces during their production. Although the primary purpose of the sizing application is to improve their handling characteristics, the sizing chemicals are selected to improve wetting and adhesion with the polymer matrix. To make the bonding even stronger, the glass fiber surface is often modified using chemical surface treatments, such as a silane coupling agent. It is reported that both sizing application and silane coupling agent improve surface energy of glass fibers [20,21]. Carbon fiber surfaces are also treated with sizing chemicals to improve their handling. Further modification of the carbon fiber surface for improved wettability and bonding is achieved by a variety of oxidation processes (see Chapter 1). Both sizing application and surface modification tend to increase the fiber surface energy and, therefore, wettability of carbon fibers in polymers [22]. Carbon fibers are often heat treated to improve their modulus. It has been shown that the heat treatment of carbon fibers reduces the surface energy [23,24] and, therefore, are not desirable if wettability is a problem.

4.4   RESIN FLOW

Proper resin flow through a dry fiber network in LCM or a stack of prepreg layers in bag molding is critical in producing void-free parts and good fiber wet-out. In thermoset resins, curing may take place simultaneously with resin flow, and if the resin viscosity increases too rapidly due to curing, its flow may be inhibited or even stopped, causing voids, poor interlayer adhesion, nonuniform resin distribution, and resin-starved areas. In LCM, restricted resin flow through dry fiber networks can create areas that are completely dry (i.e., without any resin).

4.4.1 Permeability

Resin flow through fiber networks has been modeled using Darcy's equation [25], which was derived for flow of Newtonian fluids through a porous medium. This equation relates the volumetric resin flow rate q per unit area to the pressure gradient that causes the flow to occur. For one-dimensional flow in the x-direction,

$$q = -\frac{K}{\eta}\left(\frac{dp}{dx}\right),$$
(4.21)

where q is the volumetric flow rate per unit area in the x-direction (unit: m/s); K is the permeability (unit: m²); η is the viscosity; and dp/dx is the pressure gradient (unit: N/m²/m), which is negative in the direction of flow (assumed here to be the positive x-direction).

In the context of PMCs, *permeability* is a measure of the ease with which a liquid polymer can flow through the pores in a fiber network and is determined by the following equation known as the Kozney–Carman equation:

$$K = \frac{d_f^2}{16k}\frac{(1-v_f)^3}{v_f^2},$$
(4.22)

where d_f is the fiber diameter, v_f is the fiber volume fraction, and k is the Kozney constant.

Equations 4.21 and 4.22 have been used by many investigators in modeling resin flow through prepregs in the bag molding process and during mold filling in LCM. Equation 4.21 assumes that the porous medium is isotropic, and both pore size and pore distribution are uniform. However, fiber networks are nonisotropic, and therefore, the Kozney constant k is not the same in all directions. For example, for a fiber network with unidirectional fiber orientation, the Kozney constant in the transverse direction (k_{22}) is an order of magnitude higher than the Kozney constant in the longitudinal direction (k_{11}). This means that the resin flow in the transverse direction (Figure 4.12) is much slower than that in the longitudinal direction. Furthermore, the fiber packing in a fiber network is not uniform, which also affects the Kozney constant and, therefore, the resin flow.

Equation 4.22 works well for predicting resin flow in the fiber direction. However, it is not valid for resin flow in the transverse direction, since, according to this equation, resin flow between the fibers does not stop even when the fiber volume fraction reaches the maximum value at which the fibers touch each other, and there are no gaps between them. Gebart [26] derived the following permeability equations in the

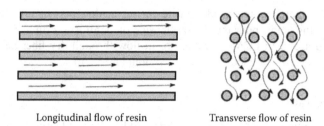

| Longitudinal flow of resin | Transverse flow of resin |

FIGURE 4.12   Resin flow in the longitudinal and transverse directions in a unidirectional fiber network.

fiber direction and normal to the fiber direction for unidirectional continuous fiber networks with regularly arranged, parallel fibers.

In the fiber direction:

$$K_{11} = \frac{2d_f^2}{C_1} \frac{\left(1 - v_f^3\right)}{v_f^2}. \tag{4.23a}$$

Normal to the fiber direction:

$$K_{22} = C_2 \left(\sqrt{\frac{v_{f,max}}{v_f}} - 1\right)^{5/2} \frac{d_f^2}{4}, \tag{4.23b}$$

where C_1 is the hydraulic radius between the fibers, C_2 is a constant, and $v_{f,max}$ is the maximum fiber volume fraction (i.e., fiber volume fraction at the maximum fiber packing).

The parameters C_1, C_2, and $v_{f,max}$ depend on the fiber arrangement in the network. For a square arrangement of fibers, $C_1 = 57$, $C_2 = 0.4$, and $v_{f,max} = 0.785$. For a hexagonal arrangement of fibers, $C_1 = 53$, $C_2 = 0.231$, and $v_{f,max} = 0.906$. Note that Equation 4.23a for resin flow parallel to the fiber direction has a similar form as the Kozney–Carman equation (Equation 4.22). According to Equation 4.23b, which is applicable for resin flow transverse to the flow direction, $K_{22} = 0$ at $v_f = v_{f,max}$, which indicates that the transverse resin flow will stop at the maximum fiber volume fraction.

It should be noted that the permeability equations assume that the fiber distribution is uniform, the gaps between the fibers are the same throughout the network, the fibers are perfectly aligned, and all fibers in the network have the same diameter. These assumptions are not valid in practice, and therefore, the permeability predictions using Equation 4.22 or 4.23 can be considered only approximate.

In general, permeability is a directional property. If the fiber network is orthotropic in which flow directions coincide with the principal directions, there will be three different permeability values, K_{11}, K_{22}, and K_{33}. For an isotropic fiber network, in which flow takes place equally in all directions, $K_{11} = K_{22} = K_{33} = K$. For a 2D fiber network, K_{11} and K_{22} are the in-plane permeabilities and K_{33} is the through-thickness permeability. If the in-plane principal permeabilities K_{11} and K_{22} for an orthotropic fiber network are known, the in-plane permeabilities in the x- and y-directions can be calculated using the following equations:

$$K_{xx} = K_{11} \cos^2 \theta + K_{22} \sin^2 \theta = \frac{K_{11} + K_{22}}{2} + \frac{K_{11} - K_{22}}{2} \cos 2\theta,$$

$$K_{yy} = K_{11} \sin^2 \theta + K_{22} \cos^2 \theta = \frac{K_{11} + K_{22}}{2} - \frac{K_{11} - K_{22}}{2} \sin 2\theta, \qquad (4.24)$$

$$K_{xy} = K_{yx} = - (K_{11} - K_{22}) \sin \theta \cos \theta,$$

where θ is the angle between the 1 and x-directions (Figure 4.13). For a unidirectional fiber preform, $\theta = 0°$ and $K_{xx} = K_{11}$, $K_{yy} = K_{22}$, and $K_{xy} = K_{yx} = 0$.

4.4.2 Measurement of Permeability

The two most commonly used experiments for measuring permeability of resin flow through a dry fiber network, such as a fabric and a random fiber mat, are the unidirectional flow experiment and the radial flow experiment (Figure 4.14). Both experiments are conducted in a flat rectangular mold with a transparent top cover of glass or acrylic so that the flow front can be viewed and video-recorded as the liquid resin flows into the dry fiber network specimen. It is important that the specimen dimensions are large compared to the fiber bundle size, the thickness of the layers, and the dimensions of the repeat pattern in the fabric so that a volume averaged flow behavior can be ensured [28]. The resin can be injected into the mold either under a constant pressure or at a constant flow rate. Pressure and

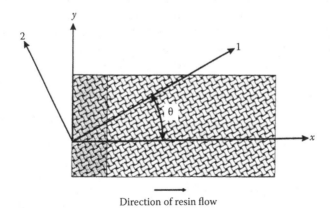

FIGURE 4.13 Resin flow in an orthotropic fiber network. (From S. G. Advani and E. M. Sozer, *Process Modeling in Composites Manufacturing*, CRC Press, Boca Raton, FL, 2000.)

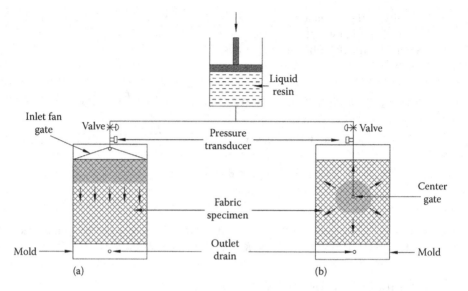

FIGURE 4.14 Permeability measurement using (a) unidirectional flow and (b) radial flow experiments.

flow sensors are placed at several locations in the mold for gathering information on pressure and flow rate.

1. Unidirectional flow experiments involve injecting the liquid resin through an inlet gate located at one end of the mold. The flow front in this case propagates in the length direction of the mold. If the injection is done under a constant pressure, the flow front will not advance at a constant rate. If injection is done at a constant flow rate, the flow front will linearly increase with time.

2. Radial flow experiments involve injecting the liquid resin through a central inlet gate. The flow front in this case propagates in a circular shape if the dry fiber network is isotropic or quasi-isotropic and in an elliptic shape if it is orthotropic. The advantage of radial flow experiments is that two components of the permeability tensor can be determined in one experiment. In-plane permeability values K_{11} and K_{22} can be calculated from the shape of the elliptic flow front and the directions of its major and minor axes.

Both permeability measurement experiments can be done under unsaturated or transient and saturated or steady-state conditions [29]. In the unsaturated flow experiments, measurements are made as the liquid resin is injected into the dry fiber network during which the instantaneous pressure varies with both distance and time. In the saturated flow experiments, the dry fiber network is completely saturated with the liquid resin, and a steady state has reached between the pressure drop and flow rate before the measurements are started.

The following equations are used to calculate the permeability K in unidirectional flow experiments under constant pressure [29]:

1. Unsaturated flow:

$$t = \frac{1}{2}\frac{\eta}{K}\frac{(1-v_f)}{\Delta P}x_f^2.$$

(4.25a)

2. Saturated flow:

$$\frac{Q}{A} = \frac{K}{\eta}\frac{\Delta P}{L},$$

(4.25b)

where K is the permeability; Q is the volumetric flow rate (unit: m^3/s); A is the cross-sectional area of the fiber network sample; L is the length of the fiber network specimen; t is the time from the start of filling; v_f is the fiber volume fraction; ΔP is the pressure drop across the length of flow; and η is the viscosity of the liquid resin.

The pressure at the mold inlet is measured using two pressure sensors positioned on the back face of the mold. The pressure at the outlet of the mold is assumed to be 1 atm. The flow rate is measured using a differential pressure cell as the resin enters the mold. Experiments are often conducted with fluids that behave as a Newtonian fluid. For such fluids, a linear relationship is obtained between the steady-state flow rate Q/A and pressure drop ΔP, which is shown in Figure 4.15. Note that Q/A in this figure represents the average velocity of fluid flow. Using Equation 4.25b, an effective value of K is determined from the slope of Q/A and ΔP.

Permeability measurements are done in both saturated and unsaturated conditions [31]. Permeability measured in a saturated condition is higher than permeability measured in an unsaturated condition. Typically, a difference of 20–30% is observed between these two measurements. It should be noted that during unsaturated flow, the resin flow through the fiber network is caused by a combination of viscous and capillary forces. At low flow rates, the capillary forces are much larger than the viscous forces, which causes the resin to fill the interbundle spaces by capillary action instead of filling the gaps between the bundles. At high flow rates, viscous forces are much higher and large gaps are filled first.

The accuracy and reproducibility of permeability measurements depend on a number of factors [28,29]. One of these factors is the edge effect or racetracking effect, which is caused by the flow of resin in the narrow gaps between the edges of the fiber network and the mold wall. Depending on the size of the gap relative to the width of the mold, the velocity of the resin flow in the gap will be much higher than the velocity of resin flow through the fiber network. Even though the pressure drop vs. flow rate may show a linear relationship, the effective permeability will be higher if there is an edge effect. In general, for the same gap length, wider molds are less sensitive to the edge effect than narrower molds.

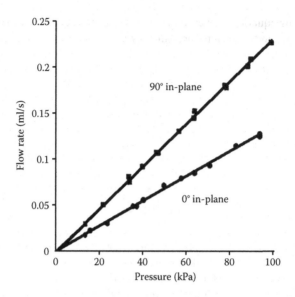

FIGURE 4.15 Resin flow rate vs. pressure drop in a saturated (steady-state) flow measurement. (R. S. Parnas, K. M. Flynn, and M. E. Dal-Favero: A permeability database for composites manufacturing. *Polymer Composites*, 1997, 18, 623–633. Copyright Wiley-VCH Verlag GmbH & Co. KGaA. Reproduced with permission.)

4.4.3 PERMEABILITY OF FIBER NETWORKS

The permeability of a fiber network is a directional property. This can be observed in Figure 4.16, which shows that permeability is an order of magnitude lower in the thickness direction than that in the plane of the fabric. Even in the plane of the fabric, there is a difference, albeit small in this case, between the 0° and 90° directions. Also to be noted in Figure 4.16 is that a linear relationship exists between the flow rate and pressure drop for steady-state flow of a Newtonian liquid.

The effective permeability of fiber networks depends on the fiber architecture, fiber volume fraction, and pore size. Figure 4.17 shows effective permeability values of a number of different fiber networks. It can be observed in this figure that random fiber mats have a higher permeability than woven fabrics. Among the random fiber mats, chopped strand mats have a higher permeability than continuous strand mats. The permeability of woven fabrics depends on their construction. For example, the woven fabric with the highest level of fiber undulation gives the largest permeability. Figure 4.18 shows the effect of fiber volume fraction on the permeability of a plain-woven fabric. The permeability decreases with increasing fiber volume fraction and decreasing porosity of the fiber network. Other factors that may influence permeability are fiber sizing, resin viscosity, flow rate, compaction pressure, and capillary pressure. Many permeability studies reported in the literature were conducted with fluids other than the liquid resin. An example of such a fluid is silicone oil. The nature of the fluid has also an effect on permeability.

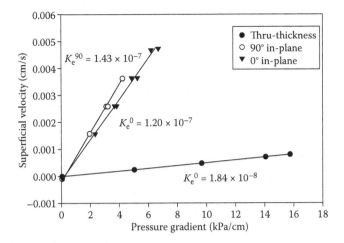

FIGURE 4.16 Effective permeability (K_e) values (in cm^2) of nonwoven ±45° fabric in the 0°, 90°, and through-thickness directions (data obtained in steady-state unidirectional flow experiments using a Newtonian fluid). (From R. S. Parnas, *Liquid Composite Molding*, Hanser, Munich, 2000.)

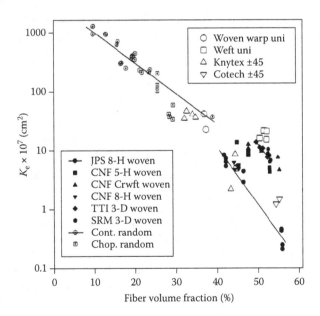

FIGURE 4.17 Effective permeability of different fiber networks. (From R. S. Parnas, *Liquid Composite Molding*, Hanser, Munich, 2000.)

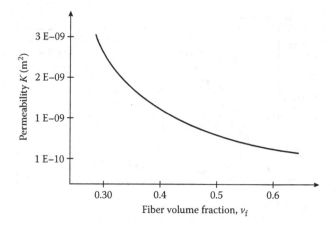

FIGURE 4.18 Effect of fiber volume fraction on the permeability of a plain woven fabric. (From S. G. Advani and E. M. Sozer, *Process Modeling in Composites Manufacturing*, CRC Press, Boca Raton, FL, 2000.)

4.4.4 CAPILLARY FLOW

Resin flow inside the fiber bundles mostly occurs by capillary action. The reason for this is that the flow channels between the filaments in fiber bundles are extremely narrow. They are typically of the order of a few microns. The flow in these micro-channels takes place by surface tension. For one-dimensional resin flow, an estimate of the capillary pressure can be obtained from the modified Young–Laplace equation [32] given as follows:

$$P_{\mathrm{c}} = \frac{F}{d_{\mathrm{f}}}\frac{(1-\varepsilon)}{\varepsilon}\,\sigma\cos\varphi, \tag{4.26}$$

where P_{c} is the capillary pressure; F is the form factor which depends on the fiber alignment and the flow direction; d_{f} is the filament diameter; ε is the porosity; σ is the surface tension of the wetting liquid; and φ is contact angle between the filament surface and the liquid.

For unidirectional fiber orientation, the form factor F is equal to 4 for flow along the fiber direction (longitudinal flow) and 2 for flow normal to the fiber direction (transverse flow). For other fiber orientations or complex fiber arrangements, F is determined by experiments [33].

It is important to note that at low injection pressures or low flow rates and high fiber volume fractions, the capillary flow becomes the major driving force for resin flow. Due to this, faster resin flow occurs in the narrow channels within fiber bundles than in the larger channels between the fiber bundles. While this results in the good wet-out of the filaments within the fiber bundles, air may be entrapped between the slower moving flow fronts between the fiber bundles, causing macrovoid formations in the interbundle areas. At high injection pressures or high flow rates, flow in the larger channels between the fiber bundles is faster than that inside the fiber bundles, and therefore, microvoids are formed within the fiber bundles.

4.5   CONSOLIDATION

The consolidation of layers in a dry fiber network or a prepreg layup requires good resin flow and compaction; otherwise, the resulting composite laminate may contain a variety of defects, including voids, interply cracks, resin-rich areas, and resin-starved areas. Good resin flow by itself is not sufficient to produce good consolidation.

Both resin flow and compaction require the application of pressure during processing in a direction normal to the dry fiber network or prepreg layup. The pressure is applied to squeeze out the trapped air or volatiles as the liquid resin flows through the dry fiber network or prepreg layup, suppress voids, and attain uniform fiber volume fraction. Initially, the fibers in the fiber network may not carry any load and the pressure in the resin is equal to the applied pressure. As the applied pressure is increased, the layers in the fiber network come closer, and eventually, multiple fiber-to-fiber contacts are established, and fibers now start to carry a portion of the applied pressure. Gutowski et al. [34] developed a model for consolidation in which it is assumed that the applied pressure is shared by the fiber network and the resin so that

$$p = \sigma_e + \bar{p}_r, \tag{4.27}$$

where p is the applied pressure; σ_e is the average effective stress on the fiber network; and $\bar{p}_r$ is the average pressure on the resin.

Assuming that the fiber segment is slightly curved and applying an elastic beam bending model between the contact points, Cai and Gutowski [35] derived the following equation for the effective stress on the fiber network as a function of fiber volume fraction:

$$\sigma_e = A \left[\frac{1 - \sqrt{\dfrac{v_f}{v_o}}}{\left(\sqrt{\dfrac{v_a}{v_f}} - 1 \right)^4} \right], \tag{4.28}$$

where A is a constant; v_o is the initial fiber volume fraction in the fiber network (before compaction); v_f is the fiber volume fraction at any instant during compaction; and v_a is the maximum possible fiber volume fraction (at the end of compaction). The constant A in Equation 4.28 is a measure of the deformability of the fiber network. Gutowski et al. [36] gives the following expression for A:

$$A = \frac{3\pi E}{\beta^4}, \tag{4.29}$$

where E is the fiber modulus and β is the fiber waviness given by the ratio of the length and height of the fiber in the fiber network (Figure 4.19). The maximum volume fraction v_a depends on a number of factors, such as the type of fiber network, stacking sequence, rate and time of compaction pressure application, and the state of lubrication.

In Equation 4.28, $\sigma_e = 0$ at $v_f = v_o$, i.e., at the initial fiber volume fraction. As the fiber volume fraction v_f increases with increasing compaction, Equation 4.28 predicts that the average effective stress σ_e on the fibers also increases and the fiber network begins to take up an increasing amount of the applied pressure. On the other hand, since the average pressure on the resin decreases with increasing compaction, the possibility of void formation increases.

A number of experiments conducted by applying compressive load on stacks of plane woven fabrics in both dry and resin-impregnated conditions show the following compaction behavior.

1. The fiber volume fraction v_f nonlinearly increases with increasing compaction pressure p, which is shown in Figure 4.20. The relationship between v_f and p can be represented by the following power law equation [37]:

$$v_f = v_{fo}p^b, \tag{4.30}$$

where v_{fo} is the fiber volume fraction in the stack at p equal to 1 Pa and β is called the stiffening index. Both are determined by fitting Equation 4.30 to the experimental data.

Experiments conducted with 3, 6, and 12 layers of NCFs, woven fabrics, and random fiber mats showed the same general trends of v_{fo} increasing and β decreasing with the number of layers in the stack. Fiber volume fraction initially increases rapidly with compaction pressure, but tends to reach

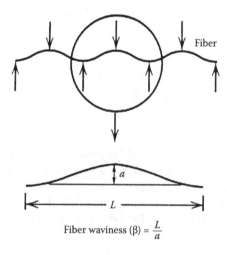

Fiber waviness $(\beta) = \dfrac{L}{a}$

FIGURE 4.19 Fiber unit cell with periodic undulation. (From Z. Cai and T. Gutowski, *Journal of Composite Materials*, 26, 1207–1237, 1992.)

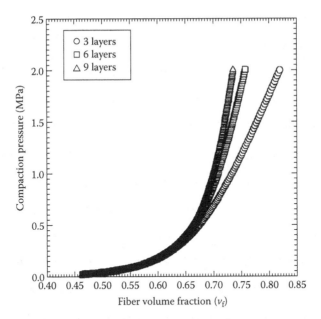

FIGURE 4.20 Fiber volume fraction as a function of compaction pressure. (F. Robitaille and R. Gauvin: Compaction of textile reinforcements for composites manufacturing. I: Review of experimental results. *Polymer Composites*, 1998, 19, 198–216. Copyright Wiley-VCH Verlag GmbH & Co. KGaA. Reproduced with permission.)

a plateau at high pressures. The plateau defines the practical limit of the maximum fiber volume fraction.

Another form of the power law equation [38] expressing v_f in terms of p is given as follows:

$$p = cv_f^m, \tag{4.31}$$

where both c and m are constants determined by fitting Equation 4.31 to the experimental data.

2. Another approach to representing the compaction behavior of fiber networks is to plot the thickness of the fiber network stack as a function of compaction pressure [39]. Typical thickness–pressure curves for a single layer of a woven fabric and a stack of 10 layers of the same woven fabric are shown in Figure 4.21. It shows that at any given compaction pressure, the thickness per layer of a multilayer stack is smaller than the thickness of a single layer.

The general trend of the effect of compaction pressure on stack thickness is shown in Figure 4.22. Three different regimes of deformation can be observed in Figure 4.22: (1) an initial linear regime at low compaction pressures, which is attributed to the compression of gaps and pores in the fiber network, fiber slippage, and rearrangement of fibers into nests; (2) a nonlinear regime at intermediate compaction pressures, which is attributed to nest formations; and (3) a final linear regime

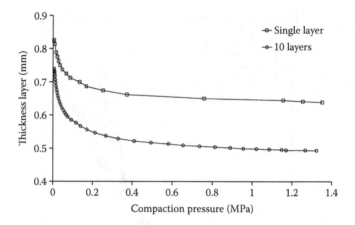

FIGURE 4.21 Effect of compaction pressure on the thickness per layer of a single layer of woven fabric and a stack of 10 layers of woven fabric. (Reprinted from *Composites Science and Technology*, 60, B. Chen and T.-W. Chou, Compaction of woven-fabric preforms: Nesting and multilayer deformation, 2223–2231, 2000, with permission from Elsevier.)

at high compaction pressures, which is attributed to the elastic deformation of the fiber cross-section and changes in the cross-sectional shape of the fiber bundle. Since prepregs are partially consolidated sheets containing B-staged resin-impregnated fibers, a stack of prepreg layers will undergo much lower reduction in thickness than a stack of dry fiber reinforcement. They will therefore require much higher pressure for consolidation. Their consolidation behavior is also highly dependent on the temperature and pressure application rates.

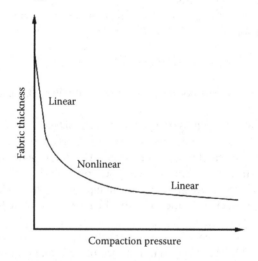

FIGURE 4.22 Typical compaction pressure vs. fabric stack thickness. (Reprinted from *Composites Science and Technology*, 60, B. Chen and T.-W. Chou, Compaction of woven-fabric preforms: Nesting and multilayer deformation, 2223–2231, Copyright (2000), with permission from Elsevier.)

The compaction of fiber stacks during mold closing in an LCM process flattens the fiber bundles and changes their shapes, reduces the intrabundle and interbundle spaces (Figure 4.23), and causes elastic deformation of the fibers. In stacks of woven, fabrics, nesting of fiber bundles and interlayer packing also occur (Figure 4.24). All these factors have a large effect on the permeability that controls resin flow through the stack and on the final microstructure of the molded composite part.

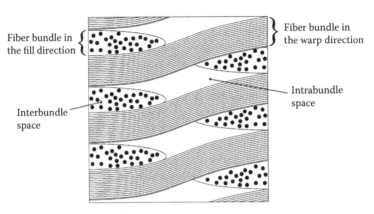

FIGURE 4.23 Intrabundle and interbundle spaces in a woven fabric.

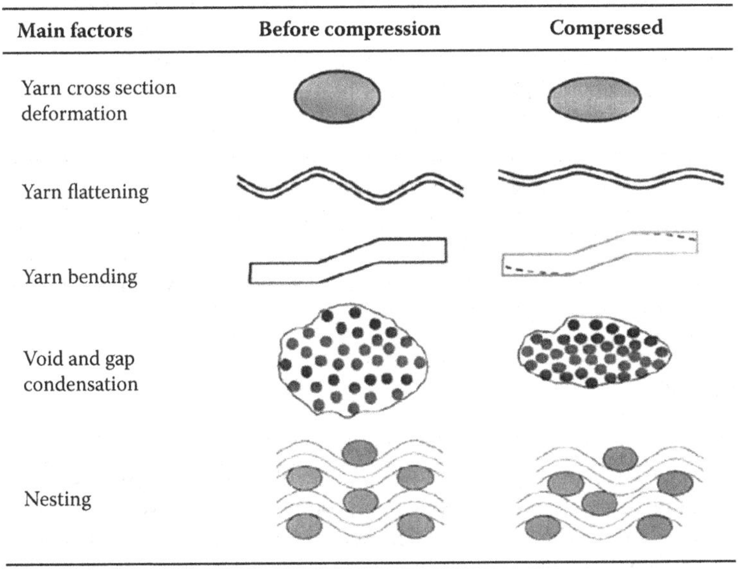

Main factors	Before compression	Compressed
Yarn cross section deformation		
Yarn flattening		
Yarn bending		
Void and gap condensation		
Nesting		

FIGURE 4.24 Factors affecting the compaction behavior of a fabric stack. (Reprinted from *Materials Science and Engineering*, A317, B. Chen, E. J. Lang, and T.-W. Chou, Experimental and theoretical studies of fabric compaction behavior in resin transfer molding, 188–196, Copyright (2001), with permission from Elsevier.)

4.6 SHRINKAGE

Shrinkage in thermosetting polymers occurs due to (1) curing shrinkage or the reduction in volume caused by curing and (2) thermal shrinkage or volumetric contraction caused by decrease in temperature from the cure temperature to room temperature. Curing shrinkage occurs because of the rearrangement of polymer molecules into a more compact mass as the curing reaction proceeds and increases with increasing degree of cure. The thermal shrinkage occurs during the cooling period that follows the curing process and takes place both inside and outside the mold until the temperature of the cured composite part reaches the room temperature.

The volumetric shrinkage of cast epoxy resins is on the order of 1–5%. For polyester and vinyl ester resins, volumetric shrinkage may range from 5% to 12%. The addition of fibers or fillers reduces the volumetric shrinkage of a resin. In the case of PMCs containing unidirectional fibers, the shrinkage in the transverse direction of fibers is higher than in the longitudinal direction. The difference in shrinkages in the longitudinal and transverse directions can give rise to residual stresses and warpage. The high shrinkage of polyesters and vinyl esters helps in releasing the part from the mold surface; however, it can contribute to dimensional variation and several molding defects, such as warpage and sink marks.

High shrinkage in polyester or vinyl ester resins can be significantly reduced by the addition of low-shrink additives (also called low profile agents), which are powders of thermoplastic polymers, such as polyethylene, polymethyl acrylate, polyvinyl acetate, and polycaprolactone. These thermoplastic powders are usually mixed in a styrene monomer during blending with the liquid resin. They do not take part in the curing reaction and reduce the shrinkage by forming a dispersed second phase in the cured resin.

Among the thermoplastic polymers used as matrix in composite parts, semicrystalline polymers, such as PEEK, exhibit a higher volumetric shrinkage than amorphous polymers, such as PEI. In both types of thermoplastics, thermal shrinkage is caused by volumetric contraction as the temperature reduces from the processing temperature to room temperature. Semicrystalline polymers undergo an additional shrinkage that occurs due to the formation of the crystalline phase. Depending on the polymer type and cooling rate, the crystallization shrinkage can be an order of magnitude higher than the thermal shrinkage. Also, the crystallization shrinkage increases with increasing degree of crystallinity.

4.7 VOIDS

Voids are formed during molding (1) due to entrapment of air and volatile gases in the resin and (2) by void nucleation on the fiber surface. Among the various manufacturing process-induced defects in composite parts, voids are considered the most critical defect influencing their mechanical properties. Voids do not affect the longitudinal tensile strength of unidirectional composites much, but many other properties, such as longitudinal compressive strength, transverse tensile strength, flexural strength, and shear strength, are reduced in the presence of voids. In general, depending on the void volume fraction, clustering, and size, they become the areas of high stress concentrations and reduce the tensile, compressive, and flexural strengths of the molded composite part. Large reductions in matrix-dominated properties, such as transverse tensile strength and interlaminar shear strength, occur with increasing void content. An example is shown in Figure 4.25 where a 50% reduction in interlaminar shear strength can be observed with void volume fraction ranging between 2% and 3%. The presence of voids generally increases the rate and amount of moisture absorption in a humid environment, which may cause reduction in matrix-dominated properties. In some composite parts, moisture absorption may result in swelling and changes in part dimensions.

The most common cause for void formation is the inability of the resin to displace air from the fiber surface during the time fibers are pulled through or coated with the liquid resin. The speed of pulling or coating rate, resin viscosity, and mechanical manipulation of fibers in the liquid resin affect air entrapment at the fiber–resin interface. Relative values of fiber and resin surface energies also play an important role, since they control the wettability of the fiber surface by the liquid resin. Surface tension is also an important parameter that controls void formation and transport. If the liquid resin has high surface tension, it becomes more difficult to eliminate voids formed within the fiber bundles.

Voids may also be caused by air bubbles, moisture, and volatiles entrapped in the liquid resin. Solvents used for resin viscosity control and chemical contaminants in the resin often remain dissolved in the resin mix and volatilize during elevated temperature curing. In addition, air pockets at ply interfaces, ply gaps and

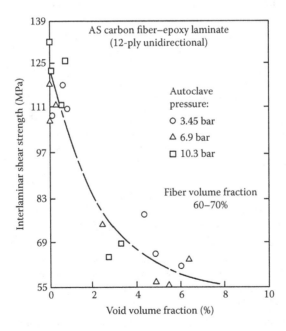

FIGURE 4.25 Effect of void volume fraction on the interlaminar shear strength of a composite laminate (as determined by short beam shear tests). (From M. J. Yokota, *SAMPE Journal*, 11, 1978.)

ply terminations, wrinkles created during layup, and draping operations can create voids.

Air and gas bubbles formed during the mixing of resin with the other ingredients, such as curing agent with epoxy and styrene with polyester, can also be sources of voids unless they are removed by (1) degassing, (2) applying vacuum during the molding process, and (3) allowing the resin mix to freely flow in the mold, which helps in carrying the air and volatiles out through the vents in the mold.

In addition, voids may be nucleated on the fiber surface, mostly from absorbed moisture in the resin and fiber surfaces. The growth of these voids may occur by diffusion of water vapor or air and agglomeration of nearby voids. They may dissolve if their solubility in the resin increases at the processing temperature and pressure is used.

4.8   RESIDUAL STRESSES AND DISTORTIONS

Residual stresses are internal stresses generated in composite parts as they are being cooled from the processing temperature to room temperature. They are present in composite parts even before any external load is applied on the parts. They can be either tensile or compressive and nonuniformly distributed. When combined with stresses due to the applied external loads, they can adversely affect the mechanical properties of the material, influence the failure modes of the composite part, and cause failure at lower than the design loads. Distortions and dimensional variations are observed in many composite parts after demolding and can cause difficulty in assembling them with other composite or metal parts.

The major sources of residual stresses are the thermal mismatch due to differential thermal contractions within a ply and between the plies in a composite laminate [42]. Most composite manufacturing processes involve cooling the material from high processing temperature to room temperature. For thermoset matrix composites, the processing temperature is the cure temperature, whereas for thermoplastic matrix composites, the processing temperature is the melt temperature or the forming temperature. Since the coefficient of thermal contraction of the polymer matrix is much higher than that of the fibers, a 3D residual stress pattern is created in the fibers as well as the matrix due to differential thermal contraction (Figure 4.26). In the longitudinal direction, residual stresses in the fibers are compressive and residual stresses in the matrix are tensile. In the hoop direction, tensile residual stresses are created in the matrix as it shrinks around the fibers. In the radial direction, residual stresses in the matrix are compressive if the fibers are close together, but are tensile if they are separated by large distances. The tensile residual stresses in the matrix can be large enough to initiate microcracks in the matrix.

Another source of residual stresses is the thermal mismatch due to the difference in thermal contractions of adjacent plies in a laminate. For example, consider a [0/90] s laminate being cooled from a cure temperature of 150 to 23°C. If the plies were not consolidated and joined, each 0° ply will contract much less in the x-direction than the adjacent 90° ply, while the reverse is true in the y-direction. However, since the plies must contract by the same amount as they cool down after curing, internal residual stresses are created to maintain the geometric compatibility between the

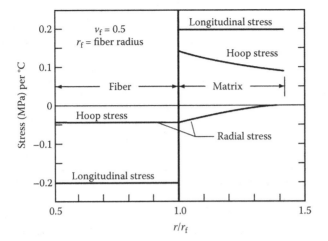

FIGURE 4.26 Residual stresses in fibers and matrix due to differential thermal contraction upon cooling from the processing temperature.

plies. As a result, residual compressive stresses are created in the fiber direction and residual tensile stresses are created transverse to the fiber direction. Thus, when such a laminate is loaded in tension in the x-direction, residual tensile stress added to the applied tensile stress in the 90° ply can initiate transverse cracks in the 90° plies in the laminate.

There are several other sources that can generate residual stresses in composite parts. Examples of these sources are ply-to-ply variations in fiber volume fraction, nonuniform distribution of fibers, fiber movement during consolidation, differential curing, and cooling rate variation through the thickness [42].

Distortions are changes in shape from the desired shape and may be observed in composite parts when pressure in the mold is released or when it is taken out of the mold. Some of the major causes of distortion are described in the following.

1. Asymmetric layup: If the construction of the laminate is not symmetric with respect to the midplane, then the laminate will either bend, twist, or both bend and twist after cool down. For example, a [0/90] laminate will bend into a circular arc (Figure 4.27) and a [45/−45] laminate will be twisted on cool down from the cure temperature. This is one of the reasons symmetric layups are preferred for most composite applications.

2. Mold surface–part surface interaction: Thin flat composite parts often form a convex curvature as they are released from the mold (tool). The reason for such distortion is the frictional interaction between the mold surface and the composite part surface [43]. If the mold material has a higher coefficient of expansion than the composite, the mold will tend to stretch the composite part when both are heating up. The amount of stretching is the highest on the composite part surface and is lower toward the interior. As a result, a stress gradient is created in the thickness direction of the laminate. When

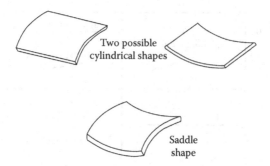

FIGURE 4.27 Different distortion modes observed in asymmetric [0/90] laminates after cooling from the processing temperature.

the stresses are released on demolding, the laminate will tend to bend as shown in Figure 4.28.

3. Spring forward: Higher through-thickness contraction compared to in-plane contraction during cooling from the processing temperature causes spring forward of curved composite parts (Figure 4.29). The higher through-thickness contraction occurs due to a higher coefficient of thermal contraction and a higher curing shrinkage in the thickness direction compared to the in-plane values. This causes the curved section to become thinner, while the arc length does not decrease much. As a result, the sector angle in the curved section increases, leading to an increase in the included angle. The

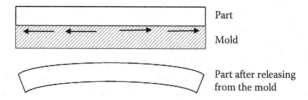

FIGURE 4.28 Bending distortion of a laminate due to mold surface–part surface interaction.

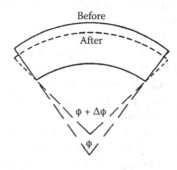

FIGURE 4.29 Change in angle due to spring forward.

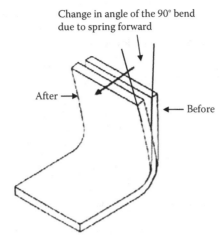

Change in angle of the 90° bend
due to spring forward

After →

← Before

FIGURE 4.30   Spring forward in an L-shaped part with a 90° bend after releasing from the mold.

magnitude of spring forward in a quasi-isotropic laminate [44], measured as the change in the sector angle, can be calculated from the following equation:

$$\frac{\Delta\phi}{\phi} = (\alpha_I - \alpha_T)\Delta T, \tag{4.32}$$

where $\Delta\phi$ is the change in angle ϕ; α_I and α_T are the in-plane and through-thickness coefficients of linear contraction of the composite part, respectively; and ΔT is the change in temperature due to cool down. The spring forward effect is very easily observed if the composite part has 90° bends [46]. An example is shown in Figure 4.30. In some cases, the spring forward effect can be reversed by reheating the part.

4.9   MANUFACTURING PROCESS-INDUCED DEFECTS AND THEIR DETECTION

A composite part may contain a multitude of surface and internal defects that are artifacts of the manufacturing process used for molding the part. A list of these defects and their probable causes is given in Table 4.4. The surface defects may be easily observed by either visual inspection or under low-magnification microscopes. The internal defects will require nondestructive inspection. Some of the internal defects may cause stress concentrations when a load is applied on the composite part and can either act as or grow into critical flaws during the service operation

TABLE 4.4

Manufacturing Process-Induced Defects in PMCs

Defect	Possible Reasons
Contamination	Foreign particles, extraneous fibers, backup film not removed properly
Undercure	Proper cure temperature and/or time not used
Variation in degree of cure	Variation in temperature distribution
Separation of plies or delamination	Poor consolidation, undercure, created during hole drilling or machining
Voids	Entrapped air, presence of moisture, excessive amount of solvent, release of gases during curing
Resin-rich or fiber-starved areas	Nonuniform resin distribution or flow
Resin-starved areas	Lack of resin flow, restricted areas
Fiber misalignment	Misoriented fibers, deviation from the preselected layup or filament winding pattern, fiber washout due to excessive injection pressure or resin flow
Broken filaments	Scratches or cuts, hole drilling
Fiber waviness or kinking	Improper tensioning during prepreg preparation, filament winding or pultrusion

of the part. This may lead to an early failure, service interruption, costly repair, or replacement.

At the end of the production process, the internal defects are detected by nondestructive tests (NDTs), and parts are either accepted or rejected on the basis of defect quality standards developed earlier at the prototype development stage. In the event of service failure, the NDT records can also serve a useful purpose in the postmortem analysis of the causes of failure. Ultrasonic and radiographic inspections are performed to detect manufacturing process-induced defects and service-generated damages in composite parts used in aircraft or aerospace applications. Other NDT methods, such as thermography, shearography, acoustic emission tests, and acousto-ultrasonic tests, are used mostly as research tools to monitor damage development during mechanical tests of composite specimens or parts [46]. A common problem with all these tests, including ultrasonic and radiography, is the lack of universal standards that can be used to distinguish between the critical and subcritical or noncritical defects. In-house standards are developed by performing both destructive tests, such as tension tests and interlaminar shear tests, and NDTs, such as radiography and ultrasonic C-scans, and comparing/corelating the data from both tests on a statistical basis.

4.9.1 ULTRASONIC INSPECTION

Ultrasonic inspection uses the energy levels of high-frequency (usually between 1 and 5 MHz) sound waves to detect internal defects in composite parts. The typical velocity of these sound waves is close to 2.8 mm/μs. The ultrasonic sound energy is generated by electrically exciting a piezoelectric transducer and is introduced into the surface of a molded composite part by means of a coupling medium, such as water. As the ultrasonic waves propagate through the material, their energy levels are attenuated by the presence of defects. Although some of the attenuated ultrasonic waves are transmitted through the part thickness, others are reflected back to the input surface. The energy levels of these transmitted and reflected ultrasonic waves are converted into electrical signals by a receiving transducer and are then compared with a preset threshold and displayed on a computer screen.

The following methods are commonly used for ultrasonic inspection of defects in PMCs.

1. Pulse-echo method: In this method, echoes reflecting from the front surface, back surface, and internal defects are picked up either by the transmitting transducer or by a separate receiving transducer. All reflected pulses are displayed as distinct peaks on the monitor (Figure 4.31). Pulse-echo depths are determined by measuring the time intervals between the front surface reflection peak and other significant peaks. Knowing the ultrasonic wave velocity in the material, these time intervals can be converted into defect location (depth) or part thickness measurements.
2. Through-transmission method: In this method, ultrasonic waves transmitted through the part thickness are picked up by a receiving transducer on the back surface of the part (Figure 4.32). Since the transmitted wave interacting with a defect has a lower energy level than an uninterrupted wave, it is displayed as a smaller peak. In contrast to the pulse-echo method, the through-transmission method requires access to both surfaces of the part.

In general, part surfaces are scanned at regular intervals by piezoelectric transducers and an ultrasonic map of the entire part is generated. The three different procedures used for data presentation are A-scan, B-scan, and C-scan. In the A-scan procedure, output signal amplitudes are displayed against a timescale (Figure 4.33) and the depths of various defect locations are determined from the positions of the signal peaks on the time sweep. The B-scan procedure profiles the top and bottom

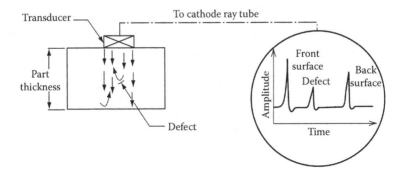

FIGURE 4.31   Pulse-echo method of ultrasonic inspection.

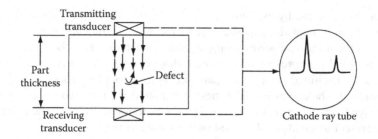

FIGURE 4.32   Through-transmission method of ultrasonic inspection.

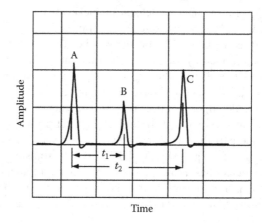

FIGURE 4.33   A-scan representation of internal defects: (a) front surface reflection, (b) reflection from a defect, and (c) back surface reflection.

surfaces of a defect (Figure 4.34). The C-scan procedure, on the other hand, displays the plan view of the defect boundaries in the material (Figure 4.35).

C-scan images of through-transmission waves are commonly used for online inspection of large molded parts. In gray-level C-scans, weaker transmitted signals are either dark gray or black. Thus, defects are identified as dark patches in a light gray background. The through-the-thickness location of any defect observed in a C-scan can be obtained by using the pulse-echo method and by recording the A-scan image of the reflected pulse.

The ultrasonic inspection is used to detect large voids, delaminations, clusters of microvoids, and foreign materials. Water is the most commonly used coupling medium for ultrasonic scanning. The composite part may be squirted with water on its surface or may be completely immersed in a water tank for more uniform coupling.

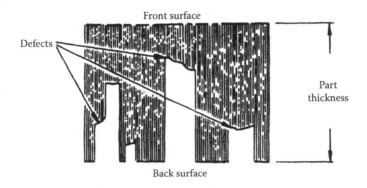

FIGURE 4.34   B-scan representation of internal defects.

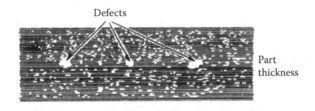

FIGURE 4.35   C-scan representation of internal defects.

4.9.2 RADIOGRAPHIC INSPECTION

In radiographic inspection, the internal structure of a molded composite part is examined by impinging a beam of radiation on one of its surfaces and recording the intensity of the beam as it emerges from the opposite surface. Conventional radiography uses X-rays (in the range of 7–30 keV) as the source of radiation and records the internal defects as shadow images on a photographic film. Gamma rays are more useful in thicker parts. Since they possess shorter wavelengths, they have greater penetrating power than X-rays. Other radiation beams, such as β irradiation and neutron radiation, are also used. Imaging techniques, such as displaying the image on a fluorescent screen (fluoroscopy) or cross-sectional scanning by computed tomography, are also available. The former is more useful for online inspection of production parts than the photographic technique.

In PMC parts, radiography is capable of detecting volumetric defects, such as voids, foreign inclusions, translaminar cracks, and nonuniform fiber distribution, as well as fiber misorientation (such as knit lines in compression-molded parts or fiber wrinkles in vacuum bag-molded parts). However, radiography is not sensitive to planar defects, such as delaminations. The volumetric defects change the intensity of the radiation beam by varying amounts and create images of different shades and contrasts on the photographic film. Thus, for example, large voids appear as dark spots and fiber-rich areas appear as light streaks on an X-ray film. The detection of microvoids and delaminations is possible by using radiopaque penetrants, such as sulfur, trichloroethylene, or carbon tetrachloride. The detection of fiber misorientation may require the use of lead glass tracers in the prepreg or SMC.

PROBLEMS

1. The degree of cure α_c for a vinyl ester resin as a function of time is given by the equation

$$\alpha_c = \frac{kt}{1+kt},$$

and the expression for k is given by the following:

$$k = 1.25 \times 10^6 \exp\left(-\frac{5000}{T}\right),$$

where k is the rate constant (min⁻¹); t is time in minutes; and T is temperature in kelvin.
   a. Determine the cure temperature needed to obtain 90% cure in 1 minute.
   b. What is the cure rate at 90% cure in letter "a"?
   c. Determine the cure temperature needed if the cure rate needs to be doubled at 90% cure. What will be the cure time now?

2. If the heat of reaction of the vinyl ester resin in Problem 1 is 450 J/g, plot the rate of heat generation as a function of degree of cure α_c at a cure temperature of 60°C. State the observation you will make from this plot.

3. In DSC experiments conducted with a thermosetting polymer, 80% cure was obtained in 3 minutes at a cure temperature of 140°C and 2 minutes at a cure temperature of 160°C. Assuming that the cure time for this polymer can be expressed by the following equation, estimate the cure time for 80% cure at 180°C.

$$t = \frac{\alpha_c^{2/3}}{k(1-\alpha_c)^{1/3}},$$

where $k = A\exp\left(\frac{-E}{RT}\right)$ and $R = 8.314$ J/mol K.

4. The following isothermal cure rate equation was found to fit the DSC data of a thermosetting polymer:

$$\frac{d\alpha_c}{dt} = k(1-\alpha_c)^2$$

where k is a temperature-dependent rate constant.
   Show that the time required to reach a degree of cure α_c is $\dfrac{\alpha_c}{k(1-\alpha_c)}$.

5. Assume that $k_1 = 0$ in Equation 4.4. Determine the cure level at which the cure rate has the maximum value.

6. In Problem 3, the temperature-dependent rate constant is found to fit the following Arrhenius equation:

$$k = A \exp\left(-\frac{E}{RT}\right)$$

where $A = 2.2 \times 10^6$ per minute, $E = 56.94$ J/mol, and T is the cure temperature in kelvin.

Calculate and compare the cure times and cure rates at 60%, 80%, 90%, and 99.9% degrees of cure at 50°C, 100°C, and 150°C.

7. The following rate constants were determined from the DSC data of a polyester resin:

Temperature (°C)	K (min^{-1})
75	0.11
80	0.20
90	0.63
95	1.10

Determine the activation energy in joules per mole (J/mol) for this resin.

8. The viscosity parameters in the Carreau equation for PEEK at 370°C are $\eta_o = 687$ Pa s, $\lambda = 0.0932$ s, and $n = 0.787$.
    a. Compare the viscosities of PEEK at 370°C and $\dot{\gamma} = 10, 100$, and 1000 s^{-1}.
    b. Which of the three viscosity parameters will change as the temperature is increased to 380°C? Will it increase or decrease?

9. The viscosities of a semicrystalline polymer are 800 Pa s and 100 Pa s at 290°C and 320°C, respectively, both measured at 500 s^{-1}. The temperature at which the polymer will be processed is 300°C. Using the Arrhenius equation, determine its viscosity at this temperature.

10. Lee et al. [6] found that the following values of the parameters in the viscosity equation given by Equation 4.15 fit the experimental viscosity data of a curing epoxy resin at $\alpha_c < 0.5$:
    $\eta_o = 7.93 \times 10^{-4}$ Pa s
    $E = 9.08 \times 10^4$ J/mol
    $K = 14.1$ Pa s
    Using these values, investigate the effect of the degree of cure α_c on the viscosity of the resin (in the range of validity of the given parameters) at 100°C, 125°C, and 150°C cure temperatures.

11. A liquid resin is being injected under an isothermal condition through a dry parallel fiber bed at a constant pressure p_i. Assuming that the flow is one dimensional (in the x-direction), show that the fill time t is proportional to x^2, where x is the fill distance.

12. Show how the flow front in a center-gated square mold may proceed (1) if the warp and fill directions of the fabric are parallel to the edges of the mold and (2) if the warp and fill directions of the fabric is at a 30° angle to the edges of the mold. Assume that the fabric is balanced, which means it has equal yarn count in the warp and fill directions.

13. The permeability values of a unidirectional continuous fiber network in the fiber direction and normal to the fiber direction are 10^{-9} and 2×10^{-10} m², respectively. Calculate the permeability values of the fiber network that is positioned at a 45° angle relative to the flow direction.

14. Three different fabrics are being considered for a composite application: (a) 8-H satin, (b) 5-H satin, and (c) 4-H (crowfoot) satin. Rank them in terms of permeability and explain the reason for the ranking.

15. What will be the effect of stitching on the permeability of a multiaxial NCF?

REFERENCES

1. J. Mijovic and H. T. Wang, Modeling of processing of composites: Part II: Temperature distribution during cure, *SAMPE Journal*, Vol. 24, pp. 42–55, 1988.
2. M. J. Marks and R. V. Snelgrove, Effect of conversion on the structure–property relationships of amine-cured epoxy thermosets, *Applied Materials and Interfaces*, Vol. 1, No. 4, pp. 921–926, 2009.
3. J. Tang and G. S. Springer, Effects of cure and moisture on the properties of Fiberite 976 resin, *Journal of Composite Materials*, Vol. 22, pp. 2–14, 1988.
4. P. J. Halley and M. F. Mackay, Chemorheology of thermosets: A review, *Polymer Engineering and Science*, Vol. 36, No. 5, pp. 593–609, 1996.
5. A. Yousefi, P. G. Lafleur, and R. Gauvin, Kinetic studies of thermoset cure reactions: A review, *Polymer Composites*, Vol. 18, No. 2, pp. 157–168, 1997.
6. W. I. Lee, A. C. Loos, and G. S. Springer, Heat of reaction, degree of cure, and viscosity of Hercules 3501-6 resin, *Journal of Composite Materials*, Vol. 16, pp. 510–520, 1982.
7. M. R. Dusi, W. I. Lee, P. R. Ciriscioli, and G. S. Springer, Cure kinetics and viscosity of Fiberite 976 resin, *Journal of Composite Materials*, Vol. 21, pp. 243–261, 1987.
8. K.-W. Lem and C. D. Han, Thermokinetics of unsaturated polyester and vinyl ester resins, *Polymer Engineering and Science*, Vol. 24, No. 3, pp. 175–184, 1984.
9. C. D. Han and K.-W. Lem, Chemorheology of thermosetting resins: IV. The chemorheology and curing kinetics of vinyl ester resin, *Journal of Applied Polymer Science*, Vol. 29, pp. 1879–1902, 1984.
10. M. R. Kamal and S. Sourour, Kinetics and thermal characterization of thermoset cure, *Polymer Engineering and Science*, Vol. 13, No. 1, pp. 59–64, 1973.
11. C.-S. Chern and G. W. Poehlein, A kinetic model for curing reactions of epoxides with amines, *Polymer Engineering and Science*, Vol. 27, No. 11, pp. 788–795, 1987.
12. J. K. Lee and K. D. Pae, Effects of hydrostatic pressure on isothermal curing of an epoxy resin-amine system, *Journal of Polymer Science Part C: Polymer Letters*, Vol. 28, pp. 323–329, 1990.
13. D.-S. Lee and C. D. Han, The effect of pressure on the curing behavior of unsaturated polyester resins, *Polymer Composites*, Vol. 8, No. 3, pp. 133–140, 1987.
14. J. Fourier, G. Williams, C. Duch, and G. A. Aldridge, Changes in molecular dynamics during bulk polymerization of an epoxide–amine system as studied by dielectric relaxation spectroscopy, *Macromolecules*, Vol. 29, pp. 7097–7107, 1996.

15. A. McIlhagger, D. Brown, and B. Hill, The development of a dielectric system for the on-line monitoring of the resin transfer moulding process, *Composites: Part A*, Vol. 31, pp. 1373–1381, 2000.
16. R. C. Chhabra and J. F. Richardson, *Non-Newtonian Flow and Applied Rheology*, 2nd Ed., Elsevier, Oxford, 2008.
17. M. R. Kamal, Thermoset characterization for moldability analysis, *Polymer Engineering and Science*, Vol. 14, No. 3, pp. 231–239, 1974.
18. K. W. Lem and C. D. Han, Rheological behavior of concentrated suspensions of particulates in unsaturated polyester resin, *Journal of Rheology*, Vol. 27, No. 3, pp. 263–288, 1983.
19. M. Roller, Characterization of the time-temperature-viscosity behavior of curing B-staged epoxy resin, *Polymer Engineering and Science*, Vol. 15, No. 6, pp. 406–414, 1975.
20. S.-J. Park and T.-J. Kim, Studies on surface energetics of glass fabrics in an unsaturated polyester matrix system: Effect of sizing treatment on glass fabrics, *Journal of Applied Polymer Science*, Vol. 80, pp. 1439–1445, 2001.
21. S.-J. Park and T.-J. Kim, Effect of silane coupling agent on mechanical interfacial properties of glass fiber-reinforced unsaturated polyester composites, *Journal of Polymer Science Part B: Polymer Physics*, Vol. 41, pp. 55–62, 2003.
22. J. D. H. Hughes, The carbon fibre/epoxy interface—A review, *Composites Science and Technology*, Vol. 41, pp. 13–45, 1991.
23. Z. Dai, B. Zhang, F. Shi, M. Li, Z. Zhang, and Y. Gu, Effect of heat treatment on carbon fiber surface properties and fibers/epoxy interfacial interaction, *Applied Surface Science*, Vol. 257, pp. 8457–8461, 2011.
24. G. Wu, Effects of carbon fiber treatment on interfacial properties of advanced thermoplastic composites, *Polymer Journal*, Vol. 29, No. 9, pp. 705–707, 1997.
25. P. Hubert and A. Pousartip, A review of flow and compaction modeling relevant to thermoset matrix laminate processing, *Journal of Composite Materials*, Vol. 17, No. 4, pp. 286–318, 1998.
26. R. R. Gebart, Permeability of unidirectional reinforcements for RTM, *Journal of Composite Materials*, Vol. 26, No. 8, pp. 1100–1133, 1992.
27. S. G. Advani and E. M. Sozer, *Process Modeling in Composites Manufacturing*, CRC Press, Boca Raton, FL, 2000.
28. R. S. Parnas, *Liquid Composite Molding*, Hanser, Munich, 2000.
29. N. K. Naik, M. Sirisha, and A. Inani, Permeability characterization of polymer matrix composites by RTM/VARTM, *Progress in Aerospace Science*, Vol. 65, pp. 22–40, 2014.
30. R. S. Parnas, K. M. Flynn, and M. E. Dal-Favero, A permeability database for composites manufacturing, *Polymer Composites*, Vol. 18, No. 5, pp. 623–633, 1997.
31. Y. Ma and R. Shishoo, Permeability characterization of different architectural fabrics, *Journal of Composite Materials*, Vol. 33, No. 8, pp. 729–750, 1999.
32. K. J. Ahn, J. C. Seferis, and J. C. Berg, Simultaneous measurement of permeability and capillary pressure of thermosetting matrices in woven fabric reinforcements, *Polymer Composites*, Vol. 12, No. 3, pp. 146–152, 1991.
33. S. Amico and C. Lekakou, An experimental study of the permeability and capillary pressure in resin-transfer molding, *Composites Science and Technology*, Vol. 61, pp. 1945–1959, 2001.
34. T. G. Gutowski, T. Morgaki, and Z. Cai, The consolidation of laminate composites, *Journal of Composite Materials*, Vol. 17, pp. 172–188, 1987.
35. Z. Cai and T. Gutowski, The 3-D deformation behavior of a lubricated fiber bundle, *Journal of Composite Materials*, Vol. 26, No. 8, pp. 1207–1237, 1992.

36. T. G. Gutowski, Z. Cai, S. Bauer, D. Boucher, J. Kingery, and S. Wineman, Consolidation experiments for laminate composites, *Journal of Composite Materials*, Vol. 21, pp. 650–669, 1987.

37. F. Robitaille and R. Gauvin, Compaction of textile reinforcements for composites manufacturing—I: Review of experimental results, *Polymer Composites*, Vol. 19, No. 2, pp. 198–216, 1998.

38. R. A. Saunders, C. Lekakou, and M. G. Bader, Compression and microstructure of fibre plain woven cloths in the processing of polymer composites, *Composites: Part A*, Vol. 29A, pp. 443–454, 1998.

39. B. Chen and T.-W. Chou, Compaction of woven-fabric preforms: Nesting and multi-layer deformation, *Composites Science and Technology*, Vol. 60, pp. 2223–2231, 2000.

40. B. Chen, E. J. Lang, and T.-W. Chou, Experimental and theoretical studies of fabric compaction behavior in resin transfer molding, *Materials Science and Engineering*, Vol. A317, pp. 188–196, 2001.

41. M. J. Yokota, In-process controlled curing of resin matrix composites, *SAMPE Journal*, Vol. 14, pp. 11–17, 1978.

42. M. R. Winsom, M. Gigliotti, N. Ersoy, M. Campbell, and K. D. Potter, Mechanisms generating residual stresses and distortion during manufacture of polymer–matrix composite structures, *Composites: Part A*, Vol. 37, pp. 522–529, 2006.

43. K. D. Potter, M. Campbell, C. Langer, and M. R. Winsom, The generation of geometri-cal deformations due to tool/part interaction in the manufacture of composite compo-nents, *Composites: Part A*, Vol. 36, pp. 301–308, 2005.

44. N. Zahlan and J. M. O'Neill, Design and fabrication of composite components; the spring-forward phenomenon, *Composites*, Vol. 20, No. 1, pp. 77–81, 1989.

45. D. W. Radford and T. S. Rennick, Separating sources of manufacturing distortion in lami-nated composites, *Journal of Reinforced Plastics and Composites*, Vol. 19, pp. 621–640, 2000.

46. R. H. Rossi and V. Giurgiutiu, Non-destructive testing of damage in aerospace com-posites, in *Polymer Composites in the Aerospace Industry*, Elsevier, Amsterdam, pp. 413–448, 2015.

5 Bag Molding Process

The bag molding process is used primarily for manufacturing laminated composite structures and components using thermoset matrix prepreg sheet or tape as the starting material. The thermoset matrix prepreg is a thin sheet containing fibers coated with a partially cured (B-staged) resin. Epoxies are the principal resins used in prepregs, although for high-temperature applications, BMIs and polyimides are also used. A typical laminate consists of multiple layers or plies laid at various fiber orientation angles in accordance with the design requirements. To manufacture a laminated composite part using the bag molding process, a stack of prepreg layers is placed on the mold or tool surface, covered with a vacuum bag, and then cured and consolidated under heat, vacuum, and pressure either inside an autoclave or in a heated platen press. Autoclaves are commonly used for vacuum bag molding of high-performance composite laminates, since temperature and pressure controls are more precise in an autoclave than in a heated platen press.

The bag molding process is the most common composite manufacturing process in the aerospace industry. It is used for manufacturing many primary and secondary structures for military and commercial aircrafts, such as wing flaps, ailerons, horizontal stabilizers, and radars. It is also used in sporting goods, motor sports, and racing yachts applications. The advantages of this process are its ability to maintain precise fiber orientation and produce high-quality composite parts with relatively low void content and high reproducibility. The drawbacks for the bag molding process are (1) it is labor intensive and time consuming and (2) autoclaves are expensive, particularly for large components. For these reasons, it is a more expensive process than the other composite manufacturing processes, such as RTM.

This chapter starts with a description of the process for making prepreg which is the starting material for the bag molding process. It then describes the bag molding process and the associated process parameters. While autoclaves are commonly used for curing bag-molded composites, out-of-autoclave curing is gaining acceptance mainly for cost-reduction purposes. This process is also described. Sandwich laminates are used for their high bending stiffness-to-weight ratios. The processing of sandwich laminates is included in this chapter.

5.1 PREPREG

Prepreg is a thin sheet (or tape) of material containing either unidirectional continuous fibers or woven fabric impregnated with a predetermined amount of partially cured thermosetting polymer matrix. Epoxy is the primary matrix material in prepreg sheets, although other thermosets, such as BMIs, are also used for high-temperature applications. There are also thermoplastic matrix prepregs that are used for manufacturing thermoplastic matrix composite parts. Thermoplastic matrix composite prepregs are described in Chapter 10.

Prepreg sheets are supplied by prepreg manufacturers in rolls (Figure 5.1). The width of prepreg sheets may vary from less than 25 to over 457 mm. Sheets wider than 457 mm are called *broadgoods*. The thickness of a single ply of a cured prepreg sheet is normally in the range of 0.125–0.25 mm. The resin content in commercially available prepregs is between 30% and 45% by weight, which, for a T-300 carbon fiber/epoxy prepreg, is equivalent to 38.6–54.5% by volume of resin.

Unidirectional fiber-reinforced epoxy prepregs are manufactured by pulling a row of uniformly spaced parallel (collimated) fiber rovings or tows through a liquid resin bath containing the epoxy resin–hardener mix dissolved in a solvent (Figure 5.2). The solvent content and resin bath temperature are adjusted to obtain the resin viscosity appropriate for good fiber wet-out and impregnation. After the resin bath, fibers impregnated with the liquid resin first pass through a metering device and then enter a heating/cooling chamber. The resin content in the prepreg is controlled by the metering device, which can be simply a set of nip rollers. In the heating/cooling chamber, heat is first applied to drive out most of the solvent from the prepreg and start the curing reaction of the resin. In the cooling chamber, the curing reaction is controlled to come to the B-stage condition. At the end of the cooling chamber, the resin is in a B-staged condition in which it is in a partially cured semisolid state at room temperature. The B-staged resin is tacky which allows it to adhere to itself and other surfaces. The prepreg sheet is then backed up with a release film or waxed paper and wound around a take-up roll. The backup material is needed so that the

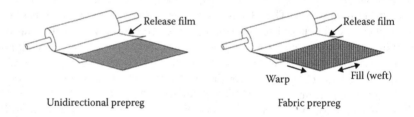

FIGURE 5.1 Schematic of prepreg rolls.

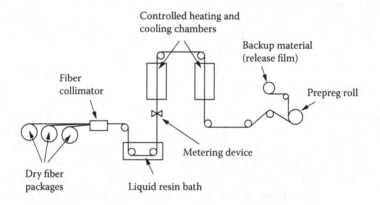

FIGURE 5.2 Prepreg manufacturing.

prepreg sheet does not stick to itself when it is being rolled, and it must be separated from the prepreg sheet before it is placed in the mold for bag molding. The prepreg roll is then placed in a polyethylene bag and stored in a freezer until the time of molding. Before molding, the prepreg roll is removed from the freezer and is allowed to reach room temperature (which may take up to 48 hours) before opening the polyethylene bag to avoid forming moisture condensation. The normal shelf life (storage time before molding) for epoxy prepregs is 6–8 days at 23°C; however, it can be prolonged 6–12 months if stored in the freezer at −18°C.

Prepregs are specified by the fiber type, fiber architecture, resin type, fiber aerial weight, resin content, and cured ply thickness (Table 5.1). The fiber aerial weight is the mass of fibers per unit area in grams per square meter (g/m^2). The resin content is the weight percentage of the resin in the prepreg, which is usually higher than the resin content in the cured composite part. This is because nearly 10 wt.% of the resin is allowed to flow out during the molding process. The excess resin flowing out from the prepreg removes the entrapped air and residual solvents, which in turn reduces the void content in the cured laminate. However, the recent trend is to employ a near-net resin content to allow only 1–2 wt.% resin loss during molding. The cured ply thickness of unidirectional epoxy prepreg is usually 0.127 mm ± 6% [1].

Two important quality characteristics of prepregs at the time of molding are tack and drapeability. *Tack* or *stickiness* is defined as the ability of a partially cured prepreg layer to adhere to the mold surface and to another partially cured prepreg layer. Prepreg with too little resin on the surfaces will have low tack and may need to be heated to increase its tack during layup. On the other hand, if the surfaces are resin rich and there is less resin inside, the prepreg may separate at the center during the layup operation. *Drapeability* is defined as the ability of the prepreg to conform to the mold surface without fiber distortion, movement, and out-of-plane buckling.

TABLE 5.1
Typical Specifications of Prepreg

Specification	Example
Fiber type	Carbon fiber
Resin type	Epoxy
Maximum service temperature—dry/wet	100°C/80°C
Available form	Unidirectional tape
Fiber aerial weight	120 g/m^2
Cure cycle—time/temperature/pressure	60 minutes/125°C/2 bar
Postcure—temperature/time	60°C/2 hours
Minimum cure temperature	100°C
Time at minimum cure temperature	10 hours
Storage life at −18°C	12 months
Outlife at 23°C	30 days
Cured ply thickness	0.17 mm
Typical applications	Aerospace structural components

Both depend on the viscosity advancement of the B-staged resin at the time of molding; in addition, drapeability depends on the selection of fiber architecture and thickness of the prepreg stack.

The advantages of using prepreg in making composite parts are: (1) it can produce composite parts with high fiber volume fraction, (2) it is easy to cut and drape, (3) it has controllable tack which allows plies to stay in place after stacking, and (4) it has low variability. The main disadvantage is that it has to be stored in a freezer; otherwise, the B-staged resin in the prepreg will continue to cure. If the prepreg is too dry and stiff at the time of molding due to cure advancement during storage, it cannot be used in bag molding and has to be discarded.

5.2 BAG MOLDING PROCESS

Figure 5.3 shows the steps in a bag molding process. Figure 5.4 shows the construction of a bag tool (mold). To prepare for the bag molding process, the hard mold surface at the bottom is thoroughly cleaned and either sprayed or coated with a mold release. It is then covered with a release film (used to prevent the molded laminate from sticking to the mold surface) and then a peel ply on which the prepreg plies are laid up one by one in the fiber orientation angles and stacking sequence as per the design requirement of the part. To prevent moisture pickup, the prepreg roll on removal

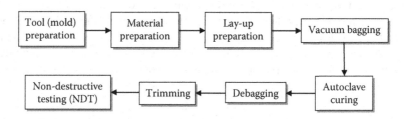

FIGURE 5.3 Steps in bag molding process.

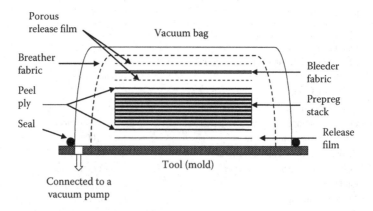

FIGURE 5.4 Schematic of a bag molding setup.

from the cold storage is allowed to warm up to the room temperature before use. Plies are trimmed from the prepreg roll in the desired shape, size, and orientation using a template and a cutting device, which can simply be a mat knife. Laser beams, high-speed water jets, or trimming dies are used for more accurate cutting. The layer-by-layer stacking operation can be performed either by hand or by a numerically controlled automatic tape-laying machine. Before laying up the prepreg layers, the backup material is peeled off from each ply. Slight compaction pressure is applied using a spatula and a roller to adhere each ply in the layup to the preceding ply. A heat gun is sometimes used to improve the tack in local areas of the prepreg layers. After laying several plies, the partial stack is debulked by applying pressure on its top surface to remove large voids and gaps between the plies and to ensure sufficient compaction. In some cases, the partial stack is placed in a vacuum bag and slightly heated in an oven (making sure that cure is not advanced) while vacuum is being applied to remove the air from between the plies. The process of debulking is repeated until the full stack is prepared.

After the layup operation is complete, a porous release cloth and a few layers of bleeder fabric are placed on top of the prepreg stack. The bleeder fabric is used to absorb the excess resin in the prepreg as it flows out during the molding process. The complete layup is covered with a release film, a caul plate, a layer of breather fabric, and then a thin heat-resistant vacuum bag, which is closed around its periphery by a sealant. The entire assembly is placed inside an autoclave (Figure 5.5) where a combination of external pressure, vacuum, and heat is applied to consolidate the separate plies into a solid laminate and cure the resin to a preselected degree of cure. The cure cycle is usually recommended by the prepreg manufacturer. Vacuum is applied to remove air and volatiles, while pressure is applied to consolidate the individual layers into a laminate. Pressure also forces the prepreg stack to fill the tight corners and conform to the curves on the mold surface.

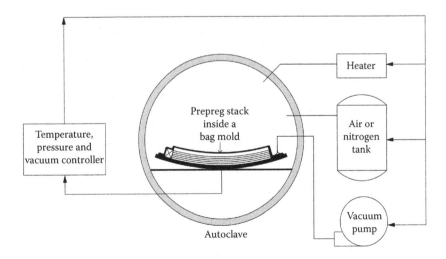

FIGURE 5.5 Bag molding inside an autocalve.

The cure cycle for a typical epoxy matrix prepreg is 5–10 hours at a cure temperature of 177°C and a pressure close to 7 bar. BMI matrix prepreg requires longer than 10 hours of cure cycle at comparable temperatures and pressures, followed by 18 hours of postcure at around 230°C. At the end of the cure cycle, the heat is turned off, but the vacuum is maintained until the temperature is reduced to 70°C or lower. After returning to room temperature, the vacuum bag is opened and the cured part is removed from the mold. The part removal from the mold may need gentle tapping by a mallet or blowing compressed air between the part and the mold surface. After the part is removed, it is nondestructively inspected for molding defects, such as voids and interply cracks. If the part passes the NDI, its edges are trimmed off any flash, which is basically the excess resin that may have flowed out and hardened around the edges. Holes are then drilled if fasteners are used for assembly.

The bag molding process requires the use of the following items, each of which plays a role in the successful molding of the part. Most of these items are consumables and are discarded after each molding cycle.

1. A mold release agent prevents sticking of the cured composite part to the mold surface and helps in its easy release from the mold. Wax is a common release agent.
2. Peel ply allows free passage of air, volatiles, and excess resin during the molding operation. It often adheres to the surface of the cured part, but it must be removed to provide a bondable or paintable surface. Nylon and thermoplastic polyester are the common peel ply materials.
3. Bleeder fabric is usually made of felt or glass fabric, and its purpose is to absorb the excess resin. The resin flow through the bleeder fabric controls the fiber volume fraction in the cured laminate.
4. Release film prevents excess resin flow. It is a perforated film that allows the passage of air and volatiles into the breather fabric and prevents the bleeder fabric from sticking to the cured part.
5. Breather fabric provides the means to apply vacuum and assists in the removal of air and volatiles from the entire assembly. Thicker breather fabrics may be needed when high autoclave pressures are used. The material for the breather fabric is usually a polyester.
6. Dams are used to limit lateral flow of the resin. They are usually a metal bar, a flexible polymer, a cork, or a rubber.
7. Caul plate is either a thin aluminum or plastic sheet placed on top of the stack up. It is used to provide a smoother surface finish to the cured part. A mold release agent is applied to the caul plate and a release film is used between the top of the stack up and the caul plate so that it can be easily separated from the part after molding.
8. Vacuum bag covers the assembly of the prepreg stack and other consumables to create a flexible containment. The vacuum bag material is usually a nylon film which can withstand temperatures of up to 180°C. Flexible films of polyethylene and rubber are also used.
9. Sealant tape seals the vacuum bag around the periphery.

5.3   LAYUP TECHNIQUES

A cured laminate consists of multiple layers (which for some applications, may be 100 or more) with fiber orientation angle from layer to layer varying from −90° to +90°. Laying up the prepreg stack in the mold requires careful attention to cutting the prepreg layers to proper dimensions and angular orientation, positioning them on the mold surface or the previous layer, and debulking the stack as it is being assembled. Debulking is necessary to reduce the gaps between the layers. To avoid errors, templates are used to draw the shapes of the prepreg pieces to be cut, and to reduce wastage, the pieces are nested as shown in Figure 5.6. Automated ply cutting machines and laser guidance systems are available to improve accuracy, reduce the amount of scrap, and reduce the laying up time.

Since prepreg sheets are flat, laying them on a flat surface or a nearly flat surface is relatively straightforward compared to laying them on a curved surface. For covering compound curved surfaces or to fit them in the corners of a mold, the prepreg layers may have to be slit or folded. These areas may become locations of weakness in the cured composite part. Considerable amount of manipulation in the forms of stretching, shearing, bending, and pressing is needed to fit them to the mold contour [3,4]. A woven fabric prepreg is generally more amenable to these manipulations than a unidirectional fiber prepreg due to the possibility of in-plane shear deformation, known as interply shear. In laying unidirectional fiber prepreg, care should be taken to make sure that the fibers are not distorted or dislocated; otherwise, the gaps created by fiber distortion and dislocation may become resin-rich areas or large voids in the cured composite part.

Many defects in bag-molded laminates relate to the ply layup and trimming operations. Close control must be maintained over the fiber orientation in each ply, the stacking sequence, and the total number of plies in the stack. Since prepreg tapes are not as wide as the part itself, each layer may contain a number of identical plies laid side by side to cover the entire mold surface. The gap between the side-by-side plies in a single layer should not exceed 0.76 mm, and the distance between any two gaps should not be less than 38 mm [5]. Care must also be taken

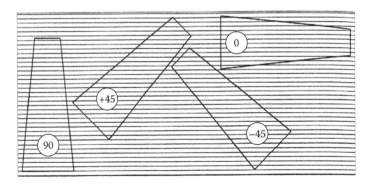

FIGURE 5.6   Nested prepreg pieces. (From M. G. Bader and C. Lekakou, *Composites Engineering Handbook* (P. K. Mallick, ed.), Marcel Dekker, New York, 1997.)

to avoid filament crossovers. Broken filaments, foreign matters, and debris should not be permitted.

Figure 5.7 shows a number of techniques used in laying up prepregs on mold surfaces containing angles, protrusions, corners, flanges, etc. For example, instead of pushing the prepeg into the corner, several short strips can be used in a stepwise fashion so that adequate uniform pressure can be applied in the corner area during cure cycling. Overlaps are used in curved areas and stepped areas to allow for slippage. Overlaps are also used on flanges; but, if possible, the overlap lengths are kept the same. If the overlap length cannot be the same, they should be staggered as shown in Figure 5.7.

In some applications, it may be necessary to taper the part from a thicker section to a thinner section. Tapering is done by terminating plies to progressively increasing lengths as shown in Figure 5.8. Ply termination, known as *ply drop-off*, causes stress concentration and increases the possibility of failure initiation by delamination during service applications. It also creates resin pockets that are usually the weakest spots in the tapered laminate. If the part contains a number of ply drop-offs, they should be staggered at distances that are three times greater than the thickness of the

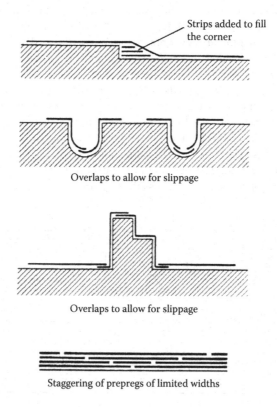

FIGURE 5.7 Techniques used for prepreg laying. (From K. Noakes, *Successful Composite Techniques*, 4th Ed., Crowood Press, Marlborough, 2008.)

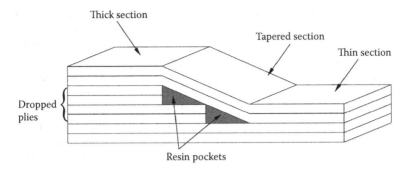

FIGURE 5.8 Tapered section in a laminate.

drop-off [6]. If the drop-off contains 45° plies, the stagger distance should be at least eight times the drop-off thickness. The plies in the tapered area should be dropped-off in decreasing order of their stiffness. For example, if the laminate contains both 0° and 90° plies, the 0° ply should be dropped off at the thicker end and the 90° ply should be dropped off at the thinner end.

5.4 AUTOMATED TAPE LAYING

Hand layup involves manually laying down each layer by hand and firmly sticking it to the previous layer or to the mold surface. This makes the production rate slow, labor intensive, and prone to human error. Although it is still the primary method of building the prepreg stack, automated tape layup machines are capable of producing prepreg layups in much shorter times and more accurately.

In automated tape laying machines, the prepreg roll is loaded on the tape-laying head which is mounted on a gantry that positions the tape-laying head in the mold. Both tape-laying head and gantry have multiple axes of movement, typically five for each, that are microprocessor operated to provide high layup flexibility and accurate feedback control of both translational and rotational motions. The tape head contains cutters, a backup paper take-up reel, a tape heating system, compaction shoes, and an optical tape flaw detector. If the prepreg lacks appropriate tack, the tape-heating system warms it up to between 25°C and 45°C to increase its tack just before it passes through the lay-down rollers. A set of compaction shoes applies pressure on the tape as it is being laid down. After a strip of tape is laid, its end is cut and the tape head moves to the next location.

The automated tape-laying machines are programmed to operate with user-defined parameters, such as mold surface topology, ply boundaries, fiber orientation angle in each ply, number of layers to be laid, and tape cutting angle. However, they are more suitable for flat and mildly contoured parts, such as wing skins and ailerons, than highly contoured parts, such as nose cones, inlet ducts, and nozzles.

5.5 CURE CYCLE

As the prepreg is heated in the autoclave, the resin viscosity in the B-staged pre-preg plies first decreases, attains a minimum, and then rapidly increases as it starts to gel, and the curing reaction proceeds toward completion. Figure 5.9 shows a typical two-stage cure cycle for a carbon fiber–epoxy prepreg. The first stage in this cure cycle consists of increasing the temperature at a controlled rate (say, 2°C/min) of up to 130°C and dwelling at this temperature for nearly 60 minutes when the minimum resin viscosity is reached. The purpose of the dwell is to equalize the autoclave temperature and prepreg stack temperature and to reduce the temperature gradient across the thickness of the stack. During this period of temperature dwell, an external pressure is applied on the prepreg stack that causes the excess resin to flow out into the bleeder fabric. The resin flow is critical since it allows the removal of entrapped air and volatiles from the prepreg stack and thus reduces the void content in the cured laminate. At the end of the temperature dwell, the autoclave temperature is increased to the cure temperature of the resin. The cure temperature and pressure are maintained for 2 hours or more until a predetermined level of cure is reached. At the end of the cure cycle, the temperature is slowly reduced while the laminate is still under pressure. The laminate is removed from the vacuum bag and, if needed, postcured at an elevated temperature in an air-circulating oven.

The flow of excess resin from the prepreg layers is extremely important in reducing the void content in the cured laminate. In a bag molding process for producing thin shell or plate structures, resin flow by face bleeding (normal to the top laminate face) is preferred over edge bleeding. Face bleeding is more effective since the resin flow path before gelation is shorter in the thickness direction than in the edge directions. Since the resin flow path is relatively long in the edge directions, it is difficult to remove entrapped air and volatiles from the central areas of the laminate by the edge bleeding process.

The resin flow from the prepreg layers significantly reduces and may even stop after reaching the gel time, which can be increased by reducing the heating rate as well as the dwell temperature (Figure 5.10). Dwelling at a temperature lower than the cure temperature is important for two reasons: (1) it allows the layup to achieve a uniform temperature throughout the thickness and (2) it provides time for the resin to

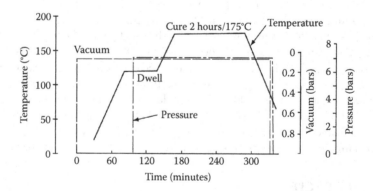

FIGURE 5.9 Typical two-stage cure cycle for a carbon fiber/epoxy prepreg.

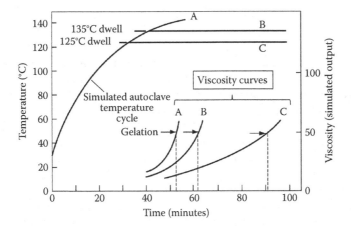

FIGURE 5.10 Effect of dwelling on gel time (shown by the horizontal arrows on the viscosity curves). (Reprinted from *Composites*, 17, D. Purslow and R. Childs, Autoclave moulding of carbon fibre-reinforced epoxies, 127–136, Copyright (1986), with permission from Elsevier.)

achieve a low viscosity. A small batch-to-batch variation in dwell temperature may cause a large variation in gel time, as evidenced in Figure 5.10 [7].

The cure temperature, pressure, and heating/cooling rate are selected to meet the following requirements:

1. The resin is uniformly cured and attains a specified degree of cure in the shortest possible time.
2. The temperature at any position inside the prepreg does not exceed a prescribed limit during the cure.
3. The cure pressure is sufficiently high to squeeze out all the excess resin from every ply before the resin starts to gel, and its viscosity starts to rapidly increase at any location inside the prepreg.
4. Heating rate is selected to reduce a large temperature gradient between the layers in the prepreg stack. Similarly, cooling rate is selected to reduce a large temperature gradient between the layers in the cured composite part.

Loos and Springer [8] developed a theoretical model for the complex thermomechanical phenomenon that takes place in a vacuum bag molding process. Based on their model and experimental work, the following observations can be made regarding the various molding parameters.

The maximum temperature inside the layup depends on (1) the cure temperature, (2) the heating rate, and (3) the initial layup thickness. The cure temperature is usually prescribed by the prepreg manufacturer for the particular resin–catalyst system used in the prepreg and is determined from the time–temperature–viscosity characteristics of the resin–catalyst system. At low heating rates, the temperature distribution remains uniform within the layup. At high heating rates and increased layup thickness, the heat generated by the curing reaction is faster than the heat transferred to the mold surface and a temperature *overshoot* may occur.

Resin flow in the layup depends on the maximum pressure, layup thickness, and heating rate, as well as the pressure application rate [9]. A cure pressure sufficient to squeeze out all excess resin from 16-ply and 32-ply layups was found to be inadequate for squeezing out resin from the layers closer to the bottom surface in a 64-ply layup. Similarly, if the heating rate is very high, the resin may start to gel before the excess resin is squeezed out from every ply in the layup. Since the compaction and resin flow progress inward from the top, the plies adjacent to the bottom mold surface may remain uncompacted and rich in resin, thereby creating weak interlaminar layers in the laminate.

Excess resin must be squeezed out of every ply before the gel point is reached at any location in the prepreg. Therefore, the maximum cure pressure should be applied just before the resin viscosity in the top ply becomes sufficiently low for the resin flow to occur. If the cure pressure is applied too early, excess resin loss would occur owing to very low viscosity in the pregel period. If, on the other hand, the cure pressure is applied after the gel time, the resin may not be able to flow into the bleeder fabric because of the high viscosity it quickly attains in the postgel period. The void content in the cured laminate depends on good resin flow, which is controlled not only by the resin viscosity, but also by the cure pressure and pressure application time. In general, the void content decreases with increasing cure pressure, but if the cure pressure is applied after the resin has reached the gel point; increasing the pressure cannot produce any significant reduction in void content [9,10]. Thus, the pressure application time is an important molding parameter in a bag molding process. In general, it decreases with increasing cure pressure as well as increasing heating rate.

The uniformity of cure in the laminate requires a uniform temperature distribution in the laminate. The time needed for completing the desired degree of cure is reduced by increasing the cure temperature as well as increasing the heating rate (Table 5.2).

TABLE 5.2

Cure Time for 90% Degree of Cure in a 32-Ply Carbon Fiber–Epoxy Laminate

Cure Temperature (°C)	Heating Rate (°C/min)	Cure Time (minutes)
135	2.8	236
163	2.8	110
177	2.8	89
177	5.6	65
177	11.1	52

Note: Based on a theoretical model developed by Loos and Springer [8].

5.6   THICK SECTION LAMINATES

Thick section laminates with thickness greater than 25 mm are used in many applications. For example, in some military applications, the laminate thickness can be as high as 300 mm. Assuming that the cured ply thickness is 0.125 mm, a 25 mm thick laminate will contain 200 plies and a 300 mm laminate will contain 2400 plies. One of the problems encountered in molding thick-section laminates is the accumulation of exothermic heat generated during curing. Since polymer matrix composites have low thermal conductivity, the cure-generated heat is not easily dissipated away from the curing laminate. As a result, the inside temperature at the center of the laminate may become much higher than the cure temperature [11,12]. The temperature overshoot may be so high that it can cause thermal degradation of the resin.

Figure 5.11 shows the simulated centerline temperatures during the cure of a 200-ply and a 500-ply AS-4 carbon fiber/3501-6 epoxy prepreg stack. The two-stage

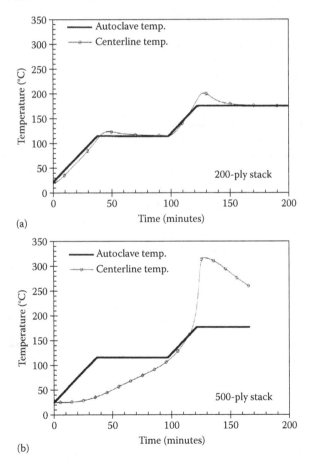

(a)

(b)

FIGURE 5.11   Centerline temperature profiles in (a) a 200-ply and (b) a 500-ply prepreg stack. (Adapted from S. R. White, *Processing of Composites* (R. S. Davé and A. C. Loos, eds.), Hanser, Munich, 2000.)

autoclave temperature is also shown in this figure. With the 200-ply stack, there is a small thermal lag during the first heating stage. The centerline temperature slightly overshoots the autoclave temperature at the end of the first heating stage, but then it rapidly increases to a peak temperature approximately 25°C higher than the autoclave temperature at the end of the second heating stage. With the 500-ply stack, the centerline temperature significantly lags behind the autoclave temperature during the first heating stage, but then it rapidly increases to a peak temperature approximately 135°C higher than the autoclave temperature during the second heating stage. As can be observed in this figure, in the initial stages of the cure cycle, the centerline temperature lags behind the autoclave temperature, but as the curing reaction begins in the center, its temperature rapidly rises to a peak value and then slowly starts to decrease. Depending on the fiber volume fraction, fiber type, and stack thickness, the peak temperature can be higher than the material degradation temperature. Both fiber volume fraction and fiber type influence thermal conductivity of the material in the thickness direction. Increasing the fiber volume fraction increases the thermal conductivity and decreases the peak temperature. Since carbon fibers have a higher thermal conductivity than glass fibers, the peak exotherm will be lower in a carbon fiber/epoxy part than that in a glass fiber/epoxy part.

Another issue with thick section molding is the pressure gradient in the thickness direction. Since pressure on the layers is not uniform, the level of compaction varies in the thickness direction. As a result, resin flow and air removal are also not uniform. It is possible that there will be excess resin in some of the plies or between the plies, and void content may become high.

Depending on the width-to-thickness ratio of the prepreg stack, edge constraints, and bleeder arrangement, the resin flow from the prepreg layers can be in the thickness direction, in the edge direction, or a combination of the two (Figure 5.12). If the resin flow is only in the thickness direction, the prepreg layers will move in the thickness direction in a sequential manner, as is shown in Figure 5.13. It will start with the top layer moving first, which will squeeze out the excess resin between the first and the second layers. These two layers will then move together toward the third layer. This sequential process will continue until the bottom of the stack is reached. When the

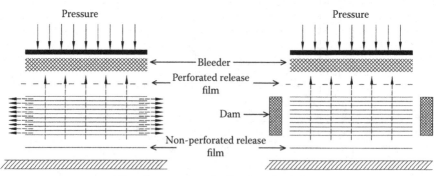

* Dotted arrows show the direction of resin flow

FIGURE 5.12 Direction of resin flow during bag molding.

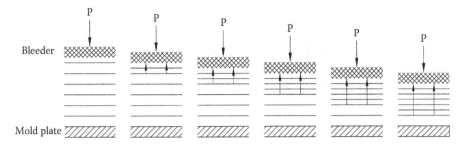

FIGURE 5.13 Sequential prepreg movement during bag molding.

resin flow is only in the edge direction, all the layers in the stack move together. As the resin flows in the transverse direction, fibers may spread out in the transverse direction and spacing between the fibers may become nonuniform.

The final thickness of the bag molded part depends on the number of prepreg layers and the degree of consolidation, which is determined by the resin flow as well as the resin impregnation of the fiber bundles in each layer.

5.7 OUT-OF-AUTOCLAVE CURING

Curing in autoclaves produces composite parts with very low void content, excellent consolidation, and uniform resin distribution. However, autoclaves are expensive and involve high operating cost. Autoclaves used for large size parts may not be efficiently utilized if they are used only for small size parts. For these reasons, out-of-autoclave curing, such as oven curing and heated blanket curing, is gaining acceptance in the composites industry [14,15]. In oven curing, heating of the vacuum bag assembly takes place in a hot air circulating oven. Vacuum is applied to remove air and volatiles, and consolidation takes place at 1 bar. Since no external pressure is applied in the oven, oven-cured parts contain higher voids and lower consolidation than autoclave-cured parts.

To produce autoclave-quality parts with very low void content using *out-of-autoclave curing*, a new generation of prepregs, called vacuum bag-only (VBO) prepregs, is used. The VBO prepregs consist of both dry and resin-impregnated areas (Figure 5.14). The dry areas act as vacuum channels that allow the migration of entrapped air toward the edges and surfaces in the early stages of curing. When the temperature is increased, resin flows into these air migration channels and infiltrates them, thus producing a void-free part.

Several other modifications are necessary for out-of-autoclave curing to work. They are as follows.

1. The resin must remain relatively viscous in the early stages of curing so that resin infiltration in the dry area is not impeded and sufficient time is available for dry areas to evacuate the air. However, its viscosity at the later stages of curing must be low enough to permit sufficient flow for full impregnation and consolidation. The VBO epoxy resins are designed to

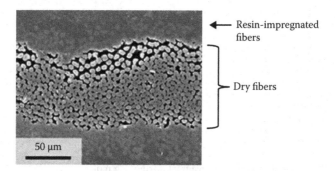

Resin-impregnated fibers

Dry fibers

50 µm

FIGURE 5.14 Dry and resin-impregnated areas in a VBO prepreg. (Reprinted from *Composites: Part A*, 70, T. Centea, L. K. Grunenfelder, and S. R. Nutt, A review of out-of-autoclave prepregs—Material properties, process phenomena, and manufacturing considerations, 132–154, Copyright (2015), with permission from Elsevier.)

cure at temperatures in the range of 80–140°C followed by in-bag or free-standing postcure at 177°C. In comparison, an autoclave curing epoxy resin is cured at 177°C followed by postcuring at about 100°C.

2. The VBO resin viscosity is generally higher than the autoclave resin viscosity, and therefore, after the cure is initiated, the time for full impregnation of the prepreg may be limited for VBO resins.

3. A minimum vacuum gauge reading of 711 mm Hg is recommended for out-of-autoclave curing. This corresponds to approximately 65 mbar of absolute vacuum, which is lower than the vacuum used for autoclave molding. This requires the use of high-quality, leak-proof vacuum bags. A vacuum leak check is recommended prior to heating up for cure to make sure that the vacuum loss is no more than 32 mbar in 10 minutes.

4. Prior to heating for curing, a vacuum hold at full vacuum is needed for an extended period. Vacuum hold time depends on the part size and complexity. It may range from 4 hours for a 0.2 m × 0.2 m part size to 16 hours for a 9 m × 8 m part size.

5. Total cycle time may range from 8 to 48 hours, depending on the part thickness, size, and complexity.

5.8 SANDWICH MOLDING

Sandwich constructions containing a lightweight, low modulus core and high modulus PMC face skins (Figure 5.15) are used in many applications. The separation of the faces by the core significantly increases the bending stiffness without adding much weight to the structure. The critical load for buckling is also significantly increased. As a result, a sandwich construction is capable of providing a higher bending stiffness and resistance to buckling per unit weight than a laminate made of only the face material.

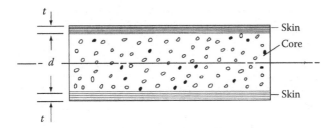

FIGURE 5.15   Sandwich construction.

To understand the advantage of using a sandwich construction, consider the equation for the bending stiffness of a sandwich beam.

$$D \cong \frac{E_s t d^2}{2}, \tag{5.1}$$

where D is the bending stiffness of the sandwich panel per unit width; E_s is the modulus of the skin material; t is the skin thickness on each side of the sandwich; and d is the core thickness.

From Equation 5.1, it can be observed that the bending stiffness D can be increased by increasing the core thickness, i.e., by using a thicker core. Since the core has a relatively low density, increasing the core thickness does not add much weight; however, if it is too thick, the core may crush or fail in shear.

Commonly used cores are aluminum honeycombs, aramid fiber-reinforced phenolic honeycombs (known as Nomex), PP honeycombs, polymeric foams (such as polyvinyl chloride and polyurethane), and balsawood. The aluminum and Nomex honeycomb cores are selected in many aerospace structures. Figure 5.16 shows two sandwich constructions with carbon fiber/epoxy in the face skins and a Nomex honeycomb core. The strength and stiffness of such cores depend on the cell size, cell wall thickness, and properties of the cell material. High core strength is desirable to resist transverse shear stresses in the core and to provide high crush resistance. High stiffness is required to resist core buckling and local wrinkling under compressive stresses.

Sandwich construction can be manufactured in two different ways: (1) by adding the core between laid up uncured prepreg layers that will form the top and bottom face skins and then vacuum bag molding in an autoclave or (2) by adhesively bonding precured face skins to the core. For adhesive bonding, a film adhesive is placed between the core and face skins, which is then cured using the normal curing procedure (Figure 5.17). In molding sandwich laminates, special attention is paid to the ends of the core. If the ends are not sealed, there will be moisture ingress into the laminate which may reduce the properties of the laminate material and, ultimately, the performance of the sandwich laminate. Several end closures are shown in Figure 5.18.

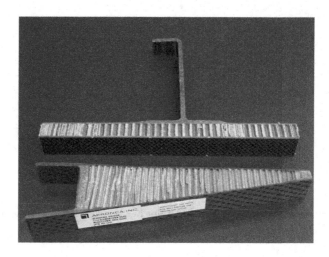

FIGURE 5.16 Examples of sandwich construction with carbon fabric/epoxy laminates in the top and bottom face skins and a Nomex honeycomb core. (Courtesy of Aeronica, Inc., Miami.)

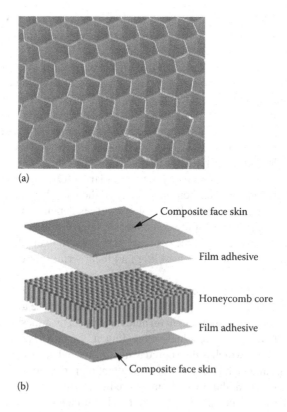

FIGURE 5.17 (a) Aluminum honeycomb core and (b) adhesive layers between the honeycomb core and face skins in sandwich molding.

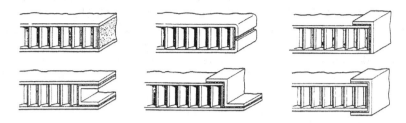

FIGURE 5.18   End closures in sandwich constructions. (From K. Noakes, *Successful Composite Techniques*, 4th Ed., Crowood Press, Marlborough, 2008.)

5.9   DEFECTS IN BAG-MOLDED PARTS

Since bag molding involves a large number of manual operations, the possibility of human error exists, and as a result, a number of defects may appear in bag-molded composite parts. Some of these defects and related quality issues are listed in Table 5.3. The most critical of these defects is the voids. It is difficult, if not impossible, to make composite parts without any voids, but since their presence reduces a number of mechanical properties, such as transverse tensile strength and interlaminar shear strength, it is important to keep the void volume fraction as low as possible. For high-performance applications, the acceptable void volume fraction is 2%. Parts with higher void volume fraction are usually rejected.

In addition to the defects listed in Table 5.3, there are two quality-related issues that are observed in many bag-molded composite parts [16]. They are (1) thickness variation and (2) warpage and distortion. Thickness variation occurs due to three

TABLE 5.3
Examples of Defects Observed in Bag-Molded Composite Parts

Defects	Possible Reasons
Part adhering to the mold surface, surface blemishes	Inadequate tool preparation, such as cleaning, failure to apply mold release or release film
Air entrapment, voids	Failure to properly debulk, inadequate pressure, inadequate vacuum, inadequate sealing, moisture in the prepreg
Reduction in mechanical properties, part warpage	Incorrect ply orientation
Interply separation	Failure to remove backup paper, inadequate pressure
High resin content, resin-rich layers	Inadequate bleeding
Uncured or unevenly cured parts, excess resin, low fiber volume fraction, voids, resin degradation, warpage	Incorrect temperature profile
Excess resin, low fiber volume fraction, voids, inadequate consolidation, thickness variation	Inadequate pressure application

reasons: (1) the top surface of the part not being in contact with a hard mold surface, (2) variation in incoming prepreg thickness, and (3) uneven consolidation. Warpage or distortion occurs for a number of reasons; the most important among them is the thermal contraction difference between the fibers and the resin. As discussed in Chapter 4, process-induced residual stresses may also exist in these parts due to thermal mismatch, interaction with the tool surface, etc.

REFERENCES

1. G. Dillon, P. Mallon, and M. Monaghan, The autoclave processing of composites, in *Advanced Composites Manufacturing* (T. G. Gutowski, ed.), John Wiley & Sons, New York, pp. 207–258, 1997.
2. M. G. Bader and C. Lekakou, Processing of laminated structures, in *Composites Engineering Handbook* (P. K. Mallick, ed.), Marcel Dekker, New York, pp. 371–479, 1997.
3. M. Elkington, D. Bloom, C. Ward, A. Chatzimichali, and K. Potter, Hand layup: Understanding the manual process, *Advanced Manufacturing: Polymer and Composites Science*, Vol. 1, pp. 128–141, 2015.
4. L. D. Bloom, J. Wang, and K. D. Potter, Damage progression and defect sensitivity: An experimental study on representative wrinkles in tension, *Composites: Part B*, Vol. 45, No. 1, pp. 449–458, 2013.
5. *Structural Composites Fabrication Guide*, Vol. 1, Manufacturing Technology Division, US Air Force Materials Laboratory, Wright-Patterson AFB, OH, 1982.
6. A. Mukherjee and B. Varughese, Design guide-lines for ply drop-off in laminated composite structures, *Composites: Part B*, Vol. 32, pp. 153–164, 2001.
7. D. Purslow and R. Childs, Autoclave moulding of carbon fibre-reinforced epoxies, *Composites*, Vol. 17, No. 2, pp. 127–136, 1986.
8. A. C. Loos and G. S. Springer, Curing of epoxy matrix composites, *Journal of Composite Materials*, Vol. 17, pp. 135–169, 1983.
9. P. Olivier, J. P. Cottu, and B. Ferret, Effects of cure cycle pressure and voids on some mechanical properties of carbon/epoxy laminates, *Composites*, Vol. 26, No. 7, pp. 509–515, 1995.
10. J. M. Tang, W. I. Lee, and G. S. Springer, Effects of cure pressure on resin flow, voids, and mechanical properties, *Journal of Composite Materials*, Vol. 21, pp. 421–440, 1987.
11. T. E. Twardowski, S. E. Lin, and P. H. Gesil, Curing in thick composite laminates: Experiment and simulation, *Journal of Composite Materials*, Vol. 27, No. 3, pp. 216–250, 1993.
12. S. Yi and H. H. Hilton, Effects of thermo-mechanical properties of composites on viscosity, temperature and degree of cure in thick thermosetting composite laminates during curing process, *Journal of Composite Materials*, Vol. 32, No. 7, pp. 600–622, 1998.
13. S. R. White, Processing-induced residual stresses in composites, in *Processing of Composites* (R. S. Davé and A. C. Loos, eds.), Hanser, Munich, 2000.
14. P. R. Ciriscioli, Q. Wang, and G. S. Springer, Autoclave curing-comparisons of model and test results, *Journal of Composite Materials*, Vol. 26, No. 1, pp. 90–102, 1992.
15. T. Centea, L. K. Grunenfelder, and S. R. Nutt, A review of out-of-autoclave prepregs—Material properties, process phenomena, and manufacturing considerations, *Composites: Part A*, Vol. 70, pp. 132–154, 2015.
16. J. Schlimbach and A. Ogale, Out-of-autoclave curing process in polymer matrix composites, in *Manufacturing Techniques for Polymer Matrix Composites (PMCs)* (S. G. Advani and K.-T. Hsiao, eds.), Woodhead, Cambridge, 2012.

6 Compression Molding

Compression molding is one of the oldest composite manufacturing processes and is now widely used for making PMC parts using sheet molding compounds (SMCs) and bulk molding compounds (BMCs) as the starting material. The most common forms of SMC and BMC contain randomly oriented discontinuous fibers dispersed in a highly viscous, but uncured thermosetting resin. In the compression molding process, the starting material is placed inside a preheated mold cavity of a matched die mold and then compressed or squeezed to flow sideways and outward as the top half of the mold moves downward relative to the bottom half. In many respects, it is a much simpler process than the bag molding process described in Chapter 5. It requires much less preparation prior to the molding operation, and it is capable of producing parts of complex shapes at a relatively rapid rate, and therefore, it is suitable for high-volume production of composite parts. Part design features, such as nonuniform thickness, ribs, bosses, flanges, holes, cutouts, curvatures, and rounded corners, can be relatively easily incorporated during the compression molding operation. Because of this, secondary postmolding operations such as drilling, machining, and joining may not be needed with compression molded parts. Very little manual labor or monitoring is needed during the compression molding operation, and therefore, the process can be automated and material handling can be done by robots. For all these reasons, compression molding has become one of the leading composite manufacturing processes in mass production industries, such as the automotive and appliance industries, where the need for fast production rates is a primary consideration in manufacturing process selection. Examples of compression-molded parts in the automotive industry include bumper beams, cross members, radiator supports, engine valve covers, hoods, and fenders. Several of these parts are shown in Figure 6.1 [1].

In this chapter, we will discuss compression molding of SMCs and BMCs. Both materials contain randomly oriented short fibers, although in some versions of SMCs, continuous fibers are also included. The matrix material is usually a thermosetting polyester or vinyl ester resin. Epoxy is also used as the matrix material; however, since the curing time of epoxy is much longer than polyesters or vinyl esters, its application in either SMC or BMC is limited. Because of lower cost, E-glass fibers are used as the reinforcement in the SMC and BMC for automotive applications. Carbon fibers can provide higher stiffness per unit mass, but they are much more expensive. Carbon fiber SMC is finding applications in a few aircraft components.

We will start this chapter by first describing the starting materials, namely, SMCs and BMCs. Then we will discuss the compression molding process, flow and cure characteristics, and other topics related to it. More information on SMCs and compression molding process can be found in references [2–4].

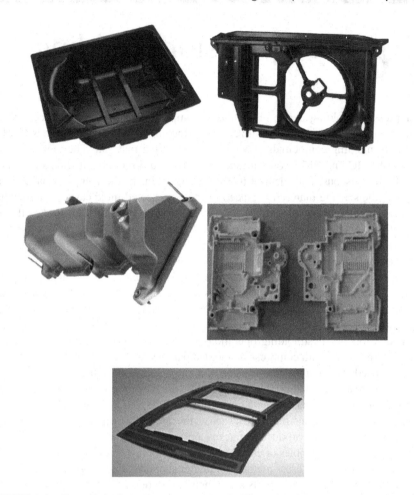

FIGURE 6.1 Examples of compression molded SMC and BMC parts. (From *Design for Success: A Design & Technology Manual for SMC/BMC*, European Alliance for SMC/BMC, Frankfurt, 2007.)

6.1 SHEET MOLDING COMPOUND

6.1.1 MATERIAL

SMC is a continuous sheet of ready-to-mold material containing fibers and mineral fillers dispersed in a highly viscous, but uncured thermosetting resin. The curing of the resin takes place during compression molding in a matched die mold. In addition to these three ingredients, SMC contains several other ingredients, such as catalyst, thickener, low-profile additive, and mold release agent. Although these ingredients are used in relatively small quantities, they serve important functions either during storage (prior to molding) or during molding. A typical composition of SMC is shown in Table 6.1.

TABLE 6.1
Typical Composition of SMCs

Ingredient	Weight Percentage	Volume Percentage
Resin (including styrene monomer)	20–27	33–42
Glass fiber	25–30	18–30
Filler	40–50	29–39
All other ingredients[a]	3–5	6–8

[a] All other ingredients include initiator, thickener, low profile agent, mold release agent, etc.

Unsaturated polyester and vinyl ester are the two most commonly used resins used in SMC. The resin is mixed with a reactive diluent that serves two purposes: (1) it reduces the viscosity of the unreacted resin and (2) it coreacts with the resin during the molding operation to form cross-links between the resin molecules. Styrene monomer, a liquid with a very low viscosity at room temperature, is the most common reactive diluent used for SMC applications. The resin is usually supplied by the resin manufacturer in a premixed condition with 30–50% by weight of styrene monomer. The viscosity of the mixture is in the range of 300–3000 mPa s.

SMCs are designated according to the form of fibers used in them (Figure 6.2). These designations are as follows.

1. SMC-R, in which R represents randomly oriented short fibers in the sheet. The short fibers are usually 25 mm long; but depending on the application, it can range from 12 to 50 mm.
2. SMC-CR, in which C represents continuous parallel fibers on one side of the sheet and R represents randomly oriented short fibers on the other side of the sheet.
3. XMC, in which X represents crisscrossed continuous fibers in the sheet.

(a)
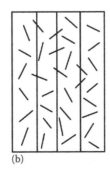
(b)

(c)

FIGURE 6.2 Various forms of SMC: (a) SMC-R, (b) SMC-CR, and (c) XMC.

In both SMC-R and SMC-CR, the fiber content (wt.%) is shown at the end of the letter designation. For example, SMC-R50 contains 50 wt.% of randomly oriented short fibers and SMC-C30R20 contains 30 wt.% continuous fibers and 20 wt.% randomly oriented short fibers. Note that the total fiber content in SMC-R50 and SMC-C30R20 is the same; however, because of the random orientation of fibers in SMC-R50, it behaves as a planar isotropic material with equal properties in all directions in the plane of the sheet, whereas SMC-C30R20 is an orthotropic material. The properties of SMC-C30R20 are direction dependent. For example, its strength and modulus have the highest value in the longitudinal (L) direction and lowest value in the transverse (T) direction of the continuous fibers (Table 6.2). Randomly oriented short fibers in SMC-C30R20 help improve the strength and modulus in the transverse (T) direction, but not to the extent that they are equal to the longitudinal values. Randomly oriented short fibers are also intermingled with X-patterned continuous fibers in XMC to improve its transverse properties. The angle between the crisscrossed continuous fibers is usually in the range of 5–7°. Larger angles are not used, since they tend to reduce the longitudinal strength and modulus of the material. Both SMC-CR and XMC are used instead of SMC-R in applications where their higher properties in the longitudinal direction can be advantageously utilized, for example, in a beam or narrow plate structures.

Fillers are used in powder or particulate form and serve several functions in a SMC. They reduce the volumetric shrinkage of the resin, improve the flow characteristics of the material during molding, and enhance the surface quality of the molded part. Calcium carbonate ($CaCO_3$) is the most commonly used filler in SMCs. Kaolin clay, talc, glass spheres (both solid as well as hollow), and alumina trihydrate ($Al_2O_3 \cdot 3H_2O$) are some of the other fillers used. Most of these fillers are very low-cost materials compared to the resin, and therefore, when added to the resin, they

TABLE 6.2
Comparison of Properties of SMC-R and SMC-CR

Property	SMC-R25	SMC-R50	SMC-R65	SMC-C20R30
Total fiber content (wt.%)	25	50	65	50
Density (g/cm³)	1.83	1.87	1.82	1.81
Tensile modulus (GPa)	13.2	15.8	14.8	21.4 (L)
				12.4 (T)
Tensile strength (MPa)	82.4	164	227	289 (L)
				84 (T)
Strain at failure (%)	1.34	1.73	1.67	1.73 (L)
				1.58 (T)
CLTE ($10^{-6}/°C$)	23.2	14.8	13.7	11.3 (L)
				24.6 (T)

Note: For SMC-C20R30, L and T are the longitudinal and transverse directions of continuous fibers in the composite, respectively.

reduce the overall material cost. Hollow glass spheres, by virtue of their low bulk density, are capable of reducing the weight of SMC. Alumina trihydrate not only acts as a filler, but also serves the purpose of a fire retardant, since the water of hydration in its molecule is easily separated at approximately 220°C, which then acts as an internal flame retardant and smoke suppressant. It is used in many electrical, appliance, and construction applications where flammability and smoke generation are important material selection criteria.

The most commonly used fiber in SMCs is E-glass owing to its low cost. Carbon fibers are used in SMCs for their higher modulus; but because of their high cost, they are limited to a few aerospace applications. The fiber content in SMC can be easily varied to match the application requirement. High fiber content, in the range of 50–70% by weight, is used in structural applications requiring high strength and modulus. Fillers are not added if the fiber content is higher than 60 wt.%, since the high viscosity of SMC at such fiber content already makes the flow of the material in the mold difficult and adding fillers further reduces its moldability. Short fibers used in SMC-R or SMC-CR are usually 25 mm long. Although longer fibers can produce higher strength in the molded composite part, material flow during molding becomes increasingly difficult with longer fibers.

The functions of the other ingredients used in a SMC are described in the following.

1. Catalyst: The role of a catalyst is to initiate the curing reaction, but only after the SMC material is placed in the preheated mold cavity. Commonly used catalysts are organic peroxides, such as TBPB, that readily decompose into free radicals at 130–160°C, which is the typical mold temperature range used in compression molding of SMC. These free radicals react with the styrene and resin molecules and start breaking up the carbon–carbon double bonds present in these molecules, thus initiating the curing reaction. The rate at which a catalyst decomposes into free radicals is one of the factors that determines the cure time in the mold and is an important consideration in selecting the catalyst.

   In some applications, a mixture of a low-temperature catalyst, such as *t*-butyl peroctoate (TBPO), is mixed with a high-temperature catalyst, such as TBPB. Since *t*-butyl peroxybenzoate (TBPB) is more reactive than TBPB and the decomposition temperature of TBPO is about 70°C, it starts the curing reaction earlier than TBPB. The heat of reaction and the corresponding temperature increase trigger the decomposition of TBPB at an earlier time than usual and accelerates its reaction with styrene and resin molecules. Such dual catalyst systems can significantly reduce both gel time and cure time, often with a slight decrease in pot life.

2. Inhibitor: An *inhibitor* (such as hydroquinone) is added only in trace amounts to prevent or reduce any curing reaction that may occur during mixing or storage prior to molding. Thus, inhibitors improve the shelf life of SMCs. On the other hand, they also tend to slow down the curing reaction at the early stages of molding and may increase the cure time.

3. Thickener: The *thickener* plays a key role in the SMC formulation. It increases the viscosity of the resin, a process known as *maturation*, during the storage

of the SMC sheet. The resin remains in an uncured state, but its viscosity increases to a very high level so that the SMC sheet is hard enough to be cut, stacked, and draped on the mold surface. In this condition, the SMC sheet is slightly tacky and very pliable. When a thickened SMC sheet is placed in a preheated mold, the resin viscosity is reduced so that the material can flow relatively easily as the molding pressure is applied.

The viscosity variation during maturation and molding is schematically shown in Figure 6.3. It is important to note that compression molding of good-quality SMC parts requires that the viscosity of the SMC material just prior to molding be within the range of 30×10^6 to 130×10^6 mPa s. If the viscosity is too low, there may be separation of fibers from the resin during molding; on the other hand, too high a viscosity may produce unfilled parts (called *short shots*). The extent of viscosity reduction during molding depends on the viscosity of the thickened SMC just before molding, which depends on the thickener type and concentration as well as maturation time.

The two common thickeners used in SMC are magnesium oxide (MgO) and magnesium hydroxide ($Mg(OH)_2$). Magnesium oxide, even at low concentrations, increases the initial resin viscosity at a higher rate than magnesium hydroxide; however, the latter tends to increase the viscosity much more slowly during maturation, thus keeping the thickened SMC in the moldable range over a longer period.

4. Low-profile additives: *Low-profile additives* are thermoplastic powders that are mixed with the resin to control the shrinkage of the SMC material during cooling from the processing temperature to room temperature. They also improve the surface quality by reducing both long-term and short-term surface waviness. Polyvinyl acetate, polymethyl methacrylate, and polystyrene are some of the thermoplastic materials used in powder form to act as low profile additives. Without the addition of low-profile additives, the linear shrinkage of molded polyester and vinyl ester SMC parts is 5–9%. With the addition of

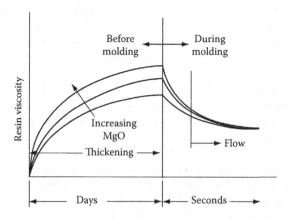

FIGURE 6.3 Resin viscosity variation in an SMC before molding (during maturation) and during flow in compression molding.

10 wt.% of polyvinyl acetate powder, the linear shrinkage reduces to less than 1%. This not only helps in dimensional control of the molded parts, but also helps reduce a number of defects, such as sink marks and voids.

5. Pigments and other additives: Pigments are added either in powder form or in a carrier resin to produce the desired color in molded SMC parts. Examples of pigments are carbon black for black color, iron oxide for red color, and titanium dioxide for white color. Some of these pigments show accelerating or inhibiting effects on the cure rate of the resins. For example, carbon black has an inhibiting effect, whereas iron oxide tends to produce an accelerating effect.

Other additives in an SMC resin paste formulation may include flame retardant, UV absorber, impact modifier, and internal mold release agent. The impact modifier is usually an elastomer, such as acrylonitrile–butadiene and styrene–butadiene copolymers. It is added in liquid form to the resin paste and is thoroughly blended in. In the cured SMC part, it is dispersed as discrete rubber particles in the matrix and provides a higher fracture toughness and impact resistance to the material.

Internal mold release agents, such as zinc stearate and calcium stearate, are added in small quantities (less than 2% by weight) and blended in with the resin paste. Their function is to prevent the cured part from adhering to the mold surface, thus facilitating part ejection from the mold.

6.1.2 PRODUCTION OF SHEET MOLDING COMPOUND

The principal steps in the production of an SMC sheet include mixing, compounding, compaction, and maturation. In the mixing step, the resin–styrene mix is blended with fillers and all the other ingredients except the fibers in a high-intensity mixer to prepare a resin paste. Typical resin paste formulations for SMC-R30, R50, and R65 are shown in Table 6.3. After mixing, the resin paste is transferred to the SMC machine where the next two steps, compounding and compaction, are carried out to

TABLE 6.3
Typical SMC Resin Paste Formulations

	Weight Percentage		
Ingredient	SMC-R30	SMC-R50	SMC-R65
Resin (Unsaturated polyester + Styrene monomer)	31	35	32
Low Profile Additive (Polyvinyl acetate	4.65	4.65	0
Filler (CaCO$_3$)	30	7	0
Thickener (MgO)	2.5	2	1.9
Catalyst (TBPB)	0.35	0.35	0.3
Mold release Agent (Zinc Stearate)	1.50	1	0.8
Inhibitor (Benzoquinone)	Trace amount (<0.005g)		
Total	70	50	35

produce SMC-R and SMC-CR sheet rolls. For XMC, compounding and compaction are performed on a filament winding machine. The XMC sheet is produced by slitting the filament-wound XMC roll along its length. Finally, the maturation step is used to increase the viscosity of the compounded material to a value suitable for handling and molding operations.

Figure 6.4 shows a schematic of an SMC machine. The resin paste is either poured or pumped onto thin polyethylene carrier films at two locations, just behind the metering blades (called the doctor blades). The carrier films are drawn from their rolls at a constant speed. The resin paste thickness carried forward by each carrier film is controlled by the gap between the carrier film and the metering blades, which is controlled by the up or down adjustment of the metering blades. Continuous fiber rovings are pulled from their roving packages and fed into the chopping station, which is located 400–800 mm above the horizontal carrier film. The function of the choppers is to cut the continuous rovings into the desired lengths and sprinkle them evenly and randomly over the horizontally moving resin paste. The amount of chopped fibers deposited on the bottom resin paste layer is controlled by changing the speed of the horizontal carrier film. For producing SMC-CR (as shown in Figure 6.4), collimated continuous fiber rovings are pulled from their packages and placed on top of the randomly oriented chopped fibers. The top carrier film coated with a second thin layer of resin paste is combined with the bottom layer of resin paste and fibers, thus sandwiching the fibers between the top and bottom layers of resin paste (Figure 6.5a). For producing SMC-R, only the randomly chopped fibers are deposited on the bottom resin paste layer, which is then sandwiched with the top resin paste layer. The sandwich construction is then pulled into the compaction section of the machine where the fibers are impregnated with the resin paste by the pressure from the compaction rollers. The gaps between the rolls progressively decrease toward the take-up section of the machine, thus improving the fiber impregnation as well as pushing the entrapped air out through the sides. Pulling the sheet assembly through a set of large rollers at the end of the compaction section further reduces the air entrapment and produces a better fiber wet-out. Finally, the SMC sheet is wound around a take-up roll, wrapped with a barrier film to prevent styrene evaporation and moisture pickup during maturation and then stored in a cold room for maturation.

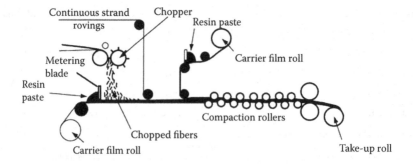

FIGURE 6.4   Process for making SMC-CR sheet.

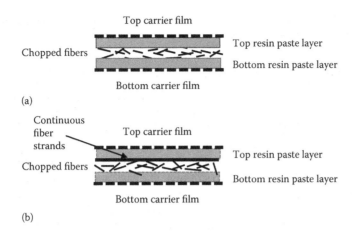

FIGURE 6.5   Construction of (a) SMC-R and (b) SMC-CR before compaction.

The SMC paste at the beginning of the SMC sheet production process has a viscosity in the range of 20,000–40,000 mPa s. Before the SMC sheet can be handled and is ready for compression molding, its viscosity is raised to between 40×10^6 and 100×10^6 mPa s in the maturation chamber, which is a temperature-controlled freezer, usually at 28–32°C. The storage time for maturation depends on the resin–thickener combination and may last up to several days. Once the proper viscosity is achieved, the SMC roll can be removed from the storage for immediate compression molding or left in the storage for several weeks. The shelf life in the storage depends on the resin–catalyst–inhibitor combination and the storage temperature. Before compression molding, the SMC roll must be placed outside the freezer for long enough time to bring it to room temperature so that there will be no moisture condensation on the roll; otherwise, the moisture will turn into steam and become entrapped in the compression molded part in the form of voids.

6.2   BULK MOLDING COMPOUND

BMC is also a ready-to-mold short fiber-reinforced thermoset used for compression molding and injection molding. It is produced in the form of a log and is supplied as a bulk material. Since it has the appearance of a dough, it is sometimes called *dough molding compound*. It is used in a variety of industrial applications, such as electrical fuses and switchgears, pump and motor housings, appliance parts, and under-the-hood automotive components, such as engine valve cover and timing belt pulley.

The commercial BMCs contain 15–20 wt.% E-glass fibers in a polyester resin. The fiber length is between 6 and 25 mm, depending on the application requirement. The resin paste formulation in a BMC is similar to that of an SMC. Chopped glass fiber strands and fillers are compounded with the resin paste in a high-intensity mixer to form a highly viscous mass, which is then extruded through a round die into the form of a continuous log. The extruded log is cut into desired lengths using a pneumatic cutter and then packed in styrene-proof bags. Thickeners are not used in the BMC resin formulation if the BMC log is molded into the finished product soon after the compounding operation.

6.3 COMPRESSION MOLDING PROCESS

Compression molding starts with the placement of a precut, assembled, and weighed amount of the starting material, also called the charge, on the bottom half of a pre-heated mold cavity (Figure 6.6) of a matched die mold. In the case of SMC, the charge contains a stack of several plies of rectangular sheets, which are die-cut from a properly matured SMC roll after thawing it from the storage temperature to room temperature and peeling off the barrier films. The basic steps that are followed in the compression molding of SMC are described in the following. In the case of BMC, precut lengths of BMC logs are placed in the preheated mold cavity; the other steps for compression molding of BMC are very similar to those for SMC.

1. Charge preparation and placement: The SMC plies are assembled and stacked outside the mold. Rectangular ply patterns are commonly used; however, depending on the part and mold design, other ply patterns, such as circular, elliptic, or oblong shapes, are also used. Ply dimensions are selected to cover 60–70% of the mold surface area. The charge is weighed before placing it in the mold. Charge–to-charge weight variation due to variation in SMC sheet weight per unit volume is made up by adding smaller

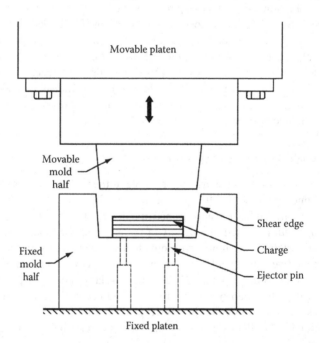

FIGURE 6.6 Compression molding process.

cut pieces from the SMC roll. The location of the charge placement influences the flow length during molding, which in turn influences the defects that may appear in the molded part, such as voids, fiber orientation pattern, and knit line formation.

2. Mold closing and compression: After placing the charge in the bottom half of the mold, the top half is quickly lowered to contact the top surface of the charge and then at a much slower rate, usually 5–10 mm/min, until the pressure on the charge increases to a preset level. As the temperature of the charge starts to increase on contact with the preheated mold surfaces, the viscosity of the SMC material starts to decrease, and with increasing pressure, it starts to flow and fill the cavity. As the material moves toward the cavity extremities, it pushes the air entrapped in the mold and in the charge (between the layers and within each layer) to escape through the vents and the shear edges of the mold. Proper flow of the material is important to reduce the air entrapment; otherwise, the molded SMC part may contain large amounts of voids, blisters, and delaminations. If the mold closes too slowly, the resin in the charge may start to gel and may not be able to flow the entire length of the cavity, causing the cavity to remain less than completely filled and creating a short shot. On the other hand, if the mold is closed too fast, there may not be enough time for the air to escape and the entrapped air will then create voids in the molded part. The length of flow in the mold controls the orientation of fibers in the molded part and, therefore, its mechanical properties and, in some cases, part distortion.

    Depending on the part complexity, part size (which controls the length of flow), and viscosity of the SMC material, the molding pressure may range from 30 to 150 bar. High pressure is required for molding parts that contain deep ribs and bosses. The mold temperature is usually in the range of 130–160°C for polyester and vinyl ester SMCs.

3. Curing: After the cavity is filled, the mold is kept closed for a predetermined length of time to assure that the desired level of curing and consolidation has taken place throughout the part. Depending on the resin–catalyst–inhibitor combination, part thickness, and mold temperature, the cure time may range from 1 minute to several minutes.

4. Part ejection and postmolding operations: After the desired degree of cure is achieved under pressure, the mold is opened and the part is removed from the bottom mold with the aid of ejector pins. The part is then cooled outside the mold, while the mold surfaces are cleaned of any remaining debris and sprayed with a mold release agent, such as zinc stearate, to prepare the mold for the next part. As the part cools outside the mold, it will continue to cure and shrink. Since there are no pressure restraints on the part during postmolding, it may exhibit distortion or generate residual stresses as a result of differential cooling rate in various sections of the part. The reason for differential cooling is nonuniform temperature distribution in the molded part, which mainly occurs due to the presence of thin

and thick sections in the part. After the part has cooled to room temperature, it may be necessary to trim its edges to remove any flash that may have formed during molding.

6.4 CURE CYCLE

The transformation of an uncured stack of SMC plies into a consolidated cured part takes place in the mold through the curing reaction, which is discussed in Chapters 3 and 4. The length of time needed to produce the desired degree of cure is called the cure cycle. In compression molding, the cure cycle occupies the major part of the molding cycle. For example, if the total molding cycle to make one part is 100 s long, about 60 s of this time is for curing. Depending on the size of the part and the capacity of the molding press, there can be multiple cavities installed in the press so that multiple parts can be compression molded within the same cycle time. Reduction in cure cycle to increase the production rate can be achieved by changing the material formulation (for example, using a fast curing resin or a highly reactive catalyst), mold temperature, and part thickness. Preheating the charge outside the mold before placing it in the mold is also an option for reducing the cure cycle.

Temperature–time curves at the outer surface and the centerline of a thick E-glass fiber–SMC molding (Figure 6.7) show that the charge surface temperature quickly reaches the mold temperature and remains relatively uniform compared to the centerline temperature. Owing to the low thermal conductivity of SMC, the centerline temperature slowly increases until the curing reaction is initiated at the midthickness of the part. Also for the same reason, the heat generated by the exothermic curing reaction in the interior of the SMC charge is not efficiently conducted away to the mold surfaces. This causes the centerline temperature to increase to a peak value which is

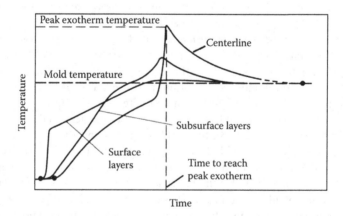

FIGURE 6.7 Temperature profiles at various locations across the thickness of an SMC part during compression molding. (P. K. Mallick and N. Raghupathi: Effect of cure cycle on mechanical properties of thick section poly/thermoset moldings. *Polymer Engineering and Science*, 1979, 19, 774–778. Copyright Wiley-VCH Verlag GmbH & Co. KGaA. Reproduced with permission.)

higher than the mold surface temperature. As the curing reaction nears completion, the centerline temperature gradually decreases to the mold surface temperature. For thin parts, the temperature rise is nearly uniform across the thickness, and the maximum temperature in the material seldom exceeds the mold temperature.

Since the surface temperature first attains the resin gel temperature, curing first begins at the surfaces and progresses inward toward the center of the thickness. The time to reach peak exotherm at the centerline of a thick section SMC part is commonly considered the minimum cure time in the mold. Nearly 90% of the curing reaction is completed at this time. The removal of the part from the mold earlier than this time can leave the center of the charge uncured, and since the pressure is relieved on mold opening, interlayer cracks may be generated near the center of the part. The mechanical properties of the molded part may also be lower if the time in the mold is less than the minimum cure time.

The time to reach peak exotherm depends on the part thickness, mold temperature, resin–catalyst–inhibitor reactivity, and thermal characteristics of the SMC material. The effect of part thickness on the time to reach peak exotherm in SMC-R65 and SMC-R72 is shown in Figure 6.8. This figure also shows the beneficial effect of preheating the charge outside the mold. Compression molding with preheating involves dielectrically heating the charge to a pregel temperature outside the mold, quickly transferring it to the mold, and finishing the curing reaction in the mold. Dielectric heating increases its temperature of the charge rapidly and uniformly throughout the volume. During molding, the thermal gradient remains nearly constant across the thickness of the preheated charge, which allows uniform curing in the thickness direction. As a result, residual curing stresses in the molded part are also reduced.

The time to reach peak exotherm can be reduced by increasing the mold temperature [5,6] (Figure 6.9); however, the peak exotherm temperature may also increase. Since peak exotherm temperature of 200°C or higher may cause burning and thermal

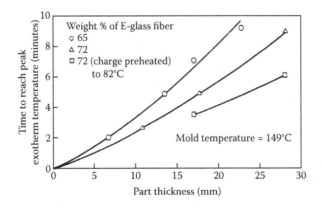

FIGURE 6.8   Effect of part thickness on the time to reach peak exotherm. (P. K. Mallick and N. Raghupathi: Effect of cure cycle on mechanical properties of thick section poly/thermoset moldings. *Polymer Engineering and Science*, 1979, 19, 774–778. Copyright Wiley-VCH Verlag GmbH & Co. KGaA. Reproduced with permission.)

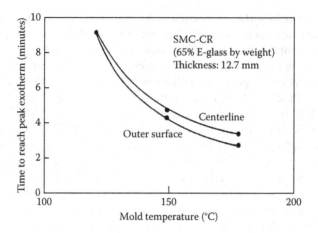

FIGURE 6.9 Effect of mold temperature on the time to reach peak exotherm. (P. K. Mallick and N. Raghupathi: Effect of cure cycle on mechanical properties of thick section poly/thermoset moldings. *Polymer Engineering and Science*, 1979, 19, 774–778. Copyright Wiley-VCH Verlag GmbH & Co. KGaA. Reproduced with permission.)

degradation of the resin, high mold temperatures in thick SMC parts should be avoided. Thermal degradation may cause a weak interlaminar zone in the interior of the molded part. Based on this observation, Panter [7] generated molding diagrams that show the cure time as a function of both mold temperature and part thickness (Figure 6.10). The safe mold temperature zone on this diagram depends not only on the mold temperature, but also on the part thicknesses.

Another parameter affecting the time to reach peak exotherm is the resin formulation or the resin–catalyst–inhibitor combination [8]. As the data in Table 6.4 show,

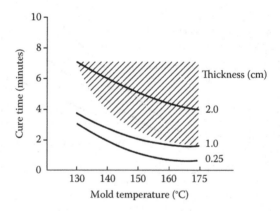

FIGURE 6.10 Molding diagram showing the effect of mold temperature and part thickness on cure time. (From M. R. Panter, The effect of processing variables on curing time and thermal degradation of compression molded SMC, *Proceedings of 36th Annual Conference, Society of Plastics Industry*, 1981.)

TABLE 6.4
Effects of Resin Formulation on the Time to Reach Peak Exotherm

Initiator		Inhibitor[b] Concentration (parts by weight)	Time to Reach Peak Exotherm Temperature (s)	
Type[a]	Concentration (parts by weight)		At Mold Temperature = 125°C	At Mold Temperature = 145°C
TBP	0.49	0	268	90
TBP	0.98	0	178	81
PDO	5.5	0	47	43
PDO	5.5	0.28	80	54
PDO	5.5	0.55	113	65
TBP + PDO	0.25 + 2.75	0.28	120	60

Note: The resin formulation contains 57 parts of unsaturated polyester in styrene, 43 parts of polymethyl methacrylate in styrene, 138 parts of $CaCO_3$ filler, and 9.6 parts of MgO (thickener).

[a] TBP: *t*-butyl perbenzoate; PDO: *t*-butyl peroxy-2-ethyl hexanoate.

[b] Inhibitor is benzoquinone.

the time to reach peak exotherm can be controlled by changing the catalyst and its concentration, adding an inhibitor and changing its concentration, and by combining two different catalysts.

The thermal characteristics of the SMC material that influence the time to reach peak exotherm are its heat capacity and thermal conductivity. In general, these properties do not strongly depend on the resin formulation; however, they can be controlled by varying the other ingredients, such as the filler content and fiber content. Increasing the filler content in SMC formulations decreases the peak exotherm temperature since it replaces part of the resin and, thereby, decreases the total amount of heat liberated. It also acts as a heat sink within the material. The time to reach peak exotherm is also reduced with increasing filler content. Thus fillers can play a significant role in reducing the cure cycle of a part. Increasing the fiber content has a similar effect.

6.5   FLOW PATTERN

The flow pattern of the SMC material during compression molding is a key factor in determining the quality of the molded part. It determines not only the extent of mold filling, but also the defects that may be caused by uneven or interrupted flow. Examples of these defects are fiber misorientation, nonuniform fiber distribution, porosity, and knit lines. As the temperature of the SMC charge increases in the mold, the effect of the thickening reaction breaks down and the resin viscosity is reduced. If the material does not attain low viscosity before gelling, its flow in the mold can be severely restricted. If premature gelation occurs before the mold is filled, the molded part will be incomplete and may contain voids and interlayer cracks.

A number of investigators have studied the basic flow behavior of SMC-R with multicolored layers in flat plaque mold cavities [9–11]. At fast mold-closing speeds, the layers flow with uniform extension (plug flow), with slip occurring at the mold surface (Figure 6.11a). The charge thickness does not influence this flow pattern at fast mold-closing speeds. At slow mold-closing speeds, on the other hand, SMC flow pattern depends very much on the charge thickness. For thick charges, the viscosity of SMC in layers adjacent to the hot mold surfaces rapidly decreases while the viscosity in the interior layers is still quite high. As a result, the outer layers begin to flow before the interior layers and may even squirt into the uncovered areas of the mold (Figure 6.11b). Thus, the outer layers in this case undergo greater extensional deformation than the interior layers, with slip occurring between the layers as well as at the mold surface. As the charge thickness is reduced, the extensional deformation becomes more uniform and approaches the same flow pattern observed at fast mold-closing speeds. For a good molded part, a rapid mold-closing speed is desirable since it avoids the possibility of premature gelation and produces the most uniform flow pattern regardless of the charge thickness.

Another aspect of flow in compression molding is whether or not the mold will be properly filled, especially if it is for a large part and contains ribs, bosses, corners, etc. Mold filling depends on the charge shape and its location in the mold. It also depends on the viscosity of the SMC as it flows in the mold and at the molding conditions being used (temperature, pressure, and mold closing speed). There are several theoretical models that provide information on the flow behavior of SMC charge in molds of simple shapes, such as a rectangular mold. In the simplest of these models, the SMC charge is assumed to be thin so that an isothermal condition exists through its thickness. The charge is centrally placed in the preheated mold.

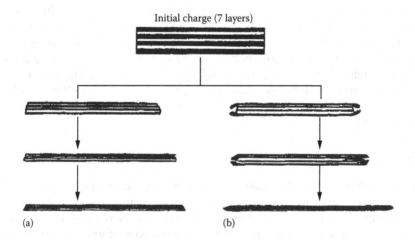

FIGURE 6.11   Stages of deformation and flow of layers in a seven-layered SMC charge at (a) fast and (b) slow mold-closing speeds. (J. D. Fan, J. M. Marinelli, and L. J. Lee: Optimization of polyester molding compound: Part I: Experimental study. *Polymer Composites*, 1986, 7, 239–249. Copyright Wiley-VCH Verlag GmbH & Co. KGaA. Reproduced with permission.)

As the mold is closed in the negative z-direction, it starts to spread in the xy plane (Figure 6.12). The following assumptions are also made.

1. There is no slip between the charge surfaces and the mold surfaces.
2. The SMC material is an incompressible, isotropic, Newtonian fluid.
3. The dominant stresses are the transverse shear stresses, and the in-plane stresses are negligible.
4. There is negligible pressure variation in the thickness direction.
5. The flow takes place under isothermal condition without any curing.

Based on the assumptions mentioned above, Tucker et al. [12] developed the following equation, which can be solved to determine the pressure distribution in the SMC charge as it flows in the xy plane:

$$\frac{\partial^2 p}{\partial x^2} + \frac{\partial^2 p}{\partial y^2} = -12\eta \left(\frac{s}{h^3} \right), \tag{6.1}$$

where p is the pressure (which is a function of x and y); η is the viscosity of the SMC material at the molding temperature; h is the instantaneous charge height; and s is the mold closing speed (assumed to be constant).

Equation 6.1 is known as the generalized Hele–Shaw (GHS) equation. It is numerically solved to determine the pressure distribution. By knowing the pressure distribution, the velocity field can be determined in the entire flow area. The thickness-averaged velocity components $\bar{u}$ and $\bar{v}$ in the x- and y-directions can then be calculated using the following equations:

$$\bar{u} = -\frac{h^2}{12\eta} \left(-\frac{\partial p}{\partial x} \right),$$
$$\bar{v} = -\frac{h^2}{12\eta} \left(-\frac{\partial p}{\partial y} \right). \tag{6.2}$$

The flow front positions and the instantaneous shape of the flowing charge can be determined after the velocity field is calculated using Equation 6.2. It is to be noted that

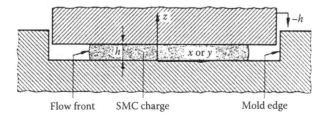

Flow front   SMC charge   Mold edge

FIGURE 6.12   Model for mold filling analysis.

Equations 6.1 and 6.2 are valid for long thin parts for which the length-to-thickness (*L/h*) ratio is very large. Figure 6.13 [12] shows the flow progression of a rectangular charge determined by using Equation 6.1. As can be observed in Figure 6.13, the predicted flow front closely matches with the experimentally observed flow front when the initial charge thickness is 3 mm, but not when the charge thickness is 9 mm.

In thick charges, the surfaces contain a thin layer of liquid resin which acts as a lubricant, and therefore, the assumption of a no-slip condition may not be valid. The velocity profiles through the thickness (1) for the no-slip assumption and (2) with slip occurring at the charge surfaces are shown in Figure 6.14. For the no-slip

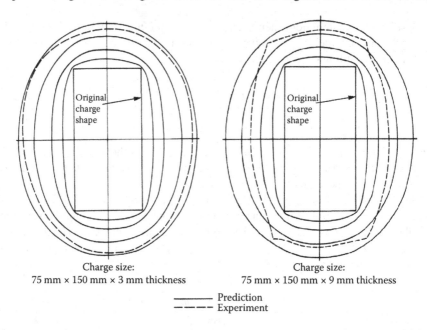

Charge size:
75 mm × 150 mm × 3 mm thickness

Charge size:
75 mm × 150 mm × 9 mm thickness

————— Prediction
— — — Experiment

FIGURE 6.13 Flow progression of a rectangular SMC charge as predicted by the GHS equation. (Reprinted from *Journal of Engineering for Industry*, 106, C. C. Lee, F. Folgar, and C. L. Tucker, Simulation of compression molding for fiber-reinforced thermosetting polymers, 114–125, Copyright (1984), with permission from Elsevier.)

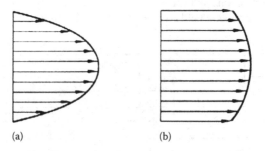

(a)   (b)

FIGURE 6.14 Velocity profiles through the thickness of an SMC charge with (a) no slip at the mold walls and (b) with slip at the mold walls.

condition, the velocity profile is parabolic with zero velocity at the top and bottom of the thickness and the maximum velocity occurring at the midthickness. On the other hand, if the thin liquid resin layers generated on the surfaces of the charge act as a lubricant and provide a condition for slippage, the velocity profile becomes more uniform, approaching the condition for plug flow or uniform extension.

Assuming that the thin liquid resin layers on the charge surfaces provide a hydrodynamic lubrication condition so that the frictional resistance to flow at the surfaces is proportional to the relative velocity between the charge and the mold surfaces, Barone and Caulk [13,14] developed the following flow equation under isothermal condition without any curing:

$$\frac{\partial^2 p}{\partial x^2} + \frac{\partial^2 p}{\partial y^2} = \eta h \left(\frac{\partial^3 u}{\partial x^3} + \frac{\partial^3 u}{\partial x \partial y^2} + \frac{\partial^3 v}{\partial x^2 \partial y} + \frac{\partial^3 v}{\partial y^3} \right) - 2k_H \left(\frac{s}{h} \right), \qquad (6.3)$$

where k_H is called the coefficient of hydrodynamic friction. The first term on the right-hand side of Equation 6.3 represents the resistance of the material to extensional deformation, and the second represents the frictional resistance at the interface of the charge and the mold surfaces. For a thin charge, the first term is negligible compared to the second term, and Equation 6.3 transforms into a simpler form:

$$\frac{\partial^2 p}{\partial x^2} + \frac{\partial^2 p}{\partial y^2} = -2k_H \left(\frac{s}{h} \right). \qquad (6.4)$$

The flow front progression predicted by Barone and Caulk's model for an elliptic charge pattern with four layers and one layer is shown in Figure 6.15. It shows that this flow model works well with both thin and thick charge patterns.

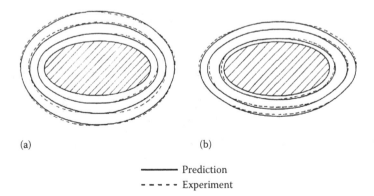

(a)   (b)

——— Prediction
- - - - - Experiment

FIGURE 6.15 Flow front progression of an elliptic SMC charge determined in molding experiments vs. prediction using Barone and Caulk model (with slip): (a) with only one layer in the charge and (b) with four layers in the charge. (M. R. Barone and D. A. Caulk: Kinematics of flow in sheet molding compounds. *Polymer Composites*, 1985, 6, 105–109. Copyright Wiley-VCH Verlag GmbH & Co. KGaA. Reproduced with permission.)

Barone and Caulk's model has been extended by several investigators who took into account the non-Newtonian viscosity characteristics of SMC, nonisothermal condition, and initiation of curing during the filling stage [15,16]. Assuming a hydrodynamic friction condition at the mold walls, Abrams and Castro [17] developed a macroscopic force balance approach to estimate the mold closing force required to fill the mold cavity. Their simplified model predicts the mold closing force that is similar to the measured values.

6.6   FIBER ORIENTATION

The ideal fiber orientation in an SMC-R composite part is random, since it produces a planar isotropic behavior with equal mechanical properties in all directions in

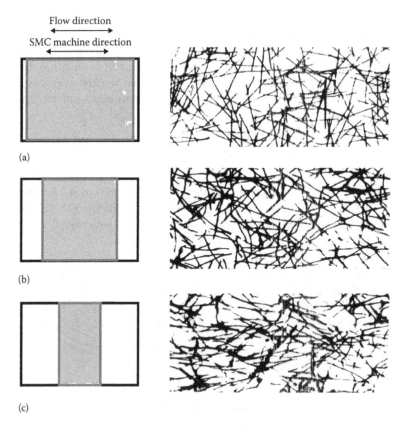

FIGURE 6.16   Fiber orientation patterns in compression molded plates with (a) 97%, (b) 67%, and (c) 32% initial mold surface coverages. (From D. L. Denton and S. H. Munson-McGee, *High Modulus Fiber Composites in Ground Transportation and High Volume Applications,* ASTM STP 873 (D. W. Wilson, ed.), ASTM, Philadelphia, PA, 1985.)

the plane of the molded part. However, the complex flow pattern in the compression molding of a part of complex shape often produces fiber orientations that are not completely random. There may be preferential orientation fibers in local areas which will make the part stronger in that direction, but weaker in other directions. Such preferential orientations are found if the flow distance is very long or is obstructed by inserts or mold edges along the flow path. The dependence of fiber orientation on flow distance can be observed in Figure 6.16 [18]. It shows the fiber orientation distributions for three different initial mold surface coverages in a compression molded flat plate. The 32% initial mold surface coverage has a much longer flow distance than the 67% initial mold surface coverage, and accordingly, it produces a much larger proportion of fibers in the flow direction. Preferential fiber orientation also occurs if the flow takes place from a thicker section to a thinner section creating a convergent flow or from a thicker section to a thinner section creating a divergent flow (Figure 6.17).

Flow-induced preferential fiber orientation causes direction-dependent variation of mechanical properties of compression-molded SMC-R parts. This can be seen in Table 6.5 in which the tensile strength and tensile modulus values are reported in the flow and normal-to-flow directions. The initial mold surface coverage in this case was 38% of the mold surface.

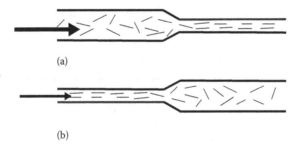

(a)

(b)

FIGURE 6.17   Fiber orientation due to (a) convergent flow and (b) divergent flow. (Arrows indicate the direction of flow.)

TABLE 6.5
Tensile Properties of SMC-R50 (with 38% Mold Surface Coverage)

Direction	Tensile Strength (MPa)	Tensile Modulus (GPa)
Parallel to flow	215	17.4
Normal to flow	75.2	10.9

6.7 DEFECTS IN COMPRESSION MOLDED SMC PARTS

Compression-molded SMC parts may contain a wide variety of surface and internal defects (Table 6.6). The surface defects usually create a poor surface appearance or unacceptable surface finish, while the internal defects (shown in Figure 6.18) may affect performance of the molded part. The origins of some of these defects are discussed next.

1. Porosity can be either a cluster of small internal voids or surface pinholes (Figure 6.18a) caused by the entrapment of air or other gases in the molded part. Air is introduced into the SMC charge due to (1) mechanical blending used for the preparation of the resin paste, (2) poor wetting of the fiber surface by the resin, (3) insufficient compaction process, and (4) layering of SMC sheets in the charge. In addition, there is a large amount of air entrapped in the closed mold. Air entrapped in the SMC prior to mold closure is squeezed into small volumes by the pressure exerted during molding. A substantial amount of these air volumes can be carried away by the material flowing toward the vents and shear edges. However, if proper venting is not provided in the mold or if the material viscosity is high during the flow in the mold, these air volumes may remain entrapped as voids in the molded part.
2. Blisters are small interlaminar cracks (Figure 6.18b) formed at the end of the molding operation due to excessive gas pressure in the interior region of the molded part. The internal gas pressure is generated during molding from unreacted styrene monomer in undercured areas of the part or from large pockets of entrapped air between the stacked layers. If this internal pressure is high, interlaminar cracks may form at the time of mold opening. The delaminated area near the surface may bulge out into a dome-shaped blister by the entrapped gas pressure. Blisters may also appear during some post-molding operations. For example, if the molded part surface is painted for appearance purposes, it may require high-temperature baking in an oven to

TABLE 6.6
List of Defects Observed in Compression Molded SMC Parts

Defect	Possible Contributing Factors
Pinhole	Coarse filler particles, filler particle agglomeration
Long-range waviness or ripple	Resin shrinkage, uneven glass fiber distribution
Craters	Poor dispersion of the lubricant (zinc stearate)
Sink marks	Resin shrinkage, uneven fiber distribution, fiber length, and fiber orientation
Surface roughness	Resin shrinkage, fiber bundle integrity, strand dimensions, fiber distribution
Dark areas	Styrene loss from the surface
Pop-up blisters in painted parts	Subsurface voids due to trapped air and volatiles
Knit lines	Divided flow front due to inserts, charge placement

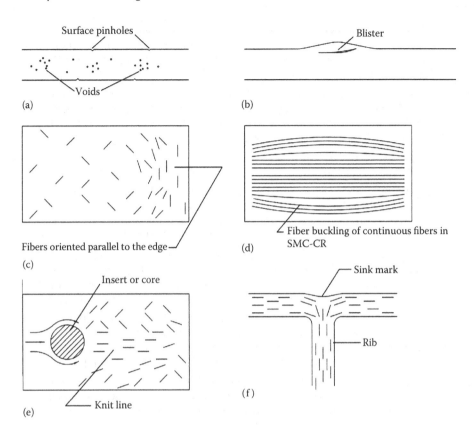

FIGURE 6.18 Defects observed in compression-molded SMC parts: (a) porosity, (b) blisters, (c) fiber mis-orientation, (d) fiber buckling, (e) knit line behind a metal insert, and (f) sink mark.

cure the surface paint, which can cause the expansion of entrapped air or gases. Griffith and Shanoski [19] suggested two possible ways of reducing blisters:

a. Minimize the entrapped air. The most effective method for minimizing the entrapped air is vacuum-assisted molding, in which air from the mold is evacuated just as the mold is closed. A second method of reducing the entrapped air is to allow more outward flow by stacking more plies over a smaller area instead of stacking fewer plies over a larger area.

b. Increase the interlaminar shear strength by changing the resin type, using coupling agents, reducing contamination between layers, decreasing the molding temperature, and assuring proper cure before the mold pressure is released.

3. In any molding operation involving long flow paths, it is extremely difficult to control the preferential orientation of fibers. With compression molding of SMC-R, abrupt changes in thickness, any obstruction in the flow path, or the presence of high shear zones can create fiber orientations that deviate

from the ideal random orientation. As a result, the molded part may become locally anisotropic with its strength and modulus higher in the direction of flow than in the transverse direction. In the compression molding of SMC-R, it is common practice to cover 60–70% of the mold surface with the charge and then use high molding pressure to spread it over the entire mold cavity. When the flow front contacts the cavity edges, discontinuous fibers in the SMC-R tend to rotate normal to the flow direction (Figure 6.18c). This results in strength reduction normal to the flow direction and makes the edges prone to early cracking.

Compression molding of SMC-CR or XMC containing continuous fibers is normally performed with 90–95% initial mold surface coverage. For these materials, flow is possible only in the transverse direction of fibers. If excessive transverse flow is allowed, continuous fibers in the surface and subsurface layers of both SMC-CR and XMC may buckle (bow out) near the end of the flow path (Figure 6.18d). In addition, the included angle between the X-patterned fibers in XMC may also increase. As a result, the longitudinal tensile strengths of SMC-CR and XMC are reduced in areas with fiber misorientation. However, since severe fiber misorientations are generally restricted to the outer layers, increasing the number of plies improves the longitudinal strength to the level observed with no misorientation.

4. Knit lines are domains of aligned fiber orientation and are formed at the joining of two divided flow fronts (Figure 6.18e), such as behind a metal insert or a core pin and where two or more separate flow fronts arising from multiple charge pieces meet. Multiple charge pieces are used for compression molding large and complex parts. Since fibers tend to align themselves along the knit line, the strength of the part in a direction normal to the knit line is reduced.

The formation of knit lines can be reduced by proper charge placement in the mold. A common location of knit lines is behind an insert or a core pin used for forming molded-in holes. Thus, if the holes are in a high-stress area, it is better to drill them in a postmolding operation, since knit lines formed behind such holes may initiate premature cracking along the knit line. Similarly, if the holes are close to the edges of the molded SMC part, it is better to drill them after molding; otherwise, cracks generated along the knit lines can easily extend to the edges of the part.

5. Warpage is critical in thin-section moldings and caused by variations in cooling rate between sections of different thicknesses or different fiber orientations. Differential cooling rates may also lead to complex residual stresses, which may ultimately reduce the strength of a molded part.

6. Nonuniform cure is critical in thick-section moldings and can create a gradient of properties in the thickness direction. Since the curing reaction is initiated at the surfaces and progresses inward, it is likely that insufficient molding time will leave the interior undercured. As a result, the interlaminar shear strength of the molded part is reduced.

The effect of various molding times on the development of through-thickness properties of a thick-section molding is demonstrated in Figure 6.19. This figure was developed by sectioning a 12 mm thick compression-molded

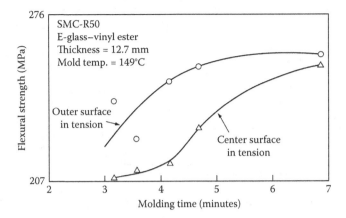

FIGURE 6.19 Effect of molding time on the flexural strength of a 12 mm thick compression-molded SMC plaque. (P. K. Mallick and N. Raghupathi: Effect of cure cycle on mechanical properties of thick section poly/thermoset moldings. *Polymer Engineering and Science*, 1979, 19, 774–778. Copyright Wiley-VCH Verlag GmbH & Co. KGaA. Reproduced with permission.)

plaque along the center plane and testing each half in three-point flexural tests, one with the outer skin in tension and the other with the exposed center in tension. For short mold-opening times, the center has a much lower strength than the outer skin, indicating that the part was removed prior to the completion of cure at the center and the material at the center was undercured. The difference in strength is reduced at higher molding times.

7. Sink marks are small-surface depressions normally observed on the surface opposite to the bases of the ribs (Figure 6.18f) that are included in the design of the part to increase its bending stiffness. The flow of material into a rib creates a fiber-rich zone near its base and resin-rich zone near the opposite surface. Since the resin-rich zone has a higher coefficient of thermal contraction, it shrinks more than the surrounding material, which contains a more uniform fiber distribution. As a result, the surface opposite to the rib is slightly depressed causing the appearance of a sink mark. If the surface contains a large number of closely spaced sink marks, it appears less glossy.

Nonuniform flow pattern of the material in the mold is generally considered the reason for the separation of resin from the fibers at or near the base of a rib. Smith and Suh [20] have shown that protruding rib corners (Figure 6.20) create less sink depths than either sharp or rounded rib corners. Their experiments also show that sink depths in 12 mm long fiber-reinforced SMC are lower than those in 25 mm long fiber-reinforced SMC. Uneven part thickness on the two sides of a rib tends to reduce the sink depth as well as shift the sink mark toward the thicker section.

Poor surface finish caused by sink marks is undesirable in highly visible exterior automotive body panels, such as a hood or an outer door panel. Short ribs are

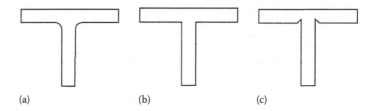

(a)   (b)   (c)

FIGURE 6.20 Rib corner designs: (a) rounded, (b) sharp, and (c) protruded. (K. L. Smith and N. P. Suh: An approach toward the reduction of sink marks in sheet molding compound. *Polymer Engineering and Science*, 1979, 19, 829–834. Copyright Wiley-VCH Verlag GmbH & Co. KGaA. Reproduced with permission.)

commonly used on the inside surface of these panels to improve their bending stiffness. However, sink marks formed on the outside surface reduce its surface finish. Although sink depths can be controlled by using longer ribs, a combination of long and short fibers in the SMC-R sheets, or a low-profile resin, sinkmarks are not completely eliminated. To mask the surface imperfections such as surface waviness, porosity, and sink marks, the outer surface is coated with a thin layer of polyester or polyester–urethane hybrid resin by a process called *in-mold coating*. Just before the completion of the cure cycle, the top mold is retracted by a small amount (0.2–0.5 mm) and the coating is injected over the top surface. The mold is then closed and the curing operation is continued to completion. Curing of the coating may require additional time, which will increase the overall mold cycle time. The second method of in-mold coating involves injecting the coating at a high pressure at the precise moment the SMC part exhibits the maximum curing shrinkage in the mold. This method does not require opening and closing the mold, but requires precise process control.

6.8 COMPRESSION MOLDING PARAMETERS

The effects of molding parameters on the strength of compression-molded SMC parts are briefly described in the following:

1. Cure time: Inadequate cure time in the mold leads to uncured or partially cured layers in the interior of a thick compression molded part, which can in turn cause strength variation across the thickness of the part.
2. Mold closing speed: Fast mold-closing speed leads to low strengths because of insufficient removal of air from the charge and subsequent void formation.
3. Molding pressure: A higher molding pressure produces better consolidation and lower void content and, therefore, higher tensile strength.
4. Charge size and weight: Charge size determines the initial mold surface coverage and the length of flow in the mold. Very long flow lengths may produce flow-induced fiber orientation that, in turn, can cause direction-dependent variation in strength and other properties. On the other hand, very limited flow may not be able to expel the air out of the charge, thus producing voids and lower strengths. To maintain a constant charge weight from one part to the next, small strips of SMC are often added to compensate

for the SMC sheet weight variation. This can also cause flow disturbances and property variation in the molded part.

5. Charge pattern and placement: The charge pattern should be as simple as possible, and the charge placement in the mold should be very consistent from one production cycle to the next so that the molded parts exhibit reproducible quality. Proper charge placement is extremely important in order to fill the mold, reduce knit line formations, and decrease air entrapment.

6.9 MOLD DESIGN CONSIDERATIONS

Since compression molding is a high pressure–high temperature process, molds used for large production volumes are constructed of alloy steels. The mold surfaces are usually heat-treated to increase their hardness and nickel plated for good surface finish. Plating also reduces frictional resistance and improves material flow. Shear edges with a minimum draft angle of 3° are provided around the perimeter of each mold half for easy opening and closing of the mold (Figure 6.21). However, the shear edges on the top and bottom mold halves are designed not to contact with each other when the mold is closed, and as a result, a very thin flash is created at the outer edges of the molded part, which needs to be trimmed after the part cools down. For easy part removal, a small draft angle (usually 3–5°) is provided on the nonappearance side of the part. Ejector pins are used on the bottom mold half to push the part out after molding. They also serve as air escape routes during material flow.

The molds are heated by pumping either hot oil or superheated steam through channels machined in the mold platens. The usual practice is to evenly space these heating channels in the platens and locate them at a uniform distance from the cavity surface. However, the heat requirement in the mold may not be uniform. Therefore, it is recommended that the heating channels be positioned such that the temperature distribution on the mold surfaces is approximately uniform. For example, the area in the mold where the charge is placed is heat deficient, whereas the area where the heated SMC material flows as the mold is closed is heat rich. In this case, the number of heating channels should be higher in the charge placement area

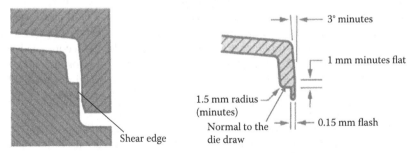

Recommendations for part edge design

FIGURE 6.21   Shear edge at the perimeter of a matched die mold. (From *Design for Success: A Design & Technology Manual for SMC/BMC*, European Alliance for SMC/BMC, Frankfurt, 2007.)

than in the rest of the mold. The temperature control of the mold is very important to avoid pregelling in hotter areas and undercuring in cooler areas.

Hydraulic presses are commonly used for compression molding operations. As the mold is closed, the upper platen of the press may not remain parallel to the lower platen due to the eccentricity of loading caused by uneven charge placement in the mold or uneven pressure distribution during flow. The clearance in the guidance system of the platens and nonuniform elastic deformation of the press frame also tend to increase the parallelism problem. Nonparallel platens cause thickness variation in the molded part, which in turn can cause fiber misorientation, flow instability, and warpage. One of the ways to correct the problem is to use four independently controlled hydraulic cylinders, one at each corner of the upper platen, instead of only one at the center as in most conventional hydraulic presses. The upper platen is maintained parallel to the lower platen during mold closing by using a control mechanism that includes position sensors mounted at each corner and hydraulic valves that are activated by the error signals from these sensors.

6.10 PART DESIGN CONSIDERATIONS

In this section, a few part design guidelines are given that are important to consider for compression-molded parts:

1. Part thickness: The part thickness is primarily determined to meet the design criteria for strength, stiffness, or both. However, if the part is too thick, it will take a longer time to cure the part in the mold. Since this will affect the production cycle time, it may be better to use ribs or other design features that will increase the load-carrying capacity or stiffness of the part without causing significant increase in cycle time. It is also important to maintain uniform thickness so that there is uniform flow, curing, and shrinkage. Uniform thickness is also important to maintain random fiber orientation and reduce the possibility of warpage due to the possibility of uneven curing in parts with nonuniform part thickness.
2. Corner radius: Many parts contain radii at the corners of their vertical and horizontal sections. In general, the inside and outside radii at a corner should be the same so that there is uniform flow at the corner. To promote uniform flow and facilitate part removal from the mold, the recommended minimum radii are 2 mm for the inside corner and 1.5 mm for the outside corner. To produce sharp corner radii on a part, corner radii on the mold surfaces also have to be sharp. Fatigue cracks are often observed at the corners of molds, particularly if they have sharp corners.
3. Ribs and bosses: Ribs are added to increase the bending stiffness of the part (instead of increasing its nominal thickness) and bosses are used to mount threaded fasteners. Properly designed ribs and bosses can reduce the depth of sink marks on the surface opposite to the ribs and bosses and improve the surface appearance. For example, sink marks are reduced if the base of the rib is thinner than the part thickness. General recommendations for rib dimensions are shown in Figure 6.22.

Rib base thickness = 0.75 h for class A; otherwise, h

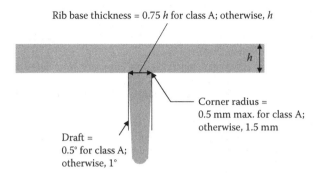

h

Corner radius =
0.5 mm max. for class A;
otherwise, 1.5 mm

Draft =
0.5° for class A;
otherwise, 1°

FIGURE 6.22 Rib design recommendations. (Class A refers to the surface quality for exterior automotive body panels.) (From *Design for Success: A Design & Technology Manual for SMC/ BMC*, European Alliance for SMC/BMC, Frankfurt, 2007.)

4. Molded-in openings: Openings, such as holes and cutouts, can be molded in compression-molded parts using inserts and cores. However, if the charge is not properly designed and located in the mold, knit lines may form at one or more areas close to the molded-in openings. Knit lines formed in areas where the stresses due to service loads are high and normal to their orientation can be a source of early cracking and failure initiation in the part. In that case, it is better to use secondary operations after molding, such as drilling and trimming, to make these openings, albeit with additional cost. To facilitate drilling, it is possible to use mash-offs (Figure 6.23) in the molded parts. The charge is placed directly over the mash-off areas to avoid the formation of knit lines. Since the mash-offs are very thin, they can be punched out or drilled out after molding the part.

5. Shrinkage: Part shrinkage is measured by comparing the long dimension of the molded part at room temperature with the long dimension of the mold at room temperature. It not only depends on the thermal contraction of the material as it cools down from the mold temperature to room temperature, but also on the curing shrinkage of the resin. Volumetric shrinkage of neat polyester and vinyl ester resins is in the range of 5–12%. When fibers and fillers are added to the resin, the volumetric shrinkage is reduced to less than 3%. Lower shrinkage can be attained by adding low-profile additives, such as polyvinyl acetate or polystyrene powder to the resin mix. Typically, to hold the dimensions of the molded part to the design, the amount of expected shrinkage is added to the mold dimension so that after cooling, the final part dimensions become equal to the desired dimensions.

Mash-off area

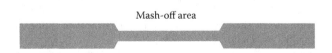

FIGURE 6.23 Mash-off at the location where a hole will be drilled in a secondary operation after compression molding.

REFERENCES

1. *Design for Success: A Design & Technology Manual for SMC/BMC*, European Alliance for SMC/BMC, Frankfurt, 2007.
2. P. K. Mallick, Sheet molding compounds, in *Composite Materials Technology* (P. K. Mallick and S. Newman, eds.), Hanser, Munich, 1990.
3. R. W. Meyer, *Handbook of Polyester Molding Compounds and Molding Technology*, Chapman & Hall, New York, 1987.
4. H. G. Kia (ed.), *Sheet Molding Compounds*, Hanser, Munich, Germany, 1993.
5. P. K. Mallick and N. Raghupathi, Effect of cure cycle on mechanical properties of thick section poly/thermoset moldings, *Polymer Engineering and Science*, Vol. 19, pp. 774–778, 1979.
6. L. J. Lee, Curing of compression molded sheet molding compound, *Polymer Engineering and Science*, Vol. 21, pp. 483–492, 1981.
7. M. R. Panter, The effect of processing variables on curing time and thermal degradation of compression molded SMC, *Proceedings of 36th Annual Conference, Society of Plastics Industry*, 16F, 1981.
8. J. D. Fan, J. M. Marinelli, and L. J. Lee, Optimization of polyester molding compound: Part I: Experimental study, *Polymer Composites*, Vol. 7, pp. 239–249, 1986.
9. L. F. Marker and B. Ford, Flow and curing behavior of SMC during molding, *Modern Plastics*, Vol. 54, pp. 64–70, 1977.
10. G. Kotsikos and A. G. Gibson, Investigation of the squeeze flow behaviour of sheet molding compounds (SMC), *Composites: Part A*, Vol. 29A, pp. 1569–1577, 1998.
11. P. T. Odenberger, H. M. Andersson, and T. S. Lundström, Experimental flow-front visualization in compression moulding of SMC, *Composites: Part A*, Vol. 35, pp. 1125–1134, 2004.
12. C. C. Lee, F. Folgar, and C. L. Tucker, Simulation of compression molding for fiber-reinforced thermosetting polymers, *Journal of Engineering for Industry*, Vol. 106, pp. 114–125, 1984.
13. M. R. Barone and D. A. Caulk, Kinematics of flow in sheet molding compounds, *Polymer Composites*, Vol. 6, pp. 105–109, 1985.
14. M. R. Barone and D. A. Caulk, A model for the flow of a chopped fiber reinforced polymer compound in compression molding, *Journal of Applied Mechanics*, Vol. 53, pp. 361–371, 1986.
15. P. Dumont, L. Orgéas, D. Faiver, P. Pizette, and C. Venet, Compression molding of SMC: In situ experiments, modelling and simulation, *Composites: Part A*, Vol. 38, pp. 353–368, 2007.
16. M. M. Shokrieh and R. Mosalmani, Modelling of sheet molding compound compression molding under non-isothermal conditions, *Journal of Reinforced Plastics and Composites*, Vol. 33, pp. 1183–1198, 2014.
17. L. M. Abrams and J. M. Castro, Predicting molding forces during sheet molding compound (SMC) compression-molding: I. Model development, *Polymer Composites*, Vol. 24, pp. 291–303, 2003.
18. D. L. Denton and S. H. Munson-McGee, Use of X-radiographic tracers to measure fiber orientation in short fiber composites, in *High Modulus Fiber Composites in Ground Transportation and High Volume Applications*, ASTM STP 873 (D. W. Wilson, ed.), ASTM, Philadelphia, PA, 1985.
19. R. M. Griffith and H. Shanoski, Reducing blistering in SMC molding, *Plastics Design and Processing*, Vol. 17, pp. 10–12, 1977.
20. K. L. Smith and N. P. Suh, An approach toward the reduction of sink marks in sheet molding compound, *Polymer Engineering and Science*, Vol. 19, pp. 829–834, 1979.

7 Liquid Composite Molding

In liquid composite molding (LCM) processes, a premixed liquid thermoset resin is injected into a dry fiber preform which is previously placed in a closed mold. As the liquid spreads through the preform, it coats the fibers, fills the space between the fibers, and expels the air trapped in the fiber layers of the preform, and finally as it cures, it transforms into the thermoset matrix [1,2]. LCM processes have several advantages over other composite manufacturing processes. The most important of them are (1) relatively fast production cycle time, (2) low molding pressure, (3) good surface finish, (4) close tolerance, and (5) design flexibility. The use of dry fiber preforms in LCM provides the design flexibility, since depending on the performance requirements, a variety of fiber architectures can be utilized in dry fiber preforms. The quality of the liquid composite-molded parts depends on both preform design and mold design, and therefore, both require a high level of attention for manufacturing parts with fewer defects and greater reproducibility.

This chapter describes two different LCM processes, namely, resin transfer molding (RTM) and structural reaction injection molding (SRIM). Even though RTM and SRIM are based on the concept of liquid resin injection into a dry fiber preform, there are differences between them. For example, RTM uses epoxy, polyester, or vinyl ester resins, whereas SRIM uses polyurethane or polyurea resins. BMI resins are used in RTM for high-temperature aerospace applications. Other differences between RTM and SRIM are listed in Table 7.1 and 7.2. Both processes have been successfully used for many years to manufacture many different types of composite parts, such as cabinet walls, chairs, bench seats, hoppers, water tanks, bathtubs, and boat hulls. They offer a cost-saving alternative to the labor-intensive bag molding process and the capital-intensive compression molding process, which is the reason they are gaining increasing acceptance in both aerospace and automotive industries.

For the aerospace industry that uses prepregs and bag molding process for manufacturing the majority of its composite parts, RTM provides the following benefits:

1. Less labor intensive than vacuum bag molding
2. Much lower capital investment, since autoclaves used in the bag molding process are much more expensive than the molding presses used for RTM
3. Lower tooling cost, since the molding pressure in RTM is much lower than that in bag molding
4. Lower material cost, since the production cost of dry fiber preforms used in RTM is much lower than that of prepregs
5. Lower material storage cost, since unlike dry fiber preforms used in RTM, prepregs must be stored in freezers; otherwise, its shelf life will be very limited

TABLE 7.1
Differences between RTM and SRIM Processes

RTM	SRIM
Resins: epoxies, polyesters, vinyl esters	Resins: polyurethanes, polyureas
Either one-component or two-component resin formulation	Two-component resin formulation
Low resin reactivity; therefore, static mixing may be sufficient for two-component formulations	Highly reactive components; therefore, requires very fast, high-pressure mixing before resin injection
Resin viscosity at the time of injection: 100–1000 cP	Resin viscosity at the time of injection: 10–100 cP
Low injection pressure	High injection pressure
Much simpler resin delivery system than SRIM and low equipment cost	Higher equipment cost due to more complex resin delivery system
Low tooling cost	High tooling cost

TABLE 7.2
Typical RTM and SRIM Process Parameters

	RTM	SRIM
Flow rate (kg/min)	2.3	55
Mixing	Static mixing	Impingement mixing
Mold pressure (MPa)	0.3	2.4
Mold temperature (°C)	25–40	95
Component viscosities (MPa s)	100–500	<200
Cycle time	10–60	2–6
Void content (vol.%)	0.1–0.5	0.5–2

Source: C. W. Macosko, *Fundamentals of Reaction Injection Molding*, Hanser, Munich, 1989.

Examples of aerospace parts made by RTM include wing structures, helicopter flaperons, various aerodynamic surfaces, and brackets.

For the automotive industry which uses injection molding and compression molding for manufacturing its composite parts, RTM is a much better option if the production volume is between 10,000 and 50,000 parts per year. It can produce composite body panels and other structural components with very good surface finish at a significantly lower manufacturing cost and a relatively fast production rate. Compared to the compression molding process, RTM has a very low tooling cost and simple mold clamping requirements. In some cases, a ratchet clamp or a series of nuts and bolts can be used to hold the two mold halves together. RTM is a low-pressure process, and therefore, parts can be resin transfer molded in low-tonnage presses. Another advantage of the RTM process is its ability to include ribs, metal or composite inserts, stiffeners, washers, etc., in the dry fiber preform, which later become integral to the molded composite part. It is also possible to mold a hollow part by encapsulating a foam core between the top and bottom preforms. Examples of automotive parts made by RTM include various body panels, pickup truck beds, and front-end structures.

7.1 RESIN TRANSFER MOLDING

In the RTM process, to manufacture a flat composite part, several layers of dry fiber network, such as continuous filament mat, chopped strand mat, woven fabrics, or other forms of fabric, are placed in the bottom half of a matching mold. The mold is then closed and clamped around the edges and a catalyzed liquid resin is injected under pressure into the mold via one or more gates (Figure 7.1). The resin injection points are usually at the lowest elevation of the mold cavity. The duration of resin injection depends on the size of the part, number of gates, viscosity, and reactivity of the resin–catalyst system. The injection pressure is in the range of 70–700 kPa. As the resin flows through the dry fiber layers and spreads inside the mold, it fills the space between the fibers, displaces the entrapped air and volatiles through the vents in the mold, and coats the fibers. Depending on the type of the resin–catalyst system used, curing is performed either at room temperature or at an elevated temperature. For small parts, curing can be performed in a hot air-circulating convection oven. For large parts, it may be necessary to use electrically heated or oil-heated molds. After the cured part is pulled or ejected out of the mold, it may require slight trimming at its outer edges to conform to the design dimensions.

For manufacturing composite parts with 3D geometry and complex shape, the starting material in an RTM process is usually a dry fiber preform that has already been given the shape of the desired product in a premolding operation. The preform does not contain any resin. An example is shown in Figure 7.2. The advantages of using a preform are good moldability with deep drawn shapes and the elimination of the trimming operation, which is often the most labor-intensive step in an RTM process.

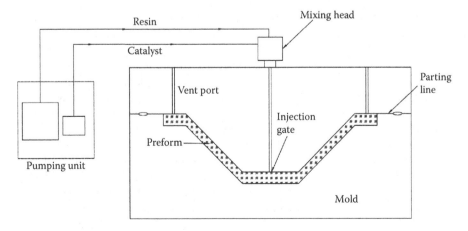

FIGURE 7.1 RTM process.

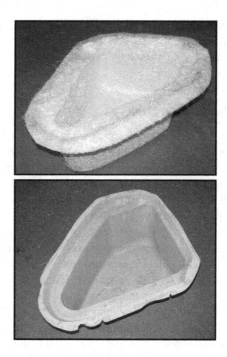

FIGURE 7.2   Random short glass fiber preform and the resin transfer molded part (before trimming).

There are several variations of the basic RTM process. They are briefly described in the following:

1. Vacuum-assisted RTM (*VARTM*), in which vacuum is used to pull the liquid resin into the preform.
2. Seemann's composite resin infusion molding process (*SCRIMP*), which is a patented process named after its inventor, William Seemann: Vacuum is also used in SCRIMP to pull the liquid resin into the dry fiber preform; but in this process, a porous resin distribution layer is placed on the top surface of the preform to uniformly distribute the resin throughout the preform. The porous resin distribution layer is selected such that it has a very low resistance to resin flow (and therefore, high permeability), and it provides the liquid resin an easy flow path to the preform. As the liquid resin is drawn in by the application of vacuum, it is quickly soaked by the resin distribution layer, and then it percolates in the thickness direction of the preform.

   Both VARTM and SCRIMP fall in the category of liquid molding processes called resin infusion under flexible tooling (*RIFT*) [3]. Unlike the RTM process which uses a matching pair of molds, RIFT uses a single-sided hard mold and a flexible vacuum bag. Instead of injecting liquid resin under pressure, it is drawn into the dry fiber preform using vacuum. The preform is placed on the hard mold surface (Figure 7.3) and covered with the vacuum bag, which is then sealed to the mold surface around its edges.

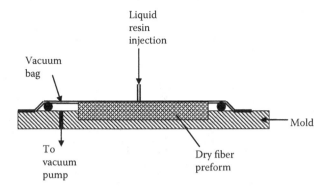

FIGURE 7.3   VARTM process.

After removing most of the air from the preform, liquid resin is introduced via a resin inlet port. The pressure during resin flow is the difference
between the atmospheric pressure and the vacuum, and because of this, the
mold filling time in VARTM and SCRIMP can be very long, especially for
large parts. Since only one side of the mold is made from a hard material,
the tooling cost is lower. By using the resin distribution layer between the
preform and the vacuum bag, the mold filling time is reduced and the resin
distribution becomes more uniform. Although both VARTM and SCRIMP
are used in manufacturing large composite parts, there are several short-
comings. For example, there may be uneven and low compaction of the
preform due to very low pressure used in these processes, which means the
fiber volume fraction in the cured part will also be uneven and low. The two
surfaces of the molded part may not have equal surface finish, since only
one surface of the part is in contact with the hard mold surface, while the
other surface is in contact with a flexible vacuum bag.

3. Compression RTM (*CRTM*) in which the upper mold is slightly opened to
   create a small gap between the top of the dry fiber preform and the bottom
   surface of the upper mold (Figure 7.4): The amount of liquid resin needed
   to fill the preform is injected into the gap. Very little infiltration of the preform by the liquid resin occurs during the injection phase. In the compression phase, the gap is closed by moving the upper mold to its final position,
   which forces the liquid resin to flow into the preform. The injection pressure
   and mold fill time are lower in CRTM than in RTM; however, the equipment cost is higher for CRTM, since it requires a molding press that can be
   closely controlled during the compression phase.
4. Resin film infusion (*RFI*), in which lightly cross-linked or partially cured
   B-staged resin films are either placed at the bottom of the dry fiber preform
   or interleaved with the layers in the dry fiber preform (Figure 7.5). Upon
   the application of heat and pressure, the resin viscosity decreases and the
   liquid resin flows through the preform and fills the cavity. Vacuum can also
   be used to assist in the resin flow. The RFI process can also be carried out
   using the vacuum bag molding technique described in Chapter 5.

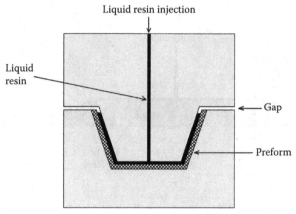

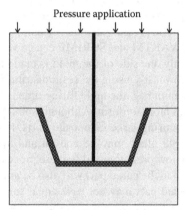

FIGURE 7.4 CRTM process.

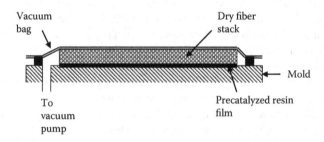

FIGURE 7.5 RFI process.

7.2   STRUCTURAL REACTION INJECTION MOLDING

Another LCM process very similar to RTM is called the SRIM. It also uses dry fiber preforms that are placed in the mold prior to resin injection. The difference in RTM and SRIM is mainly in the resin reactivity, which is much higher for SRIM resins than for RTM resins. SRIM is based on the reaction injection molding (RIM) technology [4], in which two highly reactive, low-viscosity liquid streams are impinged on each other at high speeds in a mixing chamber immediately before injecting the liquid mix into a closed mold cavity (Figure 7.6). Commercial RIM resins are mostly based on polyurethane chemistry, although polyureas and epoxies are also used.

As described in Chapter 3, the two reactive ingredients in the RIM of polyure-thane resin are diisocyanate, which is stored in liquid tank A, and a polyol blended with a chain extender, a catalyst and an internal mold release agent, which is stored in a separate liquid tank B (Figure 7.6). The storage temperature is between 30°C and 35°C. They are metered and pumped into the mixing chamber where they are sprayed on each other from two nozzles located on the opposite sides of the chamber. The impinging pressure is typically in the range of 70–200 bar. As the fine sprays of liquids A and B impinge on each other at a high speed, they are thoroughly mixed and the curing reaction begins. As soon as the amount of liquid resin required to fill the mold cavity is collected at the bottom of the mixing chamber, the spraying action is stopped and the collected liquid is injected into the mold cavity using a displace-ment piston.

The cure temperature for polyurethane resins is typically between 60°C and 120°C. The reaction rate for polyurethane resins used in RIM or SRIM is much faster than epoxy, polyester, or vinyl ester resins that are commonly used in RTM.

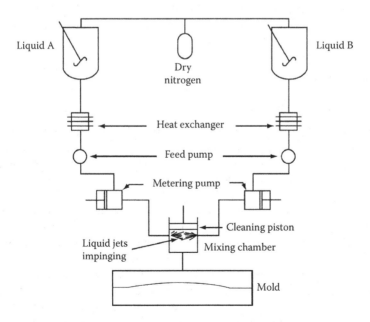

FIGURE 7.6   SRIM process.

The cure time for the SRIM resins is in the range of a few seconds compared to a few minutes for the RTM resins. Unlike the RTM resins for which curing is usually heat-activated after the mold is filled, the curing of the SRIM resins starts during impingement mixing and continues as the mold is being filled.

In both RTM and SRIM processes, the curing reaction continues as the liquid resin flows through the dry fiber layers of the preform. For producing good-quality parts, it is important that the resin completely fills the mold and wets out the reinforcement before arriving at the gel point. Therefore, the viscosity of the liquid resin for both processes must be low. However, since the curing reaction is much faster in SRIM, the initial viscosity of the SRIM resins must be lower than that of the RTM resins. The preferred room-temperature viscosity range for SRIM resins is 0.01–0.1 Pa s compared to 0.1–1 Pa s for RTM resins. Since the reaction rate of the liquid resin mix in SRIM is very high, its viscosity rapidly increases, and therefore, the mold must be filled very quickly. For this reason, the injection pressure in SRIM is several times higher than in RTM.

7.3   HIGH-PRESSURE RESIN TRANSFER MOLDING

High-pressure RTM (*HP-RTM*) is an attractive manufacturing process for large composite parts in the automotive industry where short production cycle times are needed. HP-RTM utilizes higher injection pressure than the traditional RTM, but also a fast mixing technique similar to the one used in SRIM. In the traditional RTM process, the injection pressure is in the range of 1–20 bar and the resin injection time is very long, typically 30–60 minutes. In HP-RTM, the injection pressure is in the range of 100–150 bars and the resin injection time is typically 1–5 minutes. The resins used in HP-RTM are fast-curing epoxies with cure time in the range of 1–2 minutes. The resin and hardener are mixed just outside the mold using the impingement mixing process similar to the SRIM process, and the mix is then injected at a high pressure to flow into the dry fiber preform.

7.4   PREFORMS

7.4.1   Preform Fabrication

Preforms are preshaped and preassembled layers of dry fiber network with the same shape as the composite part, but not containing the polymer matrix. They are produced separately in a premolding operation and placed in the mold prior to injecting the liquid resin into them. The use of preforms eliminates the time-consuming steps of cutting, assembling, and fitting the fiber network to the shape of the mold at the time of manufacturing the composite parts. Thus, the major benefit of using preforms is the reduction in production time and cost. The other benefits are (1) easy handling, (2) better control on fiber orientation and fiber volume fraction, (3) elimination of wrinkles, and (4) better control on part dimensions.

Preforms typically consist of multiple layers of dry fiber reinforcement that are bound together with either stiches or polymeric binders or tackifiers. The fiber architecture in the preform can be one, two or three dimensional; it can also be random. The most common types of binders for glass fiber mats are low-melting point thermoplastic polyesters and/or epoxies that are solid at room temperature [5]. The epoxy binders can be either catalyzed or uncatalyzed. Thermoplastic polyester and uncatalyzed epoxy binders are designed to be soluble in the resin, whereas catalyzed binders are designed to take part in the curing reaction and must be compatible with the resin.

There are various methods of producing random fiber preforms. One of them is the *spray-up* process, which is also called *directed fiber preforming*. In this process, fiber rovings are fed into a chopper gun where they are cut into 12–76 mm lengths. The chopped fibers are sprayed onto a perforated screen which has the shape of the part to be molded. Vacuum applied on the rear side of the screen securely holds the fibers on the screen. A thermoset resinous binder sprayed with the fibers keeps them in place and maintains the preformed shape. At the end of the spray-up operation, the screen covered with the chopped fibers is placed in an oven and heated to cure and harden the binder so that fibers stay in place. Figure 7.7 shows the schematic of a spray-up process, which has a rotating turntable with four screens.

A modified version of the spray-up process is called the *programmable powdered preform process*. In this process, the chopper gun placed on a robotic arm sprays chopped fibers along with a powdered epoxy or thermoplastic polyester binder on a screen designed in the shape of the part. The binder is typically 3–5% of the preform weight. Vacuum applied from the rear side of the screen keeps the chopped fibers and powdered binder in place. After the spray-up operation, a matched screen is pressed onto the spray-up and heat is applied to consolidate the fibers.

Another process for making random fiber preforms involves the use of a *slurry* of chopped reinforcement fibers and thermoplastic binder fibers in a carrier liquid (usually water) in a tank (Figure 7.8). A perforated screen held in a cradle is initially located at the bottom of the tank. The screen is shaped in the form of the composite part. The slurry is kept agitated by airflow to prevent the fibers from settling and to maintain a uniform dispersion of fibers in the slurry. As the platform is raised up, a reduction in atmospheric pressure develops below the screen and the resulting pressure differential forces the water to pass through the screen rapidly, depositing the fibers on the screen. The screen with the wet fiber preform is removed from the platform, covered with a

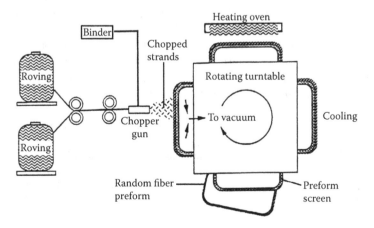

FIGURE 7.7 Spray-up process for making random fiber preforms.

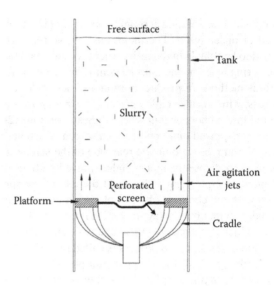

FIGURE 7.8 Slurry process for making random fiber preforms.

second perforated screen with a matching reversed shape, and then the entire assembly is placed inside a drying booth for forced air heating. The purpose of the second screen is to keep the fibers in place as they dry. After the water is removed, continued heating melts the binder fibers. Cooler air is then introduced to solidify the binder material. After cooling, the dried preform is separated from the screens.

Although the fiber spray-up and slurry process can produce preforms with complex shapes and very little waste, two major concerns with these processes are the lack of uniform fiber distribution and reproducibility. A more robust process for making preforms is stamping which can be applied to random fiber mats, woven fabrics, or their combinations. In this process, layers of fiber network are cut to the required shape and assembled into a stack, which is then preformed using a stamping operation in a press and a matched die set (Figure 7.9). Before stacking, the layers are coated with a suitable binder that helps in retaining the formed shape after stamping. To melt the binder and make stamping easier, the stack is heated before placing it in the stamping press. As the preform is cooled in the die under pressure, the binder solidifies and the individual layers are joined into a single preform. The edges of the preform are then trimmed to the final dimensions.

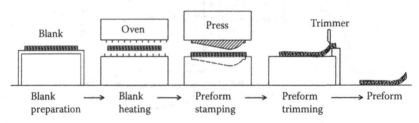

FIGURE 7.9 Preform stamping process.

FIGURE 7.10   Braided preform of a rocket motor nozzle. (Courtesy of Atlantic Research Corporation, Gainesville, Virginia.)

With woven fabrics containing bidirectional fibers, a *cut and sew* method is used in which various patterns are first cut from the fabric and then stitched together by polyester, glass, or Kevlar sewing threads into the shape of the part being produced. Braiding and textile weaving processes are also used to produce 2D or 3D preforms. Braiding is particularly suitable for producing tubular preforms such as the one shown in Figure 7.10.

An example of a 3D preform is shown in Figure 7.11 [6]. Such 3D preforms can be created using a number of weaving techniques. For example, in orthogonal weaving, the z-direction fibers are pulled through the warp and fill fibers so that there are fibers in all three mutually perpendicular directions. The fiber volume fractions in all three directions can be varied. In angular interlock weaving, the z-direction fibers are pulled through several layers of fabrics at various angles. The z-direction fibers are usually thinner so that the overall packing is tight.

7.4.2   BINDERS

The role of a binder in a preform is to keep the fibers in place and stabilize the fiber architecture in the preform against mechanical disturbances during preforming, preform placement in the mold, and liquid resin injection into the mold. Examples of mechanical disturbances are fiber displacement, preform thinning, and tearing.

Binders can be applied to the fiber network in the liquid, powder, string, or veil form. In the liquid form, the binder is dissolved in a solvent and is applied using a spray gun. Although the liquid binders give a more uniform coverage than the other forms, they have two disadvantages: (1) difficulty of removing the solvent and

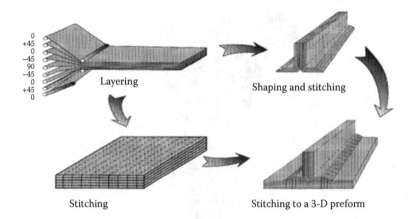

0
+45
0
−45
90
−45
0
+45
0
Layering

Shaping and stitching

Stitching

Stitching to a 3-D preform

FIGURE 7.11 A three-dimensional preform. (From M. B. Dow and H. B. Dexter, *Development of Stitched, Braided and Woven Composite Structures in the ACT Program and at Langley Research Center (1985 to 1997)*, NASA/TP-97-206234, National Aeronautics and Space Administration, Washington, DC, 1997.)

(2) possible health hazards associated with the solvent spray. Powdered binders are sprayed on the surface of the fiber network or mixed with the slurry in the slurry preforming process. After their application, they are melted either by using infrared lamps or in an oven. Upon cooling, the binder material solidifies and adheres to the fiber surfaces. The string binders are available in very thin long strands and are usually intermingled with the reinforcement fibers. The veil form of binders is very thin sheets which are inserted between the layers of fiber network so that when heated, they melt and infuse with the layers.

Some binders are selected to be soluble in the infiltrating resin. Since they are removed from the fiber surface as they dissolve in the resin, one of the criteria for selecting such binders should be that the time for their dissolution be greater than the time for mold filling; otherwise, the fiber network may be disturbed during the resin flow. Other factors to consider are the effect of binder dissolution on the resin viscosity and the effect of dissolved binder on the cure kinetics of the resin. The ingredients in some binders may act as either an inhibitor or an accelerator. If they act as an inhibitor, the cure time will be increased, which will also increase the cycle time. On the other hand, if they act as an accelerator, the reaction rate will increase, which may cause pregelling, increase the resin viscosity too fast, and reduce the mold filling rate. Another effect of binder dissolution is that it may negatively affect the properties, such as T_g and tensile modulus, of the cured resin.

7.4.3 PREFORM CHARACTERISTICS

The selection of the preform fabrication process depends on the fiber architecture, shape, and complexity of the part to be molded. The cost of preform fabrication may also be a factor. The general characteristics of preforms [1] to be considered are as follows:

1. Impregnation characteristics: The preform should be fully impregnated by the liquid resin, ideally under low injection pressure and in the shortest time possible. The impregnation characteristics of the preform depend on the permeability of the fiber architecture, fiber volume fraction, and number of layers in the preform.

2. Fiber wet-out: The liquid resin must completely wet-out the fibers in order to remove air from the fiber surface and form a strong interfacial bond with the fibers. The wet-out characteristics depends on the relative surface energies of the fibers and the liquid resin. Fiber sizing and fiber surface treatment are used to promote wettability and interfacial bonding. The selection of binder also affects fiber wet-out.

3. Conformability: The shape of the preform must be very close to the shape of the part to be molded. There should not be any gap between the mold surface and the preform; otherwise, the liquid resin will have a tendency to flow into the channel created by the gap, a process known as racetracking. The outer edges of a preform or edges around inserts are also sources of flow channeling that can create resin-rich areas. This can be prevented by cutting the preform to closely fit the edges of the mold cavity or inserts. Stamping or draping may produce preforms with wrinkles, fiber buckling, and fiber locking (due to interply shear) if the mat or the fabric has poor drapeability. They may also be spring back after stamping, which will create a narrow gap between the mold surface and the preform surface.

4. Resistance to fiber washing: Fiber washing occurs mostly near the resin entry gate, particularly at high injection pressures. The preform must have good resistance to fiber washing; otherwise, the fibers will be displaced from their intended position in the preform by the force of resin flow. This may cause changes in fiber orientation, uneven fiber distribution, blocking of the vents in the mold, etc. For random fiber mats, the resistance to fiber washing depends on the binder type and concentration. For fabric or braided preforms, the fiber packing arrangement and the degree of interlacing are important parameters in reducing fiber washout.

5. Handling characteristics: After the preform is produced, it is removed from the preforming location and moved to the molding station or to a storage, either manually or by using a robot. At the time of molding, the preform is placed in the mold and, if needed, fitted to the mold surface by tucking, pushing, pressing, etc. The mechanical actions during preform placement or fitting should not cause any fiber movement, cracks, or tear. The preform must be strong enough to resist the handling stresses.

7.5 RESIN SELECTION AND PREPARATION

The first goal in resin selection for LCM is to ensure low enough viscosity and long enough gel time for the liquid resin so that it can completely fill the mold cavity before the gel point is reached. During mold filling, it is important that the liquid resin thoroughly infiltrates the fiber preform, wets out the fibers in the preform, and pushes out air and other volatiles from the mold cavity. The second goal is to achieve full cure within a reasonable length of time. The resin characteristics, such as pot life, gel time, and cure time–temperature–viscosity relationship, depend not only on the selection of the resin and curing agent or catalyst, but also on their mix ratio. The accurate control of the mix ratio during a production run is one of the critical issues in manufacturing parts with consistent quality. Depending on the resin type, processing requirement, and desired matrix properties after curing, the resin mix may also contain an accelerator or inhibitor and various additives, such as fillers, coloring agents, fire retardants, and shrinkage control agents (also called low-profile additives). An internal mold release agent, such as zinc stearate, is also added. It helps in demolding the part and reduces the need for frequently spraying the mold surfaces with an external mold release agent.

The resin preparation, which takes place outside the mold, involves mixing various ingredients in the resin formulation. The simplest and least expensive method for storing the mixed resin is a sealed tank from which it is dispensed to the mold by applying air pressure on its free surface. The resin is kept agitated using a stirrer mixer. The resin flows from an outlet located at the bottom of the tank to the gates in the mold through flexible tubes. This method is not suitable for high-production volumes, since it cannot store the resin mix for a long time without the possibility of gelling. Since there is no metering device in the system, the amount of resin flow may not be the same from one cycle to the next, which may affect the reproducibility of the molded parts.

The most common method for storing, mixing, and dispensing resin utilizes pneumatically actuated single-acting or double-acting pumps [1]. In this method, the various ingredients are stored in separate tanks (Figure 7.12). They are drawn from their respective tanks using separate pumps and are combined in a static mixer outside the mold. Thus, the curing reaction begins just as the resin enters the mold. The advantages of this system are as follows:

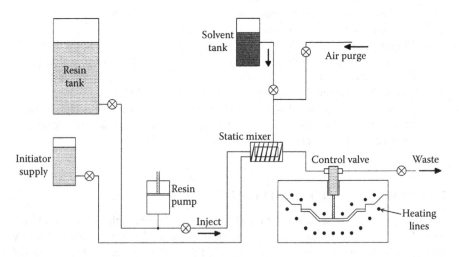

FIGURE 7.12   Resin injection system in RTM.

1. Since the resin mix is prepared on an as-needed basis and is immediately injected into the mold, it is suitable for high production volumes.
2. The mixing ratio can be controlled accurately and adjusted if needed.
3. High injection pressure can be generated.
4. Since mixing is done using a static mixer, the possibility of air bubble generation during mixing is relatively low.

After the injection is complete, the static mixer is flushed with a solvent to clean it of any residual resin that may otherwise start to cure inside the mixer.

7.6   PROCESSING STEPS IN LIQUID COMPOSITE MOLDING

The steps in LCM consists of (1) mold surface preparation, (2) preform placement, (3) mold closing and sealing, (4) resin injection and mold filling, (5) curing, and, finally, (6) mold opening and demolding. There may also be postcuring and finishing operations after the part is demolded. Postcuring outside the mold is typically done in a hot air-circulating convection oven to increase the degree of cure in the molded part. The finishing operations may include edge trimming, hole drilling, etc.

The mold surface preparation includes the steps of cleaning the mold surfaces of any residue from the previous cycle and spraying them with an external mold release agent. As the fiber preform is placed in the mold, care must be taken to ensure that the fiber network in the preform is not disturbed, no wrinkles are formed, and corners are fully draped. The demolding operation at the end of the curing step may be as simple as manually pulling the part out of the mold after the mold is opened. Ejector pins or other mechanical demolding devices can also be used. But since the part is still hot and may not have attained its full strength, care must be taken during demolding so that it is not unduly overstressed and punctured or cracked.

The most critical steps in LCM processes are resin injection and mold filling, followed by curing. They affect not only the cycle time, but also the quality of the molded part. These steps are described in the following.

7.6.1   RESIN INJECTION AND MOLD FILLING

Resin injection takes place through one or more gates that are strategically placed in the mold to reduce the mold filling time and control the occurrence of defects such as dry spots. Multiple gates are used for large parts so that the cavity can be filled in the shortest possible time. In RTM, the resin mix is transferred to the mold cavity through these gates under positive pressure using either a constant injection pressure, a constant injection flow rate, or a combination of the two (Figure 7.13). If a constant injection pressure is used, there will be a very high flow rate at the beginning of mold filling, but the flow rate will decrease with time. On the other hand, if a constant flow rate is used, the injection pressure will increase with time, and it may become so high that resin will start to leak at the mold edges or fibers will start to move from their positions (causing fiber washout). To avoid such problems, resin injection can be conducted in two steps. In the first step, the injection is started at a constant flow rate, but the pressure increase during this step is not allowed to exceed a maximum

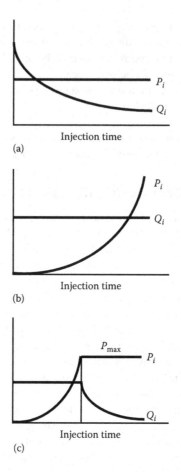

FIGURE 7.13 Resin injection strategies: (a) constant injection pressure, (b) constant injection flow rate, and (c) combination of a and b (P_i is the injection pressure and Q_i is the inlet flow rate).

value. After the maximum pressure is reached, the remaining resin injection proceeds under constant pressure.

The mold filling time depends on the size and geometry of the part, resin viscosity of the liquid resin as it flows through the fiber preform, injection pressure, and permeability of the fiber architecture. Even though the curing reaction begins at the resin mixing stage and continues as the liquid resin flows through the dry fiber preform, no significant reaction must take place during mold filling; otherwise, the resin viscosity will rapidly increase and there may be incomplete filling due to premature gelling. For a large part, the filling time can be reduced by using multiple gates, since this will reduce the flow distance of the liquid resin entering through each gate. The locations and types of gates also play important roles in mold filling. The resin viscosity can be lowered by preheating the resin and raising its temperature before it enters the cavity, but this may cause premature gelling before the cavity is filled. Increasing the injection pressure is another option, but too high a pressure can cause fiber movement and preform distortion. The third option is to select a fiber architecture with higher permeability, but before this option is exercised, it should be noted that changing the fiber architecture may cause significant changes in the mechanical properties of the cured composite.

As the liquid resin enters the mold and starts to spread, it encounters resistance to flow from the dry fiber preform. The quality of liquid composite-molded parts depends on the uniformity of resin flow through the fiber preform, since it determines the extent of mold filling, fiber surface wetting, resin distribution, and void entrapment. Several mold filling related problems are described in the following:

1. Dry spots: Incomplete mold filling causes dry spots where the resin has not infiltrated the preform. Dry spots occur if (1) the resin viscosity rapidly increases and it becomes so high that the resin flow stops before all areas of the preform are reached, (2) the resin pressure is not sufficiently high to displace the air out through the vents, and (3) vents are not located at places where the resin flow reaches last. Dry spots can also occur in locations where several flow fronts merge. Multiple flow fronts are created if there are multiple gates for resin injection or when there are inserts in the mold where a flow front is divided into two as it flows around the inserts.

2. Voids: Voids are formed between the fiber bundles due to inadequate removal of air or volatiles generated during the curing reaction. The spaces between the fiber bundles and between various layers in the preform are occupied by air that must be removed as the liquid resin flows into these spaces. Air bubbles may also be entrapped in the resin during mixing. The resin flow rate, resin viscosity, vent size, and their locations are important parameters that influence the void volume fraction in the cured composite part. Vacuum can be applied before resin injection so that much of the air is removed from the mold before the liquid resin enters the cavity.

   Voids can also form due to transverse flow of the liquid resin. In general, the resistance to resin flow is much higher in the thickness direction than in the plane of the fiber preform. As a result, the transverse flow lags behind the in-plane flow and often takes place after the in-plane flow is complete.

3. Microvoids: The main source of microvoid formation is the air entrapped in the complex fiber network in woven or stitched dry fiber preforms [7–9]. Two types of microvoids are observed: (1) intrabundle microvoids that form within the fiber bundles and (2) interbundle microvoids that form between the fiber bundles. Each fiber bundle contains a large number of filaments; the empty spaces between them are on the order of 1 μm. There are also empty spaces between the fiber bundles, but they are on the order of 1 mm. The resistance to resin flow in the empty spaces within the bundles is much higher than the resistance to resin flow in the empty spaces between the bundles. Because of this, resin flows faster between the bundles than within the bundles, but when it reaches a stitch or a transverse bundle on its path, it turns back, and as it meets the resin flowing within the bundle, an intrabundle microvoid is created (Figure 7.14a) [10]. On the other hand, capillary action may pull the liquid resin faster in the spaces within the bundles,

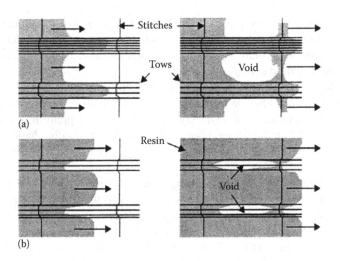

FIGURE 7.14   (a) Intrabundle and (b) interbundle microvoid formations. (From S. G. Advani and E. M. Suzer, *Process Modeling in Composites Manufacturing*, CRC Press, Boca Raton, FL, 2003.)

and when this flow front reaches a stitch or a transverse bundle, it may turn back and form an interbundle microvoid (Figure 7.14b). In general, intrabundle microvoids occur at low flow rates, and interbundle microvoids occur at high flow rates. To reduce the occurrence of microvoids, the resin flow should be controlled by adjusting the injection pressure or resin flow rate that gives sufficient time for the microvoids to shrink and collapse. Other recommendations for reducing microvoids are (1) resin degassing, (2) vacuum assistance during resin flow, (3) application of positive pressure after mold filling, and (4) injection of additional resin after the first fill.

4. Racetracking: If there is a narrow gap between the preform surface and the mold wall (Figure 7.15), the liquid resin tends to race along the channel created by the gap, since it provides a path of low resistance to resin flow. This creates a resin-rich area, which becomes an area of low strength in the cured composite part. Racetracking is usually observed at mold wall edges and corners, around ribs and inserts, and in molds of complex geometry. The possibility of racetracking can be reduced by (1) designing the preform with close fit to the mold dimensions, (2) making sure that there are no missing fibers on the outer surfaces of the preform, and (3) carefully placing the preform in the mold and properly fitting it to the mold surface. Accurate machining of the mold surfaces and corner radii is also important.

5. Fiber movement: There may be fiber movement (known as *fiber washout*) and preform distortion as the liquid resin moves through the fiber preform under pressure, especially if the viscosity rapidly increases before the mold filling, is complete. Fiber movement occurs if the fibers are not tightly woven in the woven fabrics of a fabric-based preform or if the binder that

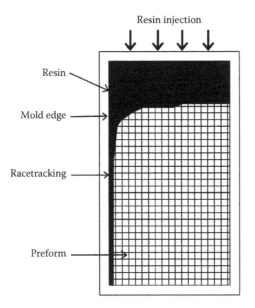

FIGURE 7.15   Racetracking along a mold edge.

holds the fibers in a random fiber mat is dissolved too quickly as it comes in contact with the resin. High initial resin viscosity and high injection pressure will also cause fiber movement.

6. Core deformation: One of the advantages of LCM processes is that hollow parts can be molded using a lightweight foam core encapsulated between two fiber preforms that are stitched together around the core (Figure 7.16). The core material is usually a low-density polystyrene or polyurethane foam. The principal purpose of the core is to provide support to the preforms as the liquid resin is injected under pressure to fill the cavity. However, if the injection pressure is too high, the core may shift, collapse, or buckle (Figure 7.16), which will cause problems in maintaining dimensional control of the part. The core material may also degas as the mold temperature is raised during the curing stage, and this may contribute to void formation in the part.

The two major process variables during mold filling are the injection pressure at the gate and the temperature of the liquid resin. Increasing the injection pressure increases the resin flow rate and, therefore, decreases the mold filling time. Increasing the liquid resin temperature at the mold filling stage will not only decrease the resin viscosity, but may also cause gelling at an earlier time. The interaction between the injection pressure and liquid resin temperature can be displayed in a mold filling diagram shown in Figure 7.17 [11]. The enclosed area in this figure can be considered the mold filling window or the moldability zone. The minimum possible gel time should be longer than the fill time so that the liquid will not gel before the mold is filled. This is shown by the left limit line in Figure 7.15. The right limit is defined by either the capacity of the injection unit or the conditions at which fiber washout

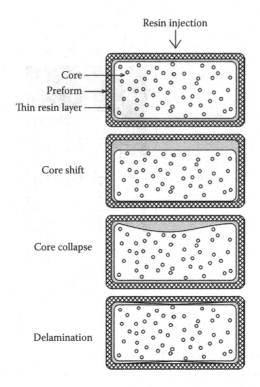

FIGURE 7.16   Foam core used for manufacturing hollow RTM parts and possible defects.

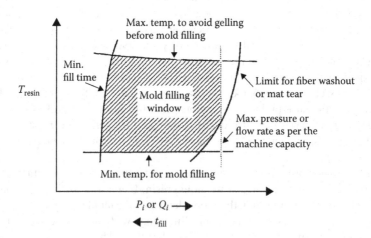

FIGURE 7.17   Mold filling diagram. (V. M. Gonzalez-Romero and C. W. Macosko: Process parameters estimation for structural reaction injection molding and resin transfer molding. *Polymer Engineering and Science*, 1990, 30, 142–146. Copyright Wiley-VCH Verlag GmbH & Co. KGaA. Reproduced with permission.)

occurs. The top limit defines the maximum temperature above which gelling will occur too fast so that the mold will not be filled. The bottom limit defines the minimum temperature below which the resin viscosity will be too high and the mold may not be completely filled.

7.6.2 CURING

Curing is the next step in an LCM process. Although curing of the resin may start at the mixing stage, the state of cure as the resin flows in the mold and fills the cavity will depend on several factors: (1) resin chemistry; (2) ingredients that influence the cure rate, such as the type and amount of the cure initiator and presence of accelerator or inhibitor; (3) resin temperature during mixing; (4) mold wall temperature; and (5) preform temperature. It is important that the mold be filled before the state of cure reaches the gel point; otherwise, the resin viscosity will rapidly increase and the resin flow will become very slow. In the extreme, the resin flow may stop before the preform is completely infiltrated. In most cases, resin mixing is done at room temperature, and the resin will enter the mold either at room temperature or in a slightly warm condition. As the mold temperature is raised either during the filling stage or after the mold is filled, the resin temperature will increase, and its viscosity will decrease. As the temperature increases further, the curing reaction will be accelerated and the resin viscosity will rapidly increase until the curing is complete.

Increasing the mold temperature in the curing step has two effects. One is the reduction in resin viscosity, which will help in fiber wetting and reducing the induction time for the curing reaction. However, if the mold temperature is too high, it can create a high peak exotherm temperature at the midthickness of the part, especially if the part is relatively thick. This can cause burning and thermal degradation of the resin.

In general, the minimum cure time in the mold is the time it takes to form 95% of the cross-linked structure of the matrix. In order to reduce the production cycle time, the curing reaction in the mold is seldom taken to full cure. After demolding the part, it can be postcured in an oven to increase the degree of cure to the desired level. If at the time of demolding the degree of cure is too low, the part may not have attained enough stiffness and may bend too much during demolding. This may warp the part or cause the development of cracks in the part.

7.7 MOLD DESIGN

Mold design is one of the key steps in manufacturing liquid composite-molded parts, since it not only determines the ease of liquid resin flow, but also has a significant influence on the quality of the cured composite part. Mold design has four major tasks: (1) selection of mold material, (2) gate locations and design, (3) vent locations and design, and (4) heating and cooling line arrangements.

1. Selection of mold material: Since the injection pressure in liquid molding processes is relatively low, the mold material commonly used for production volumes of 50,000–100,000 parts is a cast aluminum alloy. For higher production volumes, chrome-plated tool steel is selected because it has higher

hardness and abrasion resistance. The advantages of using cast aluminum compared to tool steel are its higher machinability, lower density, and higher thermal conductivity. The thermal conductivity of aluminum is in the range of 130–170 W/m K compared to 40–65 W/m K for tool steels. Higher thermal conductivity helps in quick heating during the curing step. It also helps in conducting away the heat faster during the cooling stage. Because of lower density, aluminum molds are lighter than steel molds. On the other hand, the modulus of aluminum is about one-third of the modulus of steel, which gives lower stiffness to aluminum molds. Its coefficient of thermal expansion is nearly twice that of steel, which is a disadvantage from the standpoint of holding the part dimensions to tight tolerances. Another disadvantage of aluminum is that it is a softer material with lower resistance to abrasion and scratching. In general, aluminum molds are less expensive than tool steel molds.

For low production volumes (1000 parts or less) and prototyping, the mold can be made of a glass fiber-reinforced epoxy composite shell supported on a fabricated steel frame. The composite shells are low in cost, lightweight, and easy to build (using vacuum bagging, for example). Since the glass fiber composite has a much lower thermal conductivity than metals, more heating channels are required in composite shell molds. Also, the high coefficient of thermal expansion of glass fiber composites may pose problems with dimensional control for parts molded in composite shell molds. The life of composite shell molds is limited due to their low strength, low abrasion and heat resistance, and poor damage tolerance.

Several other materials can be used for making molds for LCM. They include plaster of paris, Kirksite (which is a Zn alloy), copper, nickel alloys, and chemically bonded ceramics. They have their own advantages and disadvantages, and their use is relatively limited.

2. Gate location and design: Gate location and design not only influence the fill time, but also the amounts of voids, microvoids, and dry spots. They are usually located at the lowest points of the mold to facilitate air displacement through the vents that are located at the highest points of the mold. The common gate designs are line gates located along one edge of the mold and pin gates located at the center of the mold. For large parts, multiple gates are used to reduce the fill time. If a line gate is used, the resin flow at the gate entrance is rectilinear and air is vented at the opposite edge of the mold. If a pin gate is used, the resin flow at the gate entrance is a diverging radial flow and air is vented through all edges of the mold. If the cavity has a flat rectangular shape (Figure 7.18), the resin flow front emerging from an edge-located line gate is normal to the flow direction, whereas from a center-located pin gate, the emerging flow front is circular which radially expands until it reaches the mold wall closest to the gate, and then it gradually changes to a rectilinear flow. In general, for equal flow distance, the fill time is lower with line gates compared to pin gates, but the injection pressure is higher. The equations to calculate the fill time, injection pressure, and resin flow rate for one-dimensional flow in a flat rectangular cavity of length L and width w are given in Table 7.3.

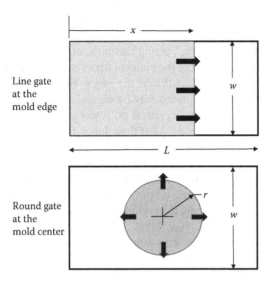

FIGURE 7.18 Resin flow in a rectangular cavity: unidirectional flow (*top*) from a line gate at the left mold edge and (*bottom*) radial flow from a round gate at the mold center. (From C. R. Rudd, A. C. Long, K. N. Kendall, and C. G. E. Mangin, *Liquid Moulding Technologies*, Woodhead, Cambridge, UK, 1997.)

TABLE 7.3
One-Dimensional Flow Equations in a Rectangular Cavity

Line Gate at One Edge of the Mold	Round Gate at the Center of the Mold
Unidirectional rectilinear flow	Radially divergent flow
Validity: $0 < x \leq L$	Validity: $r_i < r \leq w/2$
• Constant injection pressure $P_i = P_c$	• Constant injection pressure $P_i = P_c$
$$t_x = \frac{\phi\eta}{2KP_C}x^2$$	$$t_r = \frac{\phi\eta}{2KP_C}\left[r^2 \ln\left(\frac{r}{r_i}\right) - \frac{1}{2}\left(r^2 - r_i^2\right)\right]$$
• Constant flow rate $Q_i = Q_c$ $\quad\displaystyle t_x = \frac{\phi hw}{Q_C}x$ $\quad\displaystyle P_i = \frac{\eta Q_C}{hwK}x$	• Constant flow rate $Q_i = Q_c$ $\quad\displaystyle t_r = \frac{\pi\phi h}{Q_C}\left(r^2 - r_i^2\right)$ $\quad\displaystyle P_i = \frac{\eta Q_C}{2\pi hK}\ln\left(\frac{r}{r_i}\right)$

Note: x and r are the positions of the flow front; h is the cavity thickness (m); L is the cavity length (m); w is the cavity width (m); r_i is the radius of the center gate (m); K is the permeability (m²); φ is the preform porosity; η is the viscosity (Pa s); P_i is the injection pressure (Pa); and Q_i is the injection flow rate (m²/s).

3. Vent location and design: Along with gate locations, vent locations are also important considerations in mold design. Vents are usually located on the outside edges of the mold. Venting can be simply done by using a pinch-off, which involves trapping the preform between the two mold halves, thus creating a restriction to resin flow at the edges (Figure 7.19a). Although this technique is widely used, there may be excessive resin outflow, hazardous vapor escape (for example, styrene if polyester or vinyl ester resin is used), and damage to the mold ends. After demolding, the trapped material, called *flash*, needs to be trimmed off in a postmolding operation. To avoid the postmolding operation, the molds can be designed with either horizontal or vertical flash gaps (Figure 7.19b and c). A better approach is to use a flexible seal, such as a rubber O-ring (Figure 7.19d), which is compressed when the mold is closed after placing the preform. An improvement on this approach is to use a vented seal (Figure 7.19e). Venting is also possible through ejector pins if they are used for demolding (Figure 7.19f).

It is important to note that in vacuum-assisted processes, such as VARTM, the edge sealing plays a very important role in drawing liquid

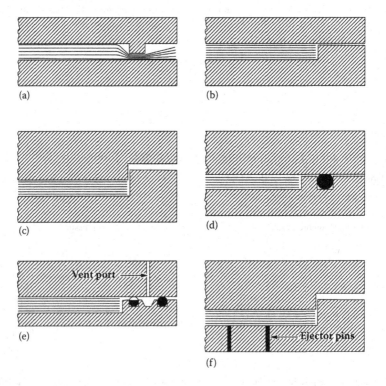

FIGURE 7.19 Examples of vent designs: (a) pinch-off, (b) horizontal flash gap, (c) vertical flash gap, (d) rubber O-ring sealing, (e) vented seal, and (f) at the ejector pins. (From C. R. Rudd, A. C. Long, K. N. Kendall, and C. G. E. Mangin, *Liquid Moulding Technologies*, Woodhead, Cambridge, UK, 1997.)

resin using vacuum. Unless it is a very tight seal, air will enter back into the cavity right after the air is evacuated by the application of vacuum.

4. Heating and cooling line arrangements: For low-volume productions and small parts, the closed mold can be placed inside a hot air-circulating oven or between the heated platens of a hydraulic press after mold filling. In both methods, the heating rate is slow and temperature rise may be uneven. A better approach is to integrate heating and cooling lines in the mold. Heating can be accomplished by circulating hot fluids, such as oil and water. Electric cartridges can also be used, but fluid lines are preferred since they can be used for both heating and cooling as needed. An integrated heating system provides a more uniform temperature distribution and a high degree of temperature control, particularly if zone heating is used for additional heat input to specific regions of the mold or to compensate for heat losses near the resin entrance gates.

An important aspect of heating the mold is the temperature distribution in the composite part, which should be as uniform as possible. High thermal gradients can cause uneven curing, which may result in uneven shrinkage. The demolded part may then show warpage and end up with high residual stresses.

REFERENCES

1. C. R. Rudd, A. C. Long, K. N. Kendall, and C. G. E. Mangin, *Liquid Moulding Technologies*, Woodhead, Cambridge, UK, 1997.
2. L. Fong and S. G. Advani, Resin transfer molding, in *Handbook of Composites* (S. T. Peters, ed.), Chapman & Hall, London, pp. 433–455, 1998.
3. J. Summerscales, Resin infusion under flexible tooling (RIFT), in *Wiley Encyclopedia of Composites*, 2nd Ed. (L. Nicolais and A. Borzacchiello, eds), John Wiley & Sons, New York, 2012.
4. C. W. Macosko, *Fundamentals of Reaction Injection Molding*, Hanser, Munich, 1989.
5. J. C. Brody and J. W. Gillespie, Jr., Reactive and non-reactive binders in glass/vinyl ester composites, *Polymer Composites*, Vol. 26, pp. 377–387, 2005.
6. M. B. Dow and H. B. Dexter, *Development of Stitched, Braided and Woven Composite Structures in the ACT Program and at Langley Research Center (1985 to 1997)*, NASA/ TP-97-206234, National Aeronautics and Space Administration, Washington, DC, 1997.
7. N. Patel and L. J. Lee, The effects of fiber mat architecture on void formation and removal in liquid composite molding, *Polymer Composites*, Vol. 16, No. 5, pp. 386–399, 1995.
8. N. Patel, V. Rohatgi, and L. J. Lee, Influence of processing and material variables on resin/fiber interface in liquid composite molding, *Polymer Composites*, Vol. 14, No. 2, pp. 161–172, 1993.
9. V. Rohatgi, N. Patel, and L. J. Lee, Experimental investigation of flow-induced micro-voids during impregnation of unidirectional stitched fiberglass mat, *Polymer Composites*, Vol. 17, No. 2, pp. 161–169, 1996.
10. S. G. Advani and E. M. Suzer, *Process Modeling in Composites Manufacturing*, CRC Press, Boca Raton, FL, 2003.
11. V. M. Gonzalez-Romero and C. W. Macosko, Process parameters estimation for structural reaction injection molding and resin transfer molding, *Polymer Engineering and Science*, Vol. 30, No. 3, pp. 142–146, 1990.

8 Filament Winding

In the filament winding process [1], a band of continuous resin-impregnated fiber rovings or tows is wound under tension around a rotating mandrel in a preselected fiber orientation pattern to produce hollow continuous FRP composite parts. The resin is usually a thermosetting polymer, such as an epoxy, vinyl ester, or polyester. After several layers are wound to build the required thickness, the part is cured on the mandrel and then the mandrel is removed. In some designs, such as a pressurized oxygen tank, the mandrel may be retained within the filament-wound composite part to meet the design requirements. Both axisymmetric and nonaxisymmetric parts can be manufactured by the filament winding process. Among the examples of axisymmetric parts are golf club shafts, automotive drive shafts, rocket motor cases, cylindrical and spherical pressure vessels (such as oxygen tanks and compressed natural gas tanks), oil and gas pipes, and large underground gasoline storage tanks. Nonaxisymmetric parts, such as helicopter blades and wind turbine blades, are also manufactured by filament winding, but they require numerically controlled machines with a sophisticated programming device to control the fiber motion and winding pattern around complex-shaped mandrels. Examples of filament-wound composite parts are shown in Figure 8.1.

There are two different types of filament winding processes: (1) wet winding, in which the fiber rovings or tows are wetted by the liquid resin mix before winding them on the rotating mandrel, and (2) dry winding, in which preimpregnated fiber rovings or tows are wound around the rotating mandrel. Among the two winding methods, wet winding is more common and has several advantages, such as lower material cost, shorter winding time, and resin formulation that can be easily changed to meet the design or processing requirements.

The principal advantages of filament winding are that it is a relatively low cost semi-continuous process for manufacturing hollow composite parts. It can produce cylindrical, spherical, conical, geodesic, and other curved shapes with highly repetitive and accurate fiber orientation. The hollow parts can be either open (e.g., pipes) or closed at the ends (e.g., oxygen tanks with hemispherical ends).

8.1 FILAMENT WINDING PROCESS

Figure 8.2 shows a schematic of the basic wet filament winding process. A large number of fiber rovings (or fiber tows) are pulled from a series of creels (fiber packages) into a liquid resin bath, which is usually a rectangular tank containing a blended mixture of liquid resin, catalyst, and other ingredients, such as pigments and UV absorbers. Fiber tension is controlled using fiber guides or scissor bars located between the creels and the resin bath. Just before entering the resin bath, the rovings are usually gathered into a band by passing them through a textile thread board or a stainless steel comb.

FIGURE 8.1   Examples of several filament-wound composite parts.

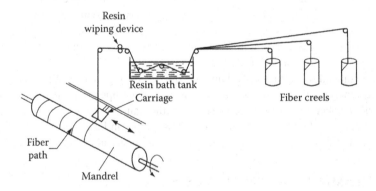

FIGURE 8.2   Filament winding process.

At the exit end of the resin bath, the resin-impregnated rovings are pulled through a wiping device that removes the excess resin from the rovings and controls the thickness of resin coating around each roving. The most commonly used wiping device is a set of squeeze rollers in which the position of the top roller is adjusted up or down to control the resin content as well as tension in the fiber rovings. Another technique for wiping the resin-impregnated rovings is to pull each roving separately through an orifice, very much like the procedure in a wire coating process. This latter technique provides a better control of resin content. However, if one or more fiber rovings break during the filament winding operation, it becomes difficult to rethread the broken roving lines through the orifices.

After the rovings are thoroughly impregnated and the excess resin is drained, they are gathered together in a flat band and positioned on the mandrel. Band formation can be achieved by using a straight bar, a ring, or a comb. The band former is usually located on a carriage, which traverses back and forth parallel to the mandrel, like a tool stock in a lathe machine. The traversing speed of the carriage and the rotational speed of the mandrel are controlled to create the desired winding angle pattern. Typical winding speeds range from 90 to 110 linear m/min. However, for more precise winding, slower speeds are recommended.

In order to completely cover the mandrel surface with resin-impregnated fibers and build thickness for the filament-wound part, the carriage moves back and forth several times parallel to the mandrel. After winding a number of layers to generate the desired thickness, the uncured filament-wound part on the mandrel is removed from the filament winding machine and is placed in an oven for curing. A shrink tape may be wound on the outside of the part to apply pressure during curing. At the end of the cure cycle, the mandrel is extracted from the cured part. In some applications, such as oxygen tanks, the mandrel is not extracted and remains inside the cured filament-wound part.

The basic filament winding process described earlier creates a helical winding pattern (Figure 8.3) and is called the *helical winding* process. The angle of the roving band with respect to the mandrel axis is called the *winding angle*, which is denoted by θ. By adjusting the carriage feed rate and the rotational speed of the mandrel, any winding angle between near 0° (i.e., *longitudinal winding*) and near 90° (i.e., *hoop winding*) can be obtained. Since the feed carriage moves forward and backward, fiber bands crisscross at plus and minus the winding angle and visually create a diamond-shaped repeat pattern. Thus, the fiber orientation in each layer of a helically wound part is $\pm\theta$. In hoop winding, the winding angle is close to 90°, but not exactly 90°. In many applications, helical and hoop winding patterns are combined on the same part to create a fiber orientation pattern such as $90/\pm\theta/90/\pm\theta/90$ through its thickness.

In another type of filament winding process, called polar winding, the carriage rotates about the longitudinal axis of a stationary (but indexable) mandrel. After each rotation of the carriage, the mandrel is indexed to advance one fiber bandwidth.

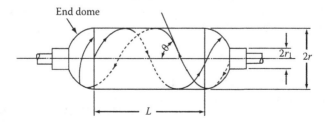

FIGURE 8.3   Helical winding pattern.

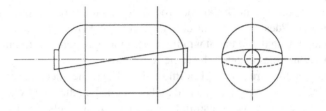

FIGURE 8.4   Polar winding pattern.

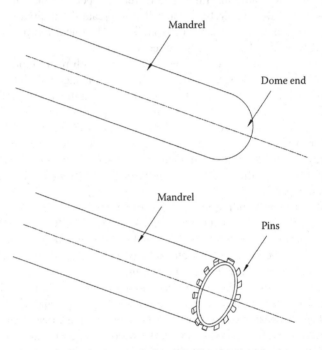

FIGURE 8.5   Dome and pins at the mandrel end for fiber band anchoring.

Thus, the fiber bands lie adjacent to each other and there are no fiber crossovers. A complete wrap consists of two plies oriented at plus and minus the winding angle on two sides of the mandrel (Figure 8.4).

It should be noted that both helical and polar winding processes require winding the fiber band around the mandrel ends. For manufacturing open-ended products, such as a pipe section or a drive shaft, pins or rounded domes (Figure 8.5) are used at the mandrel ends to anchor the fiber rovings and prevent their slippage as the carriage turns back for its reverse travel.

8.2   HELICAL FILAMENT WINDING

Conventional two-axis helical filament winding machines use a driving motor to rotate the mandrel and a chain and a sprocket to move the carriage back and forth parallel to the mandrel surface (Figure 8.6). The main sprocket is connected to the

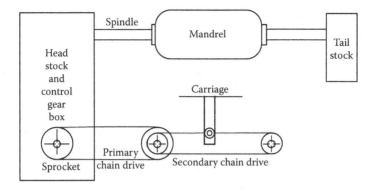

FIGURE 8.6 Conventional filament winding machine.

mandrel shaft through a set of gears so that the carriage feed can be controlled in relation to the mandrel rotation by changing the gear ratios or the sprocket size. Winding with the conventional machine also requires the carriage to travel extra lengths on both sides of the mandrel.

As the fiber band moves forward with the carriage along the length of the mandrel, it covers the mandrel only partially (Figure 8.7). To completely cover the surface of the mandrel, the carriage must reverse at the end of the forward travel and move backward until it returns to the starting location. The path the fiber band follows is called a *circuit*. It requires several circuits to completely cover the mandrel surface with the fiber band. If the fiber band is at a $+\theta$ winding angle in the forward circuit, a $-\theta$ winding angle is formed during the backward circuit. For hoop winding, θ is slightly smaller than 90°, and each full revolution of the mandrel advances the fiber band by one full bandwidth.

If θ is the desired winding angle and D is the mandrel diameter, the axial distance L_e traveled by the carriage (Figure 8.8) corresponding to one complete revolution of the mandrel can be calculated from the following equation. The distance L_e is called the *lead*.

$$L_e = \frac{\pi D}{\tan \theta}.$$ (8.1)

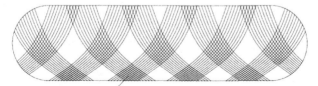

Fiber band crossovers

FIGURE 8.7 Mandrel surface coverage and diamond-shaped fiber band crossovers during helical winding.

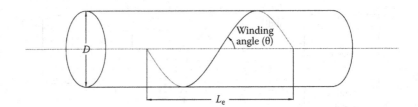

FIGURE 8.8 Axial distance traveled by the carriage in one complete revolution of the mandrel.

The number of mandrel revolutions needed for the carriage to travel a distance L parallel to the mandrel axis is

$$N = \frac{L}{L_e} = \frac{L \tan \theta}{\pi D}. \qquad (8.2)$$

Since the fiber band is at an angle θ with the mandrel axis, it covers a length greater than the bandwidth B in the circumferential direction of the mandrel (Figure 8.9). The effective bandwidth B_e in the circumferential direction is

$$B_e = \frac{B}{\cos \theta}. \qquad (8.3)$$

The number of circuits required for complete coverage around the circumference of the mandrel is given by the following equation:

$$n = \frac{\pi D}{B_e} = \frac{\pi D \cos \theta}{B}, \qquad (8.4)$$

where n is the number of circuits to completely cover the mandrel length; B_e is the effective bandwidth; and B is the bandwidth

The bandwidth B is adjusted by adding or removing the number of rovings being pulled through the carriage so that the number of circuits is an integer and there are

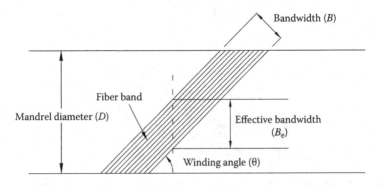

FIGURE 8.9 Effective bandwidth.

no gaps or overlaps between the bands. Another point to note is that as the carriage starts to turn back and reverse at the end of the travel, there will be a small region at each end where the winding angle will become close to 90° instead of the desired winding angle θ. In the turn-back regions at each end, there will be a slight increase in thickness due to layer buildup. This is known as *dog boning*, which can be reduced by staggering the traversing distance of the carriage in each direction. Also, during the time of carriage reversal, the mandrel rotates by several degrees before the fiber band again makes contact with the mandrel surface. This angular rotation is called the *dwell angle*.

Another term used in filament winding is called the *pattern repeat*, which is simply the number of circuits needed before a fiber band is positioned right next to itself. The pattern repeat controls the number and appearance of diamond-shaped fiber band crossovers on the surface of the filament-wound part (Figure 8.7). A low pattern repeat creates very small diamonds, while a high pattern repeat creates large diamonds.

8.3   FILAMENT WINDING MACHINES

Most filament winding machines are two-axis machines with a rotating mandrel and traversing carriage, and they produce axisymmetric parts. The versatility of the filament winding process has tremendously improved with the introduction of numerically controlled machines containing independent drives for the mandrel as well as for the carriage. With these machines, it is possible to filament wind nonaxisymmetric and complex-shaped parts, such as pipe elbows, helicopter blades, etc.

The numerically controlled machines are available with up to six axes of motion control (Figure 8.10) in which each axis is independently controlled by its own microprocessor. The six axes of motion are as follows:

1. Mandrel rotation about its own axis
2. Carriage traverse parallel to the length of the mandrel
3. Horizontal cross-feed
4. Vertical cross-feed
5. Payout eye rotation
6. Yaw

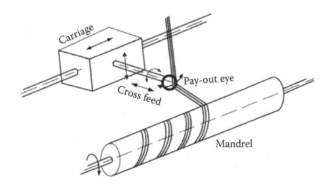

FIGURE 8.10   Schematic of a numerically controlled six-axis filament winding machine.

Various cylindrical cross-sections
that can be filament wound

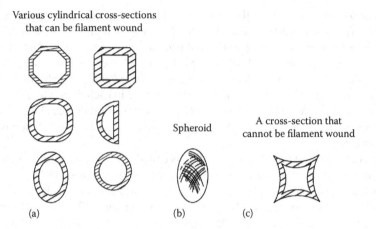

Spheroid

A cross-section that
cannot be filament wound

(a)   (b)   (c)

FIGURE 8.11 Examples of filament-wound cross sections that (a, b) can be filament wound and (c) cannot be filament wound.

The cross-feed mechanism and the rotating payout eye allow an unequal fiber placement on the mandrel. The cross-feed mechanism is mounted on the carriage and can move in and out radially, and the payout eye can be controlled to rotate about a horizontal axis. The combination of these two motions prevents fiber slippage as well as fiber bunching on mandrels of irregular shape. Although each mechanism is driven by its own motor, their movements are related to the mandrel rotation by numerical controls. Since no mechanical connections are involved, winding angles can be adjusted relatively easily without much manual operation. With conventional filament winding machines, the shapes that can be created are limited to the surfaces of revolution, such as cylinders of various cross sections, cones, box beams, or spheroids. The computer-controlled multiaxis machines can wind irregular and complex shapes with no axis of symmetry, such as the aerodynamic shape of a helicopter blade.

Figure 8.11 shows examples of cross sections that can be filament wound using conventional as well as computer-controlled machines. A round cylindrical shape with domes at each end is typical of pressure vessels and rocket motor cases. A spherical shape is selected for internal pressure containers, since this shape provides the maximum possible volume with minimum surface area. Cylindrical hollow tubes are used for shafts carrying torsional load. A $\pm45°$ winding pattern is ideal for this application.

8.4 MANDREL MATERIAL AND DESIGN

The principal design requirements for a filament winding mandrel are that (1) it must be able to resist sagging due to its own weight, (2) it must be able to withstand the applied winding tension, and (3) it must keep its form during curing at elevated temperatures. Other considerations in mandrel design are durability if it is repeatedly used for large production volumes, thermal expansion, and cost. Steel mandrels are preferred for their durability compared to aluminum or plastic mandrels, but they are heavier and may cause bending of the winding axis due to their own weight, which in turn can alter the fiber path. The difference in thermal expansions (and contractions) of the mandrel and the wound composite part is important during curing as well as during mandrel extraction. For prototyping or for low-volume productions, soluble plasters, eutectic salts, or low-melting alloys are used as the mandrel material instead of steel or aluminum.

Hemispherical domes with central end openings are commonly used at the mandrel ends for manufacturing pressure vessels. The end openings are necessary to extract the mandrel from the cured pressure vessel. Water-soluble mandrels made of fine sand particles mixed with a polyvinyl alcohol (PVA) binder are used in these applications. The mandrel is cast in a mold and cured at approximately 95°C. It can then be machined to create a smooth surface and tight tolerance. For extracting the mandrel after the composite part is cured, hot water is introduced through the end openings, which percolates through the mandrel and dissolves PVA. The resulting slurry is then poured out to create the hollow dome-ended vessel.

To facilitate mandrel extraction, collapsible mandrels, either segmented or inflatable, are used for products in which the end closures are integrally wound, as in the case of pressure vessels. Collapsible mandrels with segmented metal units or an inflatable rubber bladder are also used if the mandrel needs to be separated from the filament-wound part after curing. In segmented mandrel design, each segment is connected to a linkage mechanism that can be mechanically operated to move outward to form the mandrel surface and collapsed for mandrel removal. The inflatable bladder can be used either by itself or on the surface of a hard metal mandrel. The bladder can be inflated to apply pressure on the inside of the tube during winding as well as during curing. After the winding is complete, the wound fiber–resin system along the mandrel is placed inside a clamshell mold and cured. Pressurization of the bladder forces the winding to deform outward and press it against the mold surface. This produces a smooth outer surface of the filament-wound part after cure; however, it is difficult to accurately control the inside surface smoothness if a rubber bladder is used.

In many high-pressure gas-containing pressure vessel applications, such as pressurized oxygen tanks used by firefighters on their back, a thin metal liner is used inside the filament-wound composite vessel as a load-sharing element and for preventing gas permeation through its thickness (Figure 8.12). Thus, the liner, which serves as the winding mandrel, becomes a vital part of the pressure vessel. Since the tubular liner is very thin, it may become necessary to pressurize it so that it does not collapse during the winding operation.

To facilitate mandrel removal after curing, the mandrel surface should be highly polished. When steel mandrels are used, they can be chrome-plated. A slight taper along the length of the mandrel is also helpful for mandrel removal. The mandrel surface can also be sprayed with a mold release agent. In some cases, the mandrel surface is brushed with a gel coat (a thin layer of catalyzed resin), which creates a thin resin-rich layer on the inside of the filament-wound part.

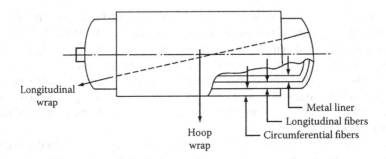

FIGURE 8.12 Construction of a filament-wound oxygen tank.

8.5 FILAMENT WINDING PROCESS PARAMETERS AND CONTROLS

The important process parameters in a filament winding operation are the winding speed, fiber delivery, fiber tension control, resin impregnation and fiber wet-out, and fiber band formation.

8.5.1 WINDING SPEED

For a circular mandrel rotating with a constant rotational speed of N_o revolutions per minute and a constant carriage feed of V_o, the winding angle is given by

$$\theta = \frac{2\pi N_o r}{V_o},\qquad (8.5)$$

where r is the radius of the filament winding part. For a thin part, r is equal to the mandrel radius. For a thick part, r is initially equal to the mandrel radius, but as the thickness builds up, r increases. Thus, as Equation 8.5 shows, a constant winding angle can be maintained in a thick part only if the ratio N_o/V_o is adjusted from layer to layer.

8.5.2 FIBER DELIVERY

Fiber rovings are pulled from fiber packages that are stacked either on a tabletop creel or on a bookrack creel with multiple shelves. As each fiber roving is unwound from its package, it is pulled through a fiber guide located at the center directly over the package. The purpose of the fiber guide is to bring the rovings together to form a fiber band in which all the rovings are running parallel. Another major function of the fiber guides is to create tension in the rovings so that no catenaries are formed and the fiber band is delivered to the resin bath under tension. It is important that each strand in the rovings is equally tensioned; otherwise, the downstream functions of resin pickup, impregnation, and fiber wet-out will be uneven and the filament-wound part may not be of high quality.

8.5.3 FIBER TENSION CONTROL

Adequate fiber tension is required to maintain fiber alignment on the mandrel as well as to control the resin content in the filament-wound part. Excessive fiber tension can cause resin squeeze-out, differences in resin content in the inner and outer layers, undesirable residual stresses in the finished product, and large mandrel deflections. Typical tension values range from 1 to 5 N per end.

Fiber tension is created by pulling the rovings through a number of fiber guides placed between the creels and the resin bath. Figure 8.13 illustrates three common types of fiber guides. Mechanical action on the fibers in the resin bath, such as looping under and over several rollers, generates additional fiber tension.

8.5.4 RESIN IMPREGNATION AND FIBER WET-OUT

In a wet filament winding process, the resin is pushed into the fiber bundle to wet out the fibers and to drive out the air that is initially entrapped between the fibers. Incomplete wetting results in the formation of voids in the filament-wound part, which has a direct effect on its quality and mechanical properties. Good fiber wet-out is essential for reducing voids in a filament-wound part and for good fiber-resin interfacial bonding. The following material and process parameters control the fiber wet-out:

1. Viscosity of the catalyzed resin mix in the resin bath, which depends on the resin type, resin bath temperature, and cure advancement in the resin bath
2. Number of strands (or ends) in a roving, which determines the accessibility of resin to each filament in the strands
3. Fiber tension, which controls the pressure on various layers already wound around the mandrel
4. Speed of winding and length of the resin bath.

There are two essential requirements for the resin used in filament winding:

1. The viscosity of the mixed resin system (which may include a solvent) should be low enough for impregnating the moving fiber strands in the resin bath, yet not so low that the resin drips and runs out easily from the impregnated fiber strands. Usually, a viscosity level of 1–2 Pa s is preferred.

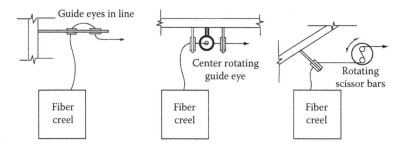

FIGURE 8.13 Fiber guides used for controlling tension in fibers in a filament winding line.

2. The resin must have a relatively long pot life so that large structures can be filament wound without premature gelation or excessive exotherm. Furthermore, the resin bath is usually heated to lower the viscosity level of the mixed resin system. Since the increased temperature of the resin bath may reduce the pot life, a resin with long pot life at room temperature is recommended for filament winding.

Two different types of resin bath designs are shown in Figure 8.14—one in which the rovings are pulled over a rotating drum and the other in which the rovings are dipped into the resin bath using three submerged rollers. In the drum-type resin bath, the rotating drum pulls up a thin layer of resin as it emerges out of the resin surface level. The resin content in this design is controlled by adjusting the gap between a stationary doctor blade and the drum. In the roller type resin tank, the rule of thumb is that the rovings should be under the resin surface level for at least 1/3–1/2 s. In a line moving at 60 m/min, this means that the length of rovings under the resin surface level should be between 33 and 50 cm. The final resin content is controlled by either a squeegee method or a stripper die method. In the squeegee

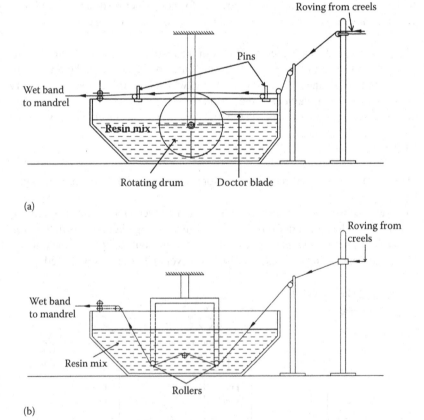

FIGURE 8.14 Resin bath designs. (a) Rotating drum and (b) submerged rollers.

method, the resin-wetted rovings are pulled through two elastomer-coated bars, and the gap between the bars is adjusted to control the resin content. In the stripper die method, the resin-wetted rovings are pulled through orifices machined on a die plate.

Proper resin content and uniform resin distribution are required for good mechanical properties as well as for weight and thickness control. Resin content is controlled by proper wiping action at the squeegee bars or stripper die, fiber tension, and resin viscosity. Dry winding in which prepregs are wound around a mandrel often provides a better uniform resin distribution than wet winding.

Fiber collimation in a multiple-strand roving is also an important consideration to create uniform tension in each strand as well as to coat each strand evenly with the resin. For good fiber collimation, single-strand rovings are often preferred over conventional multiple-strand rovings. Differences in strand lengths in conventional multiple-strand rovings can cause sagging and catenary formation in the filament winding line.

8.5.5 FIBER BAND FORMATION

Resin-wetted fiber rovings are grouped together to form a fiber band just before they are wound on the rotating mandrel. The band former unit is located on the carriage that traverses back and forth parallel to the mandrel axis. Several different band former designs are shown in Figure 8.15.

In a helical winding operation, the fiber band crisscrosses at several points along the length of the mandrel, and one complete layer consists of a balanced helical pattern with $\theta/-\theta$ fiber orientations. The thickness of a layer depends on the band density (i.e., the number of rovings per unit length of a band), the roving count (i.e., the number of strands or ends per roving), and the resin content. For the same band

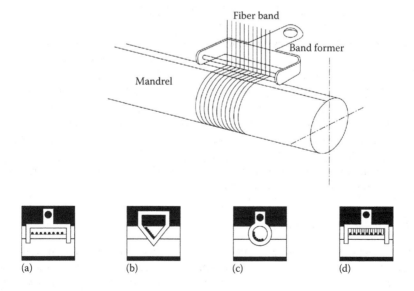

FIGURE 8.15   Fiber band former designs: (a) straight bar, (b) bent bar, (c) ring, and (d) comb .

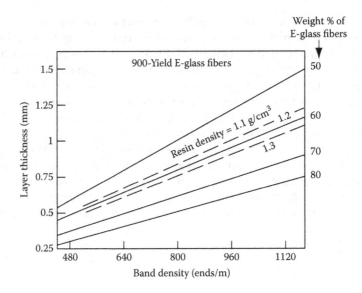

FIGURE 8.16 Effects of band density, fiber weight fraction, and matrix density on the thickness of a filament-wound E-glass fiber reinforced epoxy tube.

density, a high roving count results in larger amounts of fibers and, therefore, thicker layers. Increasing the resin content also results in thicker layers. An example is shown in Figure 8.16.

8.6 DEFECTS IN FILAMENT-WOUND PARTS

The common defects in filament-wound parts are voids, interlayer cracks, and fiber wrinkles. Voids may appear because of poor fiber wet-out, the presence of air bubbles in the resin bath, an improper bandwidth resulting in gapping or overlapping, or excessive resin squeeze-out from the interior layers caused by high winding tension. In large filament-wound parts, an excessive time lapse between two consecutive layers of windings can result in interlayer cracks, especially if the resin has a limited pot life and starts to rapidly cure before the layers are properly consolidated. Reducing the time lapse and brushing the wound layer with fresh resin just before starting the next round of winding are recommended for reducing interlayer cracks. Wrinkles result from inadequate winding tension and misaligned rovings. Unstable fiber paths that cause fibers to slip on the mandrel may cause fibers to bunch together, bridge, and improperly orient in the filament-wound part.

8.7 FIBER PLACEMENT

Fiber placement is a continuous manufacturing process in which filament winding and automated tape laying are merged to produce complex nongeodesic shapes with concave and convex surfaces. In the fiber placement process [2], preimpregnated fiber tows are pulled into a fiber delivery system under controlled tension where they are collimated into a narrow band, typically 3.2–4.6 mm in width. The resin in the preimpregnated tows can be either a thermoset or a thermoplastic. If it is a thermoset, it must have low tack so that the fiber tows can be pulled off the spool without much difficulty.

The fiber delivery system is called the fiber placement head (Figure 8.17), which is located at the end of a roll–bend–roll wrist of a robot arm. Its movement is controlled using a programmable control system that provides six independent degrees of freedom to the wrist—three for translational motions (carriage, tilt, and cross-feed) and

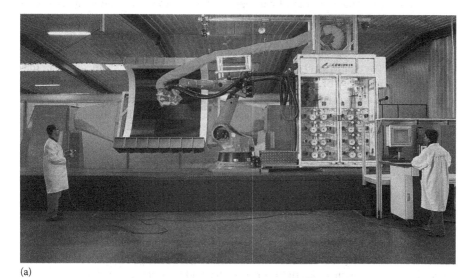

(a)

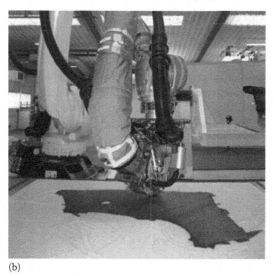

(b)

FIGURE 8.17 Fiber placement system. (a) Fiber placement machine and (b) fiber placement head. (Courtesy of Coriolis Composites, Quéven, France.)

three for rotational motions (yaw, pitch, and roll). As the fiber placement head is moved along the surface of a mandrel (or a tool), it places the fiber band on the mandrel surface under the pressure of a compaction roller. The rotational motion of the mandrel adds another degree of freedom to the system. All these motions are coordinated to ensure that the compaction roller is normal to the mandrel surface.

Unlike the filament winding process, the fiber tows in the fiber placement process are not under tension as they are placed on the mandrel surface. Instead, they are compacted by the normal pressure applied by the compaction roller. A heating arrangement attached to the fiber placement head increases the temperature of the fiber band locally and increases its tack, which helps in compaction and consolidation of the fiber band layers. The pressure applied by the compaction roller and reduced resin viscosity due to temperature increase also help in removing the trapped air between the layers. The fiber placement head includes a mechanism that can cut and restart any of the individual tows so that the width of the fiber band can be increased or decreased in increments of one tow width. This eliminates the possibility of creating large gaps or overlaps between adjacent fiber bands. Each tow can be pulled and placed on the mandrel surface at its own speed, which makes it possible for each tow to independently conform to the curved surface features of the mandrel. After finishing the fiber placement operation, the mandrel is placed in an autoclave for curing.

Even though fiber placement machines are considerably more expensive compared to filament winding machines, there are several advantages for which fiber placement is considered for applications such as aircraft fuselage shells. The key among these advantages are (1) its ability to change fiber path within each layer according to the design need, (2) its ability to start and drop off fiber bands as needed to produce tapers, open areas, tight corners, etc., and (3) significantly lower scrap and labor saving compared to hand layup.

8.8   TUBE ROLLING

Circular tubes are often manufactured using a tube rolling technique in which a precut length of a prepreg is rolled onto a removable mandrel. Several tube rolling methods are demonstrated in Figure 8.18. The uncured rolled prepreg tube is then wrapped with heat-shrinkable film or sleeve and cured in an air-circulating convection oven. As the outer wrap shrinks tightly on the prepreg tube and applies radial

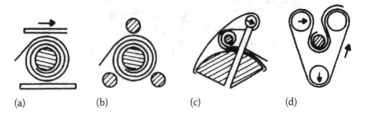

FIGURE 8.18   Schematic of several tube rolling processes: (a) between flat plates, (b) between rollers, (c) around tensioning rollers, and (d) around a continuous belt.

pressure, air entrapped between the layers of the prepreg is squeezed out through the ends. For a better surface finish, the curing operation can be performed inside a close-fitting steel tube or a split steel mold. After curing, the mandrel is removed and a hollow tube is formed.

Tube rolling is used in manufacturing bicycle frames, golf shafts, arrows, space truss, etc. Its advantages over filament winding are low tooling cost, better control over resin content, more uniform resin distribution, and faster production. However, the process is more suitable for manufacturing small diameter tubes with simple layups containing 0 and 90° layers.

REFERENCES

1. S. T. Peters and Y. M. Tarnopol'skii, Filament winding, in *Composites Engineering Handbook* (P. K. Mallick, ed.), Marcel Dekker, New York, pp. 515–548, 1997.
2. D. O. Evans, Fiber placement, in *ASM Handbook*, Vol. 21: Composites (D. B. Miracle and S. L. Donaldson), pp. 477–479, ASM International, Materials Park, OH, 2001.

9 Pultrusion

Pultrusion is a continuous manufacturing process for making long, straight structural composite parts of a constant cross-sectional area [1–3]. It can produce a wide variety of solid and hollow cross sections with wall thicknesses ranging from 2 to 30 mm. Among the common pultruded products are round and rectangular tubes, solid rods, flat and corrugated sheets, and beams of different profiles, such as angles, channels, hat sections, double web sections, and wide-flange sections (Figure 9.1). Many of these products are found in commercial and recreational applications, such as ladder rails, fence posts, tent poles, antenna rods, window frames, ski poles, and tool handles. They are also used in many structural applications, such as floor gratings, loading platforms and bridge structures, and many mass transit applications, such as overhead handrails, guardrails, and seat frames.

Pultruded composites typically contain longitudinally oriented continuous fiber rovings and several layers of random fiber mat, such as continuous filament mat (CFM) and chopped strand mat (CSM), at or near the outer surfaces (Figure 9.2). The rovings provide high modulus and tensile strength in the longitudinal direction of pultruded composite. The random mat layers, which contain randomly oriented continuous fiber strands, are added to improve its modulus and tensile strength in the transverse direction. The ratio of random fiber mat to continuous fiber rovings determines the ratio of transverse to longitudinal properties. If higher transverse properties are needed, woven roving fabrics are substituted for random fiber mats.

In typical pultruded products, the total fiber content is in the range of 50–60% by weight, and the roving to mat ratio is 2.5:1 [1]. In the majority of the pultruded products, E-glass fibers are used as the reinforcement, and either a polyester or a vinyl ester resin is used as the matrix. Carbon fibers are used as reinforcement when higher strength or modulus is required. Epoxy resins are used as the matrix in applications requiring heat resistance or fatigue resistance greater than what can be achieved using polyester or vinyl ester resins, but they require longer pultrusion dies because of their slower curing rate, and they do not easily release from the pultrusion die because of their low curing shrinkage. Thermoplastic matrix has also been used in pultruded products; however, the pultrusion process for thermoplastic matrix composites is different from that for thermosetting matrix composites.

Pultrusion is a highly automated, continuous manufacturing process with a high production rate. It requires very little manual labor, and the material scrap is very low. The tooling cost is also relatively low. All these make pultrusion a relatively low-cost production process for manufacturing structural beam sections with high fiber volume fraction. On the other hand, dimensional accuracy and tolerances of pultruded products are not as good as with other processes. The production of very thin-walled sections and complex cross-sectional profiles is also not possible with pultrusion.

FIGURE 9.1   Examples of pultruded sections.

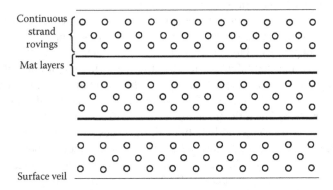

FIGURE 9.2   Fiber arrangement in a typical pultruded section.

9.1   PULTRUSION PROCESS

Figure 9.3 shows the schematic of a typical pultrusion line. Continuous fiber rovings are pulled from their creels at one end of the line, collimated, and then combined with layers of random fiber mat before entering a series of forming guides. The purpose of the forming guides is to position the rovings and mats in their respective locations in the cross section of the profile. The combination of dry rovings and mats enters the resin bath that contains a mixture of liquid resin, curing agent or catalyst, and other ingredients, such as fillers, colorant, UV stabilizer, and fire retardant. Table 9.1 gives a typical formulation of an unsaturated polyester resin mix used in a pultrusion line. The viscosity of the liquid resin mix; residence time of the fibers in the resin bath, and mechanical action, such as looping up and down around the rollers in the resin bath; are adjusted to ensure a complete wet-out of fibers with the resin. A thermoplastic (usually a thermoplastic polyester) surfacing veil is added to the fiber–resin stream just outside the resin bath to improve the surface smoothness

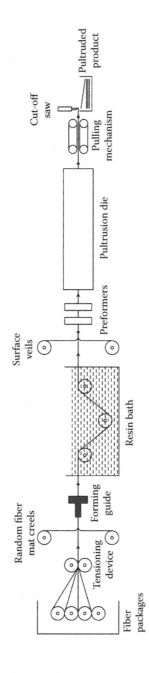

FIGURE 9.3 Schematic of a pultrusion line.

TABLE 9.1

Typical Polyester Resin Mix Formulation for Pultrusion

Ingredient	Typical Amount	Function
Unsaturated polyester resin (with 30% styrene monomer[a])	100	Becomes matrix after curing
Aluminum trihydrate	25	Filler, fire retardant, controls shrinkage, controls viscosity
Low-profile additive (a thermoplastic powder)	10	Controls shrinkage, improves surface smoothness and gloss
Pigment	2	Color
Internal mold release	1	Reduces adhesion to the die surface
UV absorbent	0.3	Weather resistance
Catalyst 1	0.5	Low-temperature reaction initiator
Catalyst 2	0.5	Higher-temperature reaction completer

[a] Styrene acts as a reactive diluent.

of the finished product. For applications requiring corrosion protection, surfacing veils containing A-glass or C-glass fibers are used. The fiber–resin stream is then pulled through a series of preformers before entering a long preheated pultrusion die. The function of the preformers is to evenly distribute the rovings, squeeze out as much excess resin as possible before entering the pultrusion die, and gradually form the shape of the cross section. Final shaping, compaction, and curing take place in the pultrusion die.

The pultrusion die is designed to ensure that the pultruded section is cured before it emerges from the die exit end. It is divided into three major sections: die entrance section, curing and consolidation section, and cooling section. The die entrance section contains a small tapered inlet zone where the fiber–resin stream coming from the last preformer is gradually pulled to enter the die. As the excess resin in the fiber–resin stream is squeezed out at the die entrance, it creates resin backflow that can generate significant increase in pressure at the die entrance [4] (Figure 9.4).

The die entrance section is usually water cooled to prevent premature gelling of the resin, but the rest of the die is heated in a controlled manner either by electric heaters or by oil heaters. Infrared heating has also been used to accelerate the curing process. The die temperature gradient along its length, die length, and pulling speed are controlled to allow the resin to completely cure as the pultruded section exits the die. A pulling system pulls the cured member out of the die. Between the die exit end and the pulling system, the cured member is cooled either naturally or by forced cooling (using either forced air or water jets) to a lower temperature so that the pulling mechanism would not cause any damage to the cured surfaces as it grips them. The pulled length of the pultruded product is then cut into desired lengths using a high-speed saw at the end of the line.

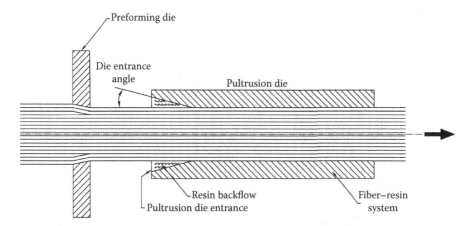

FIGURE 9.4 Resin backflow at the die entrance.

9.2 PULTRUSION PRODUCTION LINE

The common pultrusion production line has six operational areas arranged in the following order: (1) fiber storage and supply, (2) resin impregnation, (3) preforming die, (4) pultrusion die, (5) pulling, and (6) cutting. The basic functions of these operational areas are described in the following:

1. The fiber storage and supply area is the starting area of the pultrusion process. Here the rovings, mats, fabrics, and veils are stored in creels from which they are pulled in proper sequence to match the design of the pultruded section. The roving packages are usually stored on bookshelf style racks (Figure 9.5), and rovings from these packages are pulled through a series of ceramic-coated threading guides or rollers to remove any slack in them and create tension in the fiber rovings. Strips of mats, fabrics, and veils, drawn from their own rolls, join the rovings prior to entering the resin impregnation area. As all these fiber forms travel forward, it is important to maintain proper alignment to prevent any knotting or twisting and position them in the correct layering sequence as per the design requirement.
2. The resin impregnation area usually contains an open resin bath and a series of rollers submerged in the resin mix. The rollers can be arranged all in one row or divided in two rows, shown in Figure 9.6. The mechanical action on the fibers as they pass under or over these rollers helps spread them sideways, remove air entrapped between them, and squeeze resin mix into them. In some cases, there may be separate resin baths for the rovings and the other fiber forms.

   Another method of resin impregnation is direct resin injection into the fiber forms just after they enter the preforming die (Figure 9.7). The liquid

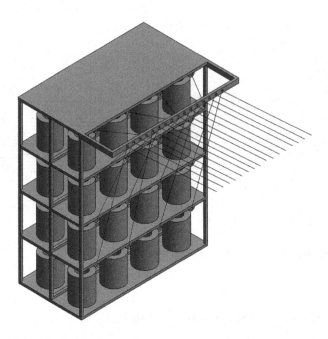

FIGURE 9.5 Bookshelf-style fiber package racks.

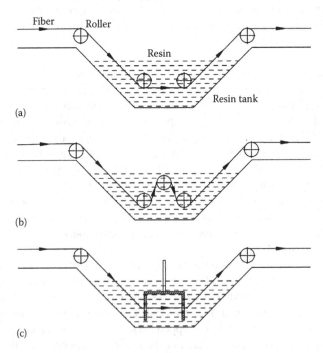

FIGURE 9.6 Various resin impregnation arrangements in a pultrusion resin bath: (a) one-row rollers, (b) two-row rollers, and (c) bracket with slotted holes.

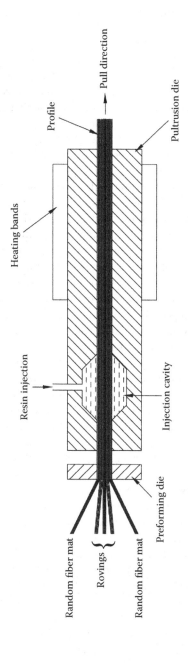

FIGURE 9.7 Resin injection process.

resin mix is injected at a high pressure, usually in the range of 5–30 bar, so that it can penetrate into the interstices of the fiber forms. There are two advantages to the direct injection process. One, it is easier to handle the fiber forms and position them more accurately in the preforming die when they are not wet with the resin mix as in the open resin bath method. Second, since the resin mix injection takes place inside a closed chamber, there is very little release of styrene or other volatile components into the production plant, and so it is more environmentally safe. One limitation of this method is that fiber forms entering the injection chamber are tightly compressed which makes resin impregnation difficult, particularly for thick sections with high fiber volume fractions.

3. The preforming die serves the purpose of preshaping the cross section of the pultruded profile using the resin-impregnated fiber forms before they enter the pultrusion die. It is usually a plate with an opening with the shape of the cross section to be formed. Depending on the complexity and dimensions of the cross-sectional shape, there may be a series of preforming dies that are sized and positioned to progressively create the shape, apply pressure on the fiber forms to compact them to the desired thickness, and remove excess resin. The opening in the last preforming die is slightly larger than the opening of the pultrusion die. For tubular shapes, a mandrel is used inside the preforming dies to form the internal shape. The mandrel is extended as a cantilever through the pultrusion die and is held in place in correct alignment and position with respect to the pultrusion die.

4. The pultrusion die is where final shaping and curing take place. It is typically 1 m long and is made of two or more segments so that the die can be easily opened for cleaning and repair. If short die lengths are used, the pulling speed will be slower and the production rate will also be lower. The pultrusion process requires that the curing of the entire section be complete within the die length. This is the reason for slower pulling speed with shorter die lengths.

Since pultrusion runs are continuous and the fibers used in pultruded products are very abrasive, pultrusion dies are made of hardened tool steels or high alloy steels. The interior surface is highly polished and hard-chrome plated to improve wear resistance and reduce friction. The die is usually built by assembling two or more segments, and therefore, the outer surface of the pultruded product will show shallow longitudinal parting lines. An example of a four-segment die for pultruding a channel section is shown in Figure 9.8. In this case, there will be four longitudinal parting lines, two on each flange of the pultruded channel section. It is important to note that the die should not have sharp corners, since they will be readily filled with resin and have to be cleaned frequently. A minimum radius of 0.5 mm is recommended and the corners should not be located at the parting lines. Another important consideration in die design is to account for curing shrinkage, which depends on the resin type, fiber volume fraction, fiber orientation, and part thickness. There is also the possibility of warpage and spring back, which also needs to be taken into account in the die design. For example, for the open channel section shown in Figure 9.8, flanges will tend

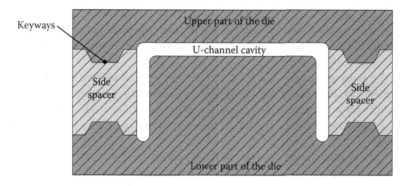

FIGURE 9.8   Four-segment die for pultruding a channel section.

to bend inward or *toe-in* as the section emerges from the die exit. This can be reduced by designing the die with a small *toe-out*. Postsizing the section immediately after the die exit while it is still hot and pliable can also reduce the spring back problem.

Die heating is one of the most critical considerations in the pultrusion process design, since it controls not only the rate of curing, but also the temperature profile and location of the exotherm. Typically, a pultrusion die is heated by two heating bands that are independently controlled. The temperature is measured using thermocouples placed at various locations along the length of the die. When the resin-impregnated fiber forms enter the first heating zone in the die cavity, the resin is in a liquid state. As the resin absorbs heat and its temperature starts to increase, its outward thermal expansion causes the pressure from the die wall to increase. There is also a slight decrease in its viscosity due to temperature rise. As the assembly moves forward and absorbs more heat, the curing reaction begins and the resin starts to change from a viscous liquid to a gel. With continued curing in the second heating zone, the gel starts to transform into a solid. Curing shrinkage at this stage causes the pultruded section to retract from the inside surface of the die and the pressure from the die wall starts to decrease. The liquid-to-gel and gel-to-solid transformations of the resin are depicted in Figure 9.9.

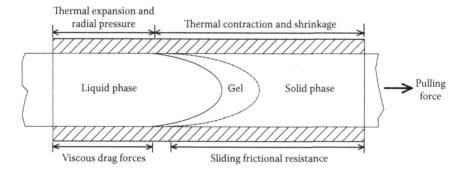

FIGURE 9.9   Transformation of resin from liquid phase to solid phase in a pultrusion die.

Figure 9.10 shows the location of the peak exotherm measured by using a thermocouple inserted into the moving material and plotting the temperature as a function of distance from the die entrance. Figure 9.10 also shows the die temperature profile.

5. The pulling area has the function of clamping and pulling the cured pultruded product out of the curing die in a continuous and steady speed: Two different pulling systems are used—either a set of continuous belt pullers or chain and sprocket system containing a series of pulling blocks (Figure 9.11).

The pulling force depends on the following factors:

a. Pulling and collimation resistance in the creels, resin baths, and preforming dies
b. Frictional forces arising from bulk compaction of the fiber forms and the resin mix in the preforming dies and at the pultrusion die entrance
c. Viscous drag force of the resin-impregnated fiber forms sliding against the die surface while the resin is in the liquid phase
d. Thermal expansion of the material as it heats up in the die and applies normal pressure on the die wall
e. Resin shrinkage and release of contact from the die surface, which may cause reduction in pulling resistance
f. Sliding frictional resistance in the die, especially in the solid-phase section of the die

The relative magnitudes of these resistances to pulling depend on the fiber and filler content, roving guidance system, resin viscosity, resin reaction rate, resin shrinkage behavior, die surface condition, and die design. A high pulling force is not desirable, since it will require a large pulling

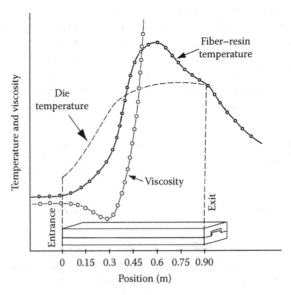

FIGURE 9.10 Temperature and resin viscosity profiles in a pultrusion die.

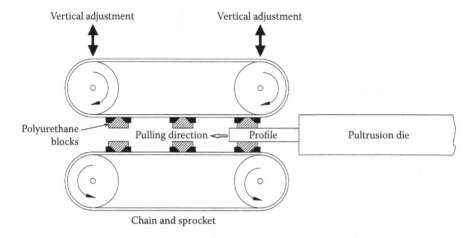

FIGURE 9.11 Pulling mechanism using a continuous chain and sprocket system

system and the cost will also increase. High clamping force and pulling force may also cause damage to the cured pultruded member as it is being pulled.

6. Cutting of pultruded products into desired lengths for storage and shipping is done using a high-speed saw mounted on a platform that moves at the same speed as the pulling speed of the pultruded product. Carbide or diamond-tipped saw blades are used for cutting E-glass and carbon fiber pultrusions.

9.3 MATERIALS FOR PULTRUDED PRODUCTS

Many pultruded products are used in commercial and engineering applications where functional characteristics such as electrical insulation (for example, in household ladders) or chemical resistance (for example, in chemical scrubbers) are important material selection considerations in addition to strength, stiffness, and cost. In the vast majority of these products, E-glass fiber is used which not only is the least expensive of the reinforcement fibers available, but also provides high strength and good electrical insulation. It is also available in a variety of forms, such as rovings, yarns, fabrics, CFM, and CSM.

The primary reinforcement form in pultruded sections is the continuous fiber rovings that are oriented in the length direction of the pultruded products. This gives high strength and stiffness in the length direction; but strength and stiffness in the transverse direction are low. Random fiber mats are added to improve the transverse properties. If higher transverse properties are needed, 0/90° or ±45° fabrics are selected instead of random fiber mats. Carbon fibers offer much higher modulus than E-glass fibers, but they are significantly more expensive. They are selected in applications that require very high stiffness, low weight, and low thermal expansion, for example, in space antenna booms.

The addition of random fiber mats in pultruded products contributes to the improvement of transverse properties [5]. However, if the overall volume fraction of fibers remains constant, the addition of increasing numbers of random fiber layers will reduce the volume fraction of longitudinal rovings, and the longitudinal properties will also be reduced (Table 9.2). Since random fiber mats have a relatively open arrangement of fibers with large gaps in between them, their use on the outer surfaces increases porosity and voids in the final product. The potential for moisture absorption also increases with greater proportion of random fiber mats in the construction of a pultruded section.

The outside surface layer in the construction of pultruded sections usually contains a thin surfacing veil which is a fabric made of very fine thermoplastic polyester or nylon filaments. Since the surfacing veil is placed on the exterior surface, it tends to produce a relatively smooth resin-rich surface and help reduce die wear from the abrasive action of glass or carbon fiber layers. Since many pultruded products, such as handrails and street signposts, are used outdoors, they are subject to weathering effect, such as rain erosion on the surface. The weathering effect is reduced if the surface is resin rich due to the presence of a surfacing veil.

The commonly used resins in pultruded products are isophthalic polyesters and vinyl esters. Typical applications of polyesters are ladder rails, handrails, fence posts, roof support beams, electrical poles, air conditioning ducts, etc. Vinyl esters are selected where higher chemical (corrosion) resistance is required, such as oil well sucker rods and beams for pollution control devices. Epoxies are selected for applications where high fatigue resistance is needed, such as a bridge floor. Epoxies also give higher heat resistance and creep resistance. A comparison of pultrusion process parameters based on these three resin types is given in Table 9.3. Phenolic resins are finding some applications where their fire resistance and low smoke production are considered to be distinct advantages. However, curing of phenolics takes place by a condensation reaction that produces water vapor and creates voids, porosities, and delaminations in the product.

TABLE 9.2

Effect of Random Fiber Mats on the Tensile Properties of a Pultruded Section

Total fiber content (wt.%)	70	60	50	40	30
Roving (wt.%)	38.8	28.8	18.8	18.8	16.1
Mat (wt.%)	31.2	31.2	31.2	20.8	13.9
Roving-to-mat ratio	1.24:1	0.92:1	0.60:1	0.90:1	1.16:1
Roving end count	79	58	38	29	33
Number of mat layers	3	3	3	3	2
Mat weight (g)	42	42	42	42	28
Longitudinal tensile strength (MPa)	373.1	332.4	282.1	265.5	217.2
Transverse tensile strength (MPa)	86.9	93.1	94.5	84.8	67.6
Longitudinal tensile modulus (GPa)	28.8	23.6	18.4	17.1	15.4
Transverse tensile modulus (GPa)	8.34	9.31	8.55	7.1	5.24

Note: Materials: continuous E-glass fiber rovings, random E-glass fiber mats, and polyester resin.

TABLE 9.3

Comparison of Pultrusion Process Parameters Based on the Resin Type

Process Parameter	Polyesters and Vinyl Esters	Epoxies
Common reaction initiator	Organic peroxides	Amines
Viscosity	Low (500–2000 MPa s)	High (>3000 MPa s)
Cure rate	Fast	Slow
Gel time	Short (s)	Long (min)
Gelation characteristics	Gelation occurs before exotherm	Gelation occurs after exotherm
Conversion at gelation	10–30%	>50%
Resin mix temperature	25–50°C	50–55°C
Mold temperature	115–177°C	150–230°C
Shrinkage (% volume)	6–12	1–6
Mold release effectiveness	Good	Fair
Typical processing rates	0.6–1.5 m/min	0.07–0.1 m/min

The selection of resin and curing reaction initiator controls the viscosity, pot life, curing characteristics, and shrinkage. For resins used in pultrusion, long pot life and short cure time are desired. Since the die wall temperature is higher than the temperature of the material entering the die, the initial heat conduction is from the die wall toward the inside of the cross section. The center of the cross section does not reach the die wall temperature until it has traveled some distance from the die entrance. The curing reaction begins at the surface of the cross section, which first reaches the activation temperature of the reaction initiator (catalyst in case of polyesters and vinyl esters or the curing agent in case of epoxies). The heat generated by the curing reaction helps promote further reactions, which causes the temperature to rise and curing reaction to proceed inward of the cross section. Eventually, curing starts at the center of the cross section and the center temperature starts to increase. Because of the exothermic heat generation and poor heat dissipation, the center temperature exceeds the die wall temperature.

The rate of temperature increase in the material after it enters the die depends on the resin-initiator combination. If the initiator is too active, the material near the die wall will cure very fast and there will be a nonuniform degree of cure across the cross section. On the other hand, if the initiator is very slow, there will be a low degree of cure and it may not be complete across the cross section. With polyester and vinyl ester resins, a combination of a low-temperature initiator and a high-temperature initiator is often used. The low-temperature initiator is activated soon after the fiber–resin system enters the die and initiates the cross-linking reaction close to the die entrance zone. After the material has travelled some distance in the die, the high-temperature initiator is activated and continues to sustain the cross-linking reaction to its completion.

Since pultrusion is a relatively low-pressure process, the surface appearance of pultruded products may not be very smooth. The surfaces tend to be fiber rich, and they may show fiber pattern or even contain exposed fibers. To make the surfaces resin rich and smoother, resins with low curing shrinkage are selected.

The curing shrinkage of polyester and vinyl ester resins is nearly twice that of epoxies. Experiments with polyesters and epoxies have shown that polyesters continue to undergo thermal expansion after gelling, which is followed by high shrink rate after cure [3]. Epoxies, on the other hand, begin to shrink well before gelation and continue to shrink at a slow rate until they are fully cured. Thus, with polyesters, normal pressure from the die wall rapidly decreases after gelling, while with epoxies, the decrease in normal pressure is gradual. The difference in the rate of decrease in normal pressure affects the rate of reduction in sliding friction force in the solid-phase section of the die and, therefore, the pulling force.

9.4 PULTRUSION PROCESS PARAMETERS

The major process parameters in a pultrusion process are pulling speed, die temperature settings, and pulling force. If resin preheating is used before entering the die, then preheating temperature also becomes an important process control parameter. The selection of these parameters determines the quality and structural performance of pultruded products, since they influence fiber wet-out, fiber alignment, entrapped air displacement before curing, and curing characteristics of the resin in the die.

The most important factor controlling the structural performance of pultruded products is the fiber wet-out, which depends on the initial resin viscosity, residence time in the resin bath, bath temperature, and mechanical action applied to fibers in the bath. For a given resin viscosity, the degree of wet-out is improved as (1) the residence time is increased by using slower line speeds or longer baths, (2) the resin bath temperature is increased (which reduces the resin viscosity), and (3) the degree of mechanical working on the fibers is increased by pulling the fibers around properly placed rollers. Since each roving pulled through the resin bath contains a large number of fiber bundles, it is extremely important that the resin penetrates inside the roving and coats each bundle uniformly. Resin penetration takes place through capillary action as well as lateral squeezing between the bundles. Lateral pressure is applied at the end of the resin bath by the squeeze-out bushings, in the preformers, at the die entrance, as well as in the initial length of the die. Generally, slower line speed and lower resin viscosity improve resin penetration by capillary action, and a faster line speed and higher resin viscosity improve resin pickup due to increased viscous drag force [6]. The fiber and resin surface energies are also important parameters in improving the resin coating on fiber rovings. Fiber surface modification, for example, by the application of silane coupling agents on E-glass fibers, also helps in fiber wet-out.

The unfilled resin viscosity in commercial pultrusion lines may range from 0.4 to 5 Pa s. Note that if fillers are added to the resin mix, the viscosity of the resin mix becomes higher than the unfilled resin viscosity. Resin viscosity higher than 5 Pa s may result in poor fiber wet-out, slower line speed, and frequent fiber breakage at the resin squeeze-out bushings. On the other hand, very low resin viscosity may cause high resin loss by dripping from the fiber–resin stream after it leaves the resin bath. To improve fiber wet-out, resin viscosity can be lowered by increasing the resin bath temperature; however, if it reduces to 0.2 Pa s or lower, the fiber–resin stream must be cooled at the resin bath exit to increase the resin viscosity and prevent excessive resin loss by dripping.

As the fiber–resin stream enters the preheated die, heat flow from the die wall decreases the resin viscosity, which aids in continued wet-out of the fibers. The temperature of the fiber–resin stream increases much faster near the die wall than near the center of the cross section. When the temperature reaches a value at which the initiator (or a catalyst) is activated, the curing reaction begins, which, in turn, generates heat that further increases the temperature. The curing reaction proceeds from the material near the die wall toward the material near the center. The accumulation of exothermic heat helps accelerate the temperature rise at the center; however, since the resin–fiber stream is in motion, the center reaches the cure initiation temperature further downstream from the die entrance. Eventually, when the curing at the center begins, its temperature may reach a level higher than that of the material near the die wall, especially since the thermal conductivity of the resin–fiber stream is relatively low. In the remaining length of the die, heat conduction takes place from the material to the die wall, which helps in decreasing the temperature of the cured section as it emerges from the die.

With the beginning of the curing reaction within a short axial distance from the die entrance, the resin viscosity starts to rapidly increase. If the die temperature is not gradually increased at the die entrance zone, a cured resin skin may quickly form on the die walls. The separation of the uncured material from the skin results in poor surface finish of the pultruded product. One method of reducing this problem is to preheat the fiber–resin stream just outside the die, which reduces the temperature gradient at the die entrance zone.

The curing reaction continues at an increasing rate as the fiber–resin stream continues to move forward toward the exit end of the die. Heat generated by the exothermic curing reaction increases the temperature in the fiber–resin stream, which, in turn, increases the reaction rate even more. The location of the exothermic peak along the length of the die depends on the speed of pulling the fiber–resin stream in the die. Figure 9.12 shows the measured die temperature profile along the length of a die and the centerline temperature of the fiber–resin stream at five different pulling

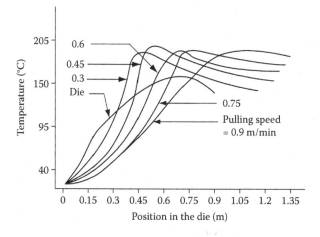

FIGURE 9.12 Temperature profiles of the fiber–resin stream at different pulling speeds.

speeds. Initially, the centerline temperature is lower than the die wall temperature, but it soon crosses over, reaches a peak, and then starts to decrease. The initial lag is due to the low thermal conductivity of the fiber–resin stream. The higher centerline temperature after the crossover is due to the exothermic heat generation. As the curing reaction nears completion, the exothermic temperature decreases and the cooling period begins. The rate of heat transfer from the cured material into the die wall is increased due to lower die wall temperature near the die exit zone. As the pulling speed increases, the crossover distance increases and is shifted toward the die exit end. The exothermic peak temperature is also higher due to lower residence time in the die at higher pulling speeds. If the pulling speed is too high, the material remains in the die for a shorter period and may not be able to conduct away the reaction-generated heat to the die wall before leaving the die.

The pulling speed also affects the degree of cure in the pultruded section. The higher the pulling speed, the lower the degree of cure at the centerline. Moreover, the difference in the degree of cure between the centerline and the surface increases with increasing pulling speed. A more uniform degree of cure distribution across the part thickness is obtained at a lower pulling speed.

Unlike many other molding processes, no external pressure is applied in a pultrusion process. However, volumetric expansion of the resin as it is heated in the die entrance zone can create a significant pressure on the material. Experiments performed by Sumerak [7] have demonstrated that the pressure in the die entrance zone is in the range of 1.7–8.6 MPa. When the curing reaction begins, the polymerization shrinkage reduces the pressure to near-zero values at approximately midlength of the die. High internal pressure at the die entrance before curing helps consolidate the material. Although increasing the internal pressure may result in higher pulling force, it improves the fiber–resin consolidation, which improves the mechanical properties of the pultruded product.

Depending on the part complexity, resin viscosity, and cure schedule, the line speed in a commercial pultrusion line may range from 50 to 75 mm/min on the low side and from 3 to 4.5 m/min on the high side. High line speed usually shifts the location of the peak exotherm toward the exit end of the die and increases the pulling force. Although the production rate increases with increasing line speed, the product quality may deteriorate owing to poor fiber wet-out, unfinished curing, and roving migration within the cross section. If high production output is desired, it is sometimes better to use multiple production lines instead of using high line speeds in one production line.

The control of the pulling force and the design of the fiber guidance system are extremely important, since they influence not only the fiber wet-out, but also the fiber alignment. Some of the defects found in pultruded products, such as fiber bunching, fiber shifting, wrinkles, and folding of mats or woven rovings, are examples of fiber misalignment that can reduce the structural properties of pultruded products. The two most important process parameters that control the pulling force are die temperature and pulling speed (Figure 9.13). Increasing the die temperature near the die entrance reduces the resin viscosity, which, in turn, reduces the viscous drag force. Although thermal expansion is also increased, its effect on the pulling force is smaller than the effect of viscosity reduction. As a result, pulling force decreases with increasing temperature. On the other hand, it also decreases with increasing pulling speed.

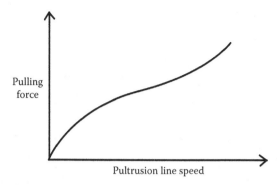

FIGURE 9.13 Effect of pultrusion line speed on the pulling force.

9.5 DESIGN GUIDELINES FOR PULTRUDED PRODUCTS

From a processing standpoint, there are several simple rules of design that can produce better pultruded products. They are given in the following:

1. Thickness: The wall thickness of a pultruded cross section should be as uniform as possible, since it allows uniform curing and cooling through the thickness and reduces the possibility of differential shrinkage and distortion by warpage. If the design requires high bending stiffness, it is better to use a deeper section with uniform thinner walls or include ribs instead of making the section thicker. This is because the thicker the section, the longer the curing time. Curing a thick section within the die length may require a very slow pulling speed, which will reduce the production rate. If a cross section has both thin areas and thick areas, the pulling speed is determined by the curing rate of the thick areas.

2. Corner radius: To prevent corner cracking, there should not be any sharp corners in the design of a pultruded section. They are not only the sources of stress concentration, but they are also more difficult to fill with fibers. For these reasons, a minimum corner radius of 1.5 mm is recommended [2]. The corners should also have a uniform thickness (Figure 9.14), which helps in reducing the possibility of resin buildup at the corners that can later crack or flake off. Uniform thickness also produces a more even distribution of fibers and more uniform curing shrinkage at the corners, which is important in maintaining consistent properties throughout the section.

3. Shrinkage: Pultruded sections typically shrink by 2–4% by volume during curing, which must be taken into account in their design. Shrinkage can be controlled by adding fillers and/or low-profile additives in the resin mix; but they cannot be completely eliminated. Higher shrinkage is observed in thicker sections. Nonuniform shrinkage causes distortion by warpage of the cross section in which the walls of a closed section may be pulled inward, or unsupported vertical walls of an open section may bend toward each other,

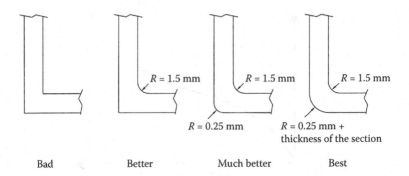

FIGURE 9.14 Recommended corner design of an L-shaped pultruded section. (From T. F. Starr (ed.), *Pultrusion for Engineers*, Woodhead, Cambridge, UK, 2000.)

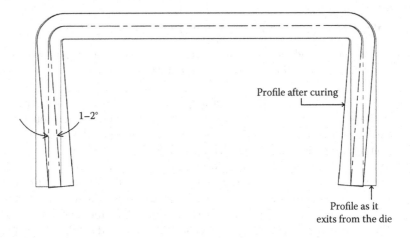

FIGURE 9.15 Warpage in a pultruded section after curing.

as shown in Figure 9.15. A uniform wall thickness helps in reducing warpage. Symmetric designs and symmetric fiber form placement on the cross section also helps in controlling warpage.

9.6 PROCESSES RELATED TO PULTRUSION

The basic pultrusion process is capable of manufacturing continuous straight members with constant cross-sectional shape. Several different manufacturing processes have been developed that are variations of the basic pultrusion process. Two of these processes are briefly described in the following.

9.6.1 PULFORMING

Pulforming [8] combines pultrusion and compression molding to manufacture curved members. It also has the ability to change the cross-sectional dimensions along their lengths. Unlike the basic pultrusion process, which uses a single die of constant cross section, pulforming uses a series of die segments that are mounted on a rotating turntable at radial distances that match the radius of curvature of the curved member. The dies are open on their outboard faces. Their other three faces are designed to match the cross-sectional shape variation needed along the length of the pultruded member. The resin-impregnated fiber preform enters the first die cavity just like in the basic pultrusion process, but is pulled into other die cavities by the rotation of the turntable. A stainless steel belt rotating with the turntable is used to close the open outboard faces and apply compressive force on the material moving through them. The application of the compressive force creates the desired cross-sectional shape change along the length of the curved member. After the dies are filled and the cross-sectional shapes have formed, they are heated to cure the resin. After the curing is complete, the stainless steel belt is pulled away. At this stage, ejector pins located in the die segments push the cured curved pultruded member out.

9.6.2 Pull Winding

Pull winding combines pultrusion and filament winding to manufacture round and square tubes containing filament-wound fibers in the interior layers of the cross section and longitudinal fibers on the inside and outside layers. The winding angle in the filament wound layers is typically between 80° and 87°. The longitudinal fibers provide resistance to axial loading and bending, and the filament-wound fibers provide resistance to hoop tension due to internal pressure loading.

REFERENCES

1. J. E. Sumerak, The pultrusion process for continuous automated manufacture of engineered composite profiles, in *Composites Engineering Handbook* (P. K. Mallick, ed.), Marcel Dekker, New York, pp. 549–577, 1997.
2. T. F. Starr (ed.), *Pultrusion for Engineers*, Woodhead, Cambridge, UK, 2000.
3. B. A. Wilson, Pultrusion, in *Handbook of Composites* (S. T. Peters, ed.), Springer, New York, pp. 488–524,1998.
4. K. S. Raper, J. A. Roux, T. A. McCartty, and J. G. Vaughn, Investigation of the pressure behavior in a pultrusion die for graphite/epoxy composites, *Composites: Part A*, Vol. 30, pp. 1123–1132, 1999.
5. D. J. Evans, Classifying pultruded products by glass loading, *Proceedings of the 41st Annual Conference*, Society of Plastics Industry, 1986.
6. N. Dharia and N. R. Shott, Resin pick-up and fiber wet-out associated with coating and pultrusion processes, *Proceedings of the 44th Annual Conference*, Society of Plastics Engineers, 1986.
7. J. E. Sumerak, Understanding pultrusion process variables, *Modern Plastics*, Vol. 62, No. 3, pp. 58–64, 1985.
8. W. B. Goldsworthy, Pulforming, in *Wiley Encyclopedia of Composites* (L. Nicolais and A. Borzaccchiello, eds.), John Wiley & Sons, Hoboken, NJ, 2012.

10 Forming of Thermoplastic Matrix Composites

As matrix material in PMCs, thermoplastic polymers are used in much larger volumes than thermosetting polymers. However, most of the thermoplastic matrix composites used today are injection-molded short E-glass fiber-reinforced thermoplastics. They are used in numerous household, industrial, commercial, and automotive applications. Fibers in these composites are typically 1–3 mm in length, and they are randomly oriented. Injection-molded thermoplastic matrix composites with fiber lengths of up to 12 mm, known in the plastics industry as long fiber thermoplastics or LFTs, are also available and finding increasing number of applications in the automotive industry. Increased fiber length in LFT improves the modulus and strength, but they are still much lower than the modulus and strength of continuous fiber-reinforced thermoplastics. On the other hand, the processing of continuous fiber-reinforced thermoplastics is significantly more difficult and expensive than injection molding short or long fiber-reinforced thermoplastics due to the difficulty of incorporating continuous fibers in thermoplastic polymers. For this reason, even though continuous fiber-reinforced thermoplastics possess several advantages over continuous fiber-reinforced thermosets, such as lower processing time, higher fracture toughness, long shelf life, and recyclability, their applications are still very limited.

The difficulty of incorporating continuous fibers in liquid thermoplastic polymers is due to their high viscosity, which is in the range of 10–1000 Pa s. When continuous fibers are incorporated in thermosetting polymers, they are in their uncured prepolymer form with viscosity in the range of 0.05–1 Pa s. Because of such low viscosity, thermosetting prepolymers can flow relatively easily into the interior of fiber bundles and coat the filaments relatively easily. The high viscosity of thermoplastic polymers in their liquid form restricts the flow needed for good fiber wetting and resin infiltration into the inside of fiber bundles [1]. Processing temperatures (see Table 10.1) and pressures for thermoplastic matrix composites are also higher than those for thermoset matrix composites. For example, autoclave processing of PEEK-based composites may require a temperature of up to 390°C and pressure up to 20 bar compared to 177°C and 3–6 bar, respectively, for epoxy-based composites.

Many of the processes used for producing continuous fiber-reinforced thermosets can also be applied for producing continuous fiber-reinforced thermoplastics, but with some modifications. Examples of these processes are filament winding and pultrusion [2,3]. Even though filament winding or pultrusion equipment is basically the same for both thermosetting and thermoplastic matrix composites, there are differences in the fiber impregnation techniques. Unlike the thermosetting polymers for which dry fiber tows are pulled through the liquid resin for fiber impregnation, the starting material for filament winding and pultrusion of thermoplastic

TABLE 10.1

Melt Processing Temperatures of Several Thermoplastic Polymers

Polymer	T_g (°C)	T_m (°C)	Melt Processing Temperature (°C)
PEEK	143	332–343	360–400
PPS	85	285	300–345
PEI	217	–	327–443
PAI	275		>350
PP	–10	168–176	260
PA-6	50–60	215–230	250–290

Note: In comparison, the processing (cure) temperature for most epoxies is between 120°C and 180°C.

matrix composites is usually a thin prepreg tape of a unidirectional or bidirectional fiber-reinforced thermoplastic. The prepreg tape is pulled through a heating zone where the thermoplastic in the tape is transformed from the solid state to the liquid state. The heat source is either infrared radiation, hot gas torch, flame, or laser. The higher the temperature, the lower the viscosity and better the consolidation of the layers; however, care has to be exercised so that the maximum temperature does not exceed the degradation temperature of the polymer. For filament winding, the heated tape is wound around the mandrel under roller pressure to improve the consolidation between the successive layers. The mandrel can also be heated to improve consolidation and make the through-thickness temperature distribution in the layers more uniform compared to a nonheated mandrel. For pultrusion, the heated tape is first pulled through the preforming dies and then through the pultrusion die to form the cross-sectional profile of the pultruded composite part.

In this chapter, we will consider the shape-forming processes for continuous fiber-reinforced thermoplastic matrix composites. In these processes, the starting material is a continuous fiber-reinforced thermoplastic prepreg tape or sheet. The fibers in the prepreg can be unidirectional, bidirectional, or randomly oriented. The forming process includes three steps: (1) melting the thermoplastic polymer in the prepreg, (2) shape forming and consolidation under the application of pressure, and (3) cooling to room temperature. If the thermoplastic in the prepreg is a semicrystalline polymer, melting will require heating it to temperatures that are 20–50°C above its crystalline melting point T_m. If it is an amorphous polymer, which does not have any crystalline phase, melting will require heating it up to 50–150°C above its glass transition temperature T_g. The liquid polymer temperature at the beginning of the shape-forming and consolidation steps is called the forming temperature, which is denoted as T_f.

This chapter begins with an overview of the processes used for incorporating continuous fibers in thermoplastic polymers and making prepreg sheets. It then describes the important steps in the shape forming of these prepreg sheets into composite parts. It also describes three different shape-forming processes that are used in making thermoplastic matrix composite parts.

10.1 INCORPORATION OF CONTINUOUS FIBERS IN THERMOPLASTIC MATRIX

Several processes have been developed to incorporate continuous fibers into thermoplastic matrix, and some of them are now commercially used to produce continuous fiber-reinforced thermoplastic matrix composite prepreg [1,4–6]. The differences in the processing of thermoset and thermoplastic matrix composite prepregs are listed in Table 10.2. One key difference is that unlike the thermoset matrix prepregs, thermoplastic matrix prepregs can be stored for an unlimited time without any special storage facility, such as a freezer. Another difference is that thermoplastic matrix prepregs are not very tacky, and therefore, they can be rolled or stacked without any separator or backup material between them. However, the lack of tackiness has one drawback—the prepreg layers do not adhere well to each other when they are being stacked.

1. Melt impregnation: In the melt impregnation process, collimated continuous fibers are pulled through a die attached at the end of an extruder, which delivers a thin sheet of liquid polymer to the die at a high pressure

---

TABLE 10.2

Comparison of Thermosets and Thermoplastics for Prepreg Processing and Composite Part Manufacturing

Processing Characteristic	Advantage for Prepreg Processing	Reason
Viscosity	Thermosets	Lower viscosity
Solvents	Thermosets	Greater choice
Handling	Thermosets	More flexible
Tack	Thermosets	Higher tack
Storage	Thermoplastics	No chemical reaction
Quality control	Thermoplastics	Fewer variables
Composite Part Manufacturing		
Layup	Thermosets	Ease of handling
Degassing	Thermoplastics	Fewer volatiles
Temperature changes	Thermoplastics	Fewer changes
Maximum temperature	Thermosets	Lower
Pressure changes	Thermoplastics	Fewer changes
Maximum pressure	Thermosets	Lower
Cycle time	Thermoplastics	Lower
Postcure	Thermoplastics	Not required
Repair	Thermoplastics	Can be remelted
Postforming	Thermoplastics	Can be remelted

Source: J. D. Muzzy and A. O. Kays: Thermoplastic vs. thermosetting structural composites. *Polymer Composites.* 1984. 5. 169–172. Copyright Wiley-VCH Verlag GmbH & Co. KGaA. Reproduced with permission.

(Figure 10.1). To expose the filaments in the fibers to the polymer melt, the fiber tows are spread by an air jet just before they enter the die. As the fibers coated with the liquid polymer exit from the die, a cold air jet rapidly cools them, and the polymer transforms from its liquid state to its solid state. After the liquid polymer solidifies, the prepreg tends to be stiff and tack-free. This may cause difficulty in draping the prepreg on the mold surface and sticking the prepreg layers to each other as they are stacked before molding the composite part.

Melt impregnation is used for both semicrystalline thermoplastics, such as PEEK, PPS, and PP, and amorphous thermoplastics, such as PEI. For good and uniform polymer coating on the fiber surface, the viscosity of the liquid polymer should be as low as possible. Low viscosity can be achieved by increasing the melt temperature, but if very high melt temperatures are used, there may be thermal degradation of the polymer. Hot melt impregnation is not suitable for polymers with degradation temperature close to their melt temperature. A high shear rate in the extruder can help in reducing the viscosity.

2. Solution impregnation: In solution impregnation, the thermoplastic polymer is dissolved in a suitable solvent to create a homogeneous solution of low viscosity. The fibers are coated with the solution by either dipping them in it or slowly pulling them through it and then drying at an elevated temperature to remove the solvent (Figure 10.2). Since some of the solvent may be left on the fibers even after drying, solution-impregnated prepreg

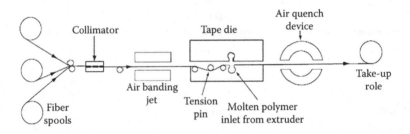

FIGURE 10.1 Melt impregnation process.

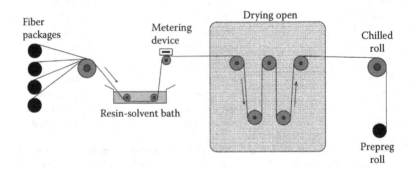

FIGURE 10.2 Solution impregnation process.

is usually tackier and more drapeable than hot melt-impregnated prepreg. However, solvent removal from the prepreg during the final forming process can become a critical issue. If the solvent is not completely removed, it may create undesirable voids and surface imperfections in the composite part. The residual solvent may also have a plasticizing effect and reduce the properties of the composite.

In general, amorphous polymers, such as PEI, are more suitable for solution impregnation, since they dissolve much more easily in solvents than semicrystalline polymers, such as PEEK, PPS, and PP. The choice of the solvent depends on the solubility parameter of the polymer in the solvent. The chemical structure and molecular weight of the polymer have a great influence on the solubility parameter. The solvent temperature also affects the solubility. In general, a low-boiling point solvent is preferred, since it is often difficult to remove a high-boiling point solvent from the fiber bundles.

3. Powder coating: In the powder coating process, shown in Figure 10.3, charged and fluidized thermoplastic powders are deposited on the fiber surface using a combination of mechanical and electrostatic processes. Before entering the fluidized bed, an air jet is used to spread the filaments in the fiber bundles and expose them to the charged powders that are typically between 15 and 150 μm in diameter. Immediately after powder deposition, the fibers are pulled through an infrared oven tunnel where the powders are melted and sintered to produce a continuous coating of the thermoplastic polymer on the fiber surface.

4. Film stacking: This process is primarily used with woven fabrics or random fiber mats, which can be interleaved with unreinforced thermoplastic polymer films (Figure 10.4). The layup is then heated between the hot platens of a press to force the thermoplastic into the reinforcement layers and, thus, form a prepreg sheet.

5. Fiber mixing: In this process, thermoplastic fibers are mixed with reinforcement fibers, such as E-glass and carbon, by commingling, cowrapping, or directly coweaving (Figure 10.5). The commingled and cowrapped fibers can be subsequently woven, knitted, or braided into 2D and 3D hybrid fabrics. The thermoplastic fibers in these fabrics are melted and spread to coat the reinforcement fiber surface during the heating and consolidation stages of the forming process.

The principal advantage of using fabrics with mixed fibers is that they are highly flexible and can be easily draped on mold surfaces with complex contours and shapes. Therefore, they are more suitable for manufacturing thermoplastic matrix composite parts of complex shapes. Prepregs made by hot melt impregnation or solution impregnation are stiffer, less drapable, and more suitable for relatively flat surfaces. However, fiber mixing is possible only if the thermoplastic is available in filamentary form. PEEK and PPS, for example, are available as monofilaments of diameter between 16 and 18 μm. They are commingled with carbon filaments, which have a diameter in the range of 10–15 μm. A commercially available fabric, known as Twintex®, consists of commingled E-glass and PP filaments.

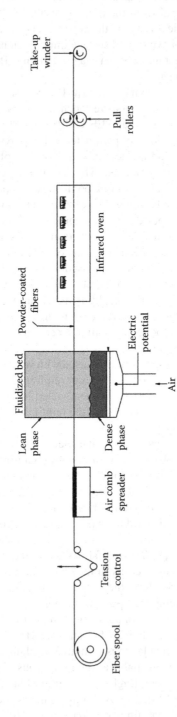

FIGURE 10.3 Powder coating process.

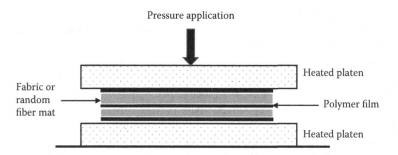

FIGURE 10.4 Film stacking process.

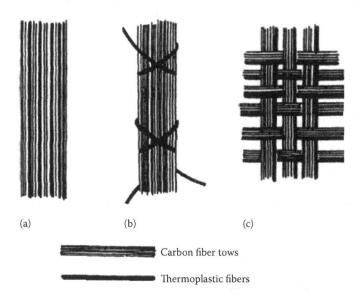

(a)   (b)   (c)

Carbon fiber tows

Thermoplastic fibers

FIGURE 10.5 Fiber mixing: (a) commingled, (b) cowrapped, and (c) coweaved fibers.

10.2 CONSOLIDATION OF THERMOPLASTIC MATRIX COMPOSITES

The consolidation of thermoplastic matrix composites starts with heating a stack of prepreg layers to the melt temperature either in the mold or outside the mold. The melt temperature is selected such that the thermoplastic transforms into a low-viscosity liquid and is able to flow under the consolidation pressure. Low viscosity is also needed to remove air from both inside the fiber bundles and between the prepreg layers and to improve fiber wetting by the liquid polymer. Sufficient time needs to be given under pressure to achieve good fiber wetting and consolidation of the layers in the stack. After the consolidation is complete, the stack is cooled down to room temperature either in the mold or outside the mold. The consolidated stack, which is now in a sheet form, is later used as the starting material for shape forming.

A simple method of consolidating prepreg layers is to place the stack between two heated platens in a press (Figure 10.6), apply the consolidation pressure for several minutes, and then cool it down to room temperature. A picture frame mold can be used to prevent the lateral spreading of the prepreg layers and the bleeding of the liquid polymer from the edges of the stack. The stack can also be preheated outside the mold in a hot air-circulating oven or in an infrared oven before placing it in the mold. Depending on the preheating temperature, this step will eliminate or reduce the heating time in the mold.

The consolidation of prepreg layers takes place by the diffusion of polymer molecules across the interfaces of successive layers in the stack. However, before diffusion can occur, uniform and intimate contact between the successive layers must be established. Since the prepreg surfaces are rough and do not possess sufficient tack to smoothly join at their interfaces during stacking, there will be numerous gaps between the successive layers of the stack (Figure 10.7). As the temperature and pressure are increased and the polymer matrix starts to soften and deform, these gaps start to reduce in size and the degree of

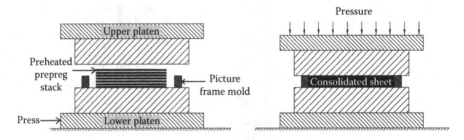

FIGURE 10.6   Consolidation of thermoplastic prepreg layers using press heating.

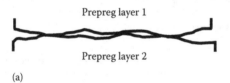

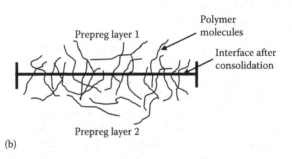

FIGURE 10.7   (a) Gaps and partial contacts between prepreg layers before consolidation and (b) molecular bridges formed across the interface after consolidation.

contact starts to increase. The time to establish intimate contact rapidly decreases as either the consolidation pressure or the processing temperature is increased (Figure 10.8) [8]. The intimate contact time is almost an order of magnitude higher for cross-plied prepreg layers than for unidirectional prepreg layers. Also, the intimate contact time is significantly higher if the prepreg layers contain an amorphous polymer instead of a semicrystalline polymer, which is attributed to higher melt viscosity of amorphous polymers.

With the melting of the polymer matrix and continued application of pressure, polymer molecules start to diffuse from one layer to the next and form molecular bridges across their interfaces. The diffusion process is both time and temperature dependent. If sufficient time is given for full consolidation to occur, the interfacial strength can approach the strength of the polymer. Pressure is also an important factor, since it brings the layers close to each other and establishes the contact across their interfaces; however, if the pressure is too high, it can squeeze out some of the liquid polymer from the prepreg. The process by which consolidation occurs is called *autohesion*, which simply means self-adhesion between two thermoplastic layers due

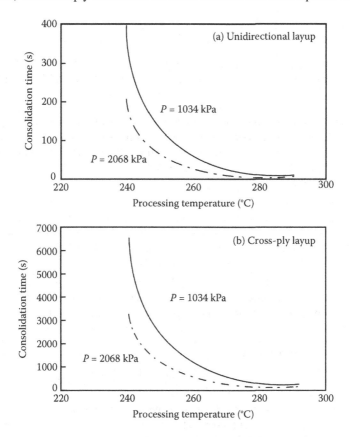

FIGURE 10.8 Effects of processing temperature and consolidation pressure on consolidation time (material: T300/P1700 prepreg). (From Loos, A. C. and M.-C. Li, Consolidation during thermoplastic composite processing, in *Processing of Composites* (R. S. Davé and A. C. Loos, eds.), Hanser, Munich, 2000.)

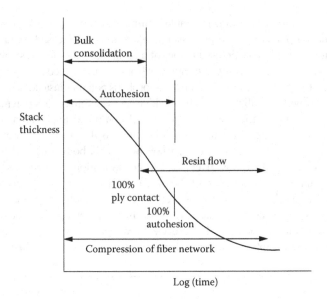

FIGURE 10.9 Decrease in prepreg stack thickness with consolidation time. (J. D. Muzzy and J. S. Colton: *Advanced Composites Manufacturing.* 1997. Copyright Wiley-VCH Verlag GmbH & Co. KGaA. Reproduced with permission.)

to molecular diffusion across the interface between them [9]. The time required for autohesion decreases as the consolidation temperature is increased. During the consolidation process, there may be further impregnation of the fibers by the liquid polymer. Furthermore, the thickness of the consolidated stack decreases with increasing time as demonstrated in Figure 10.9.

If commingled or other forms of mixed reinforcing and polymer fibers are used, fiber impregnation takes place during the consolidation process [10]. As the temperature is raised, the polymer fibers start to melt and form pools of liquid polymer around the reinforcing fibers. When the pressure is applied, the reinforcing fibers move closer to each other as well as to the liquid polymer pools. The separated liquid polymer pools also come closer and coalesce around the reinforcing fibers. With increasing pressure, the liquid polymer flows inside the reinforcing fiber bundles and impregnates the individual filaments. This is shown schematically in Figure 10.10.

10.3 SHAPE FORMING

Shape forming starts with a flat sheet of material, called the *blank* that can be either an unconsolidated or a preconsolidated stack of prepregs. Although preconsolidation is an additional step before shape forming, a preconsolidated stack is preferred for the following reasons: (1) it eliminates the time needed for assembling multiple layers of prepreg before starting the forming operation, (2) it reduces the consolidation time in the mold, (3) it requires lower forming temperature, (4) it reduces the temperature gradient between the layers, and (5) it produces fewer voids and lower porosity, which gives higher mechanical properties.

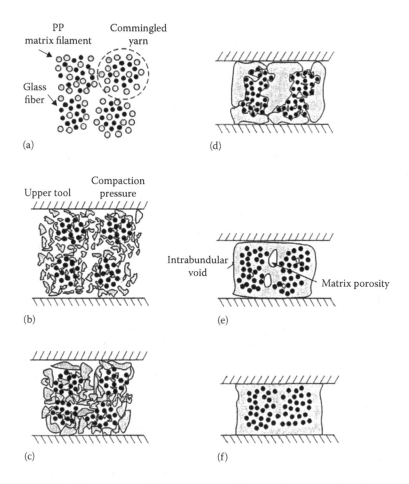

FIGURE 10.10 Fiber impregnation during consolidation of commingled glass and PP fibers. (a) Dry commingled bundles, (b) molten matrix pooled at edges of reinforcing bundles, (c) bundles move closer together, bridging of bundles with matrix, (d) individual matrix pools coalesce around dry fiber bundles, (e) flow of matrix to impregnate bundles, and (f) in-plane matrix flow on microscale, dissolution of trapped porosity. (From M. D. Wakeman and J. A. E. Manson, Design and Manufacture of Textile Composites (A. C. Long, ed.), Woodhead, Cambridge, UK, 2005.)

To produce a 3D shape from a flat blank, it is heated to the forming temperature T_f that will make it flexible enough to be pressed into the mold surface using reasonable pressure and without causing fiber distortion or tearing. Heating can be done either in the mold or outside the mold. At the forming temperature, the matrix in the blank is in a liquid state. However, if the selected forming temperature is too high, thermal degradation of the polymer matrix may occur. The application pressure not only deforms the blank into the shape of the mold, but also helps in the additional consolidation of the layers. At the end of forming, the formed part is cooled down to the demolding temperature while it is still under pressure. The demolding temperature is usually a temperature lower than the T_g of the polymer or its heat deflection temperature. After demolding, the part cools down to room temperature without any pressure acting on it. The cooling rate and pressure application during cooling in the mold determine spring forward, warpage, and residual stresses in the formed part. Very high cooling rates can create residual stresses due to differential cooling between the surfaces and the interior of the part. If the matrix is a semicrystalline polymer, the degree of crystallinity is also affected by the cooling rate. For semicrystalline polymers, different cooling rates at the surfaces and the interior can create different degrees of crystallinity and morphological characteristics of the crystalline phase across the part thickness, both of which will affect their properties and performance under load.

10.3.1 Temperature and Pressure Profiles

Shape forming can be done under either isothermal or nonisothermal conditions [2]. If a blank with an unconsolidated prepeg stack is used, consolidation takes places during the forming operation. Even with a preconsolidated stack, additional consolidation may occur during the forming operation. The temperature and pressure profiles for these two conditions are shown in Figure 10.11.

1. Isothermal forming process: In the isothermal process, the blank is placed in the mold and heated to the forming temperature T_f. After the desired temperature is reached, the forming pressure is applied. Both forming temperature and pressure are held constant during the forming and consolidation phases. As the cooling phase begins, the pressure is initially maintained constant for consolidation to complete. After the temperature has sufficiently decreased, the pressure is released. At the end of the cooling phase, the temperature is rapidly decreased to bring the part to room temperature.
2. Nonisothermal forming process: In the nonisothermal process, the blank is heated to the forming temperature T_f outside the mold, usually in a hot air-circulating oven or an infrared oven. The preheated blank is quickly transferred to the mold, which is held at a temperature much lower than T_f. As the blank immediately starts to cool on contact with the mold surfaces, the forming pressure is applied on the prepreg for forming and consolidation. While the pressure is maintained constant during the shape forming phase, the temperature continues to decrease. After the temperature has sufficiently decreased, the pressure is released and the temperature is rapidly decreased to bring the part to room temperature.

The product cycle time is longer in the isothermal process than that in the nonisothermal process. In the isothermal process, the mold cools down to room temperature along with the part as the forming cycle ends, and therefore, in the next cycle, it needs to be heated up again to the forming temperature. In the nonisothermal process, heating is done outside the mold and cooling starts almost immediately after the preheated blank is placed in the mold. The mold heating time at the beginning of each cycle makes the cycle time longer for the isothermal process. On the other hand, the isothermal process produces parts with lower void content because of longer consolidation time.

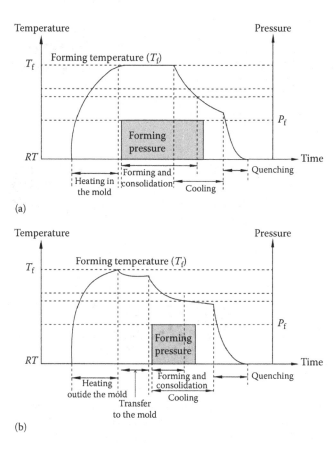

FIGURE 10.11 Temperature and pressure profiles during (a) isothermal and (b) nonisothermal forming processes. (From R. Brooks, *Composites Forming Technology* (A. C. Long, ed.), Woodhead, Cambridge, UK, 2007.)

10.3.2 DEFORMATION DURING SHAPE FORMING

During the shape forming process, the flat blank of a multilayered prepreg stack is deformed to the shape of the mold to produce the composite part. Since the continuous fibers are practically inextensible, there is very little stretching of the individual plies in the fiber direction. Instead, the principal deformation mechanisms that help form the shape are interply slip, intra-ply shearing, and inter-ply rotation [11,12]. While forming shapes with single curvature is facilitated by interply slip, forming shapes with double curvatures requires all three deformation mechanisms. Two additional mechanisms that also occur during shape forming are resin percolation and transverse resin flow. These mechanisms are illustrated in Figure 10.12.

1. Interply slip or interlaminar slip occurs through the slippage or sliding of adjacent layers relative to each other as the flat prepreg stack is deformed over a singly curved surface, e.g., a corner radius in the mold. The interply

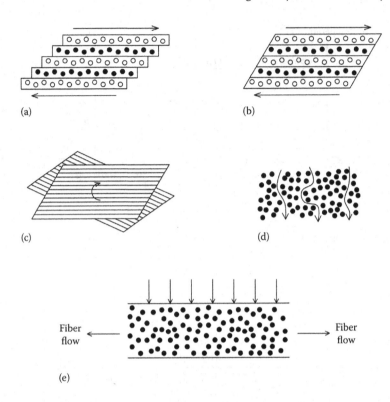

(a)   (b)

(c)   (d)

Fiber   Fiber
flow   flow

(e)

FIGURE 10.12 Deformation mechanisms during shape forming and consolidation of thermoplastic matrix composites. (a) Interply slip, (b) intra-ply slip, (c) interply-rotation, (d) resin percolation around fibers, and (e) transverse flow of fibers under pressure.

slip is helped by the liquid polymer which acts as a lubricant between the adjacent layers. It is also the principal mechanism that prevents the buckling and wrinkling of the prepreg layers during the shape change (Figure 10.13). The minimum shear stress needed to induce interply slip decreases with increasing temperature and increases with increasing slip velocity [13]

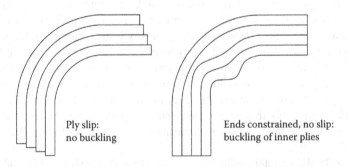

Ply slip: Ends constrained, no slip:
no buckling   buckling of inner plies

FIGURE 10.13 Interply slip between prepreg layers (*left*) and wrinkling due to lack of interply slip (*right*).

(Figure 10.14). It is also a function of fiber orientations or fiber arrangements in the adjacent layers. For example, the minimum shear stress needed
for interply shear of 45/45 layers is significantly higher than 0/45 layers.

2. Intra-ply shearing occurs when the fibers within each ply move past each other
by in-plane shearing. This mechanism is more common when the prepreg stack
is deformed over a doubly curved surface, such as a spherical dome. For fabrics
with 2D and 3D architectures, intra-ply shearing takes the form of a *trellis effect*
in which the angle between the initially orthogonal fiber tows is decreased. The
shear stress needed to induce intra-ply shearing also decreases with increasing
temperature and increases with increasing forming speed, and it is significantly
higher than the shear stress needed for inter-ply slip [13].

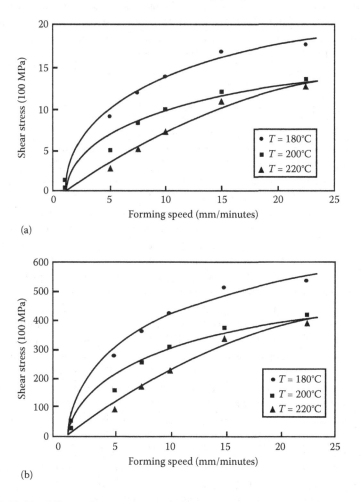

(a)

(b)

FIGURE 10.14 Effects of temperature and forming speed on (a) interply and (b) intra-
ply shear stresses. (Reprinted from *Composites: Part A*, 29A, J. Krebs, K. Friedrich, and
D. Bhattacharyya, A direct comparison of matched-die versus diaphragm forming, 183–188,
Copyright (1998), with permission from Elsevier.)

3. Interply rotation is also a shearing mechanism that creates rotational slip of one layer relative to its adjacent layer. The viscosity of the liquid polymer between the layers controls the resistance to interply rotation. Since viscosity decreases with increasing temperature, the shear stress needed for interply rotation also decreases with increasing temperature.

4. Resin percolation or flow of the liquid polymer between the fibers and along the layers in the prepreg stack is fundamental to fiber impregnation, consolidation, and evacuation of air from the fiber bundles in each layer as well as between the layers. Resin percolation is also needed for uniform distribution of the matrix in the composite part.

5. Transverse flow is a process by which the prepreg stack spreads out in the transverse direction of fibers due to the pressure gradient that develops during the processing. There is very little axial flow, since the resistance to flow in the axial direction of fibers is much higher than that in the transverse direction. The stack thickness decreases as both fibers and matrix flow together in the transverse direction as shown in Figure 10.12e.

10.3.3 COOLING AND SOLIDIFICATION

As the composite part is being cooled from the forming temperature to room temperature, the polymer matrix transforms from a liquid state to a solid state. The specific volume vs. temperature diagrams in Figure 10.15 shows the cooling paths of amorphous and semicrystalline polymers. For amorphous polymers, such as PEIS, the liquid-to-solid transformation takes place at a temperature much higher than the T_g. As the temperature decreases below T_g, the material changes from a soft rubbery solid to a hard glassy solid. For semicrystalline polymers, such as PEEK and PP, the liquid-to-solid transformation takes place at the crystallization temperature T_c, which is lower than the crystalline melting point T_m. The crystallization temperature and the cooling rate determine the degree of crystallinity and the spherulitic morphology of the crystalline phase. The degree of crystallinity and spherulite size decrease with increasing cooling rate. The effect of cooling rate on the degree of crystallinity is shown in Figure 10.16. In general, the lower the crystallization temperature, the smaller the spherulite size. The properties of semicrystalline polymers after solidification depend on these two parameters. For example, both modulus and tensile strength increase with increasing degree of crystallinity and decreasing size of spherulites. However, impact strength and fracture toughness are lower with higher degree of crystallinity and increasing size of spherulites [14].

In consolidation experiments with eight-ply carbon fiber/PEEK laminates, Manson et al. [15] observed that as the cooling rate was increased from 0.3°C/s to 120°C/s, the degree of crystallinity of the PEEK matrix decreased from 26% to 12%. They also noted that annealing the laminates at an elevated temperature after consolidation can significantly increase the degree of crystallinity. For example, annealing at 300°C for approximately 30 minutes increased the degree of crystallinity from 12% to 37%. The increase in crystallinity after annealing was higher with higher annealing temperature. Laminates annealed at 300°C showed close to 10% higher degree of crystallinity than the laminates annealed at 177°C.

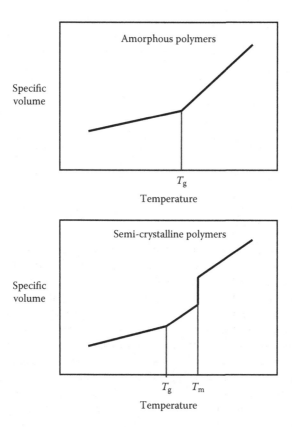

FIGURE 10.15 Specific volume vs. temperature diagrams of amorphous and semicrystalline thermoplastic polymers.

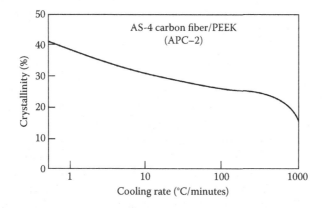

FIGURE 10.16 Effect of crystallinity on the degree of crystallinity in a carbon fiber/PEEK composite.

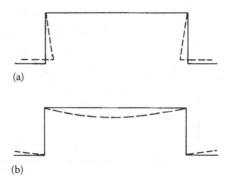

(a)

(b)

FIGURE 10.17 Spring forward of a channel section (a) and a circular hat section (b).

During solidification, it is important to maintain the pressure on the formed part until it cools down to below the glass transition temperature T_g of the polymer matrix. This will suppress the shrinkage in the mold and help control the dimensional stability of the composite part. However, residual stresses are generated in the part during cooling. One source of these residual stresses is the difference in the thermal contractions of the fibers and the matrix. Difference in cooling rates at the center and surfaces of the part and the resulting thermal gradient is another source of residual stresses. During cooling, the surface temperature decreases faster than the center temperature, thus creating a temperature gradient in the thickness direction. The difference between the surface and center temperatures is much greater for a thick laminate than for a thin laminate. As a result of the temperature gradient and difference in cooling rates, a residual stress pattern evolves along the thickness of the formed laminate. After the part is removed from the mold, it will continue to shrink, and as some of the residual stresses are relieved, the part may start to warp. Both amorphous and semicrystalline polymers shrink on cooling due to thermal contraction. Semicrystalline polymers show additional shrinkage due to crystallization. Thus, shrinkage of semicrystalline polymers is higher than that of amorphous polymers. The difference in in-plane and out-of-plane (through-thickness) thermal contractions can give rise to distortion, known as spring forward [16], of the formed shape (Figure 10.17).

10.3.4 VOIDS

Voids in thermoplastic matrix composites are formed during prepregging and consolidation. One reason for the presence of voids is the high viscosity of the liquid polymer and its inability to force all the air out from inside the fiber bundles as well as between the fiber bundles during the impregnation process. The viscosity can be decreased by increasing the forming temperature as long as it is lower than the thermal degradation temperature of the polymer. Consolidation pressure has a large influence on the void content. For example, in the autoclave consolidation of a commingled carbon fiber/PEEK fabric, it was found that the void content reduced from 13 to 1 vol.% when the consolidation pressure was increased from 0.3 to 2 MPa [17]. Other reasons for void formation are inadequate or uneven forming pressure distribution in the mold, short consolidation time, and forming temperature lower than needed for good liquid polymer flow. It is also possible that the liquid thermoplastic polymer contains dissolved air trapped between its molecules [10]. The diffusion of the dissolved air out of the composite part requires time that depends on the forming temperature and the diffusion distance. Thus, if the consolidation time is not long enough for the dissolved air to diffuse out, it will remain in the composite part and appear as voids. Another reason is the tendency for the preconsolidated stack to open up or loft when it is heated and trap air between the layers. Voids can also occur from crystallization shrinkage during cooling, particularly if the forming pressure is released before the crystallization is complete.

10.4 FORMING PROCESSES

Several different forming processes have been developed for shape forming thermoplastic matrix composite parts. Among them are diaphragm forming, thermostamping, and compression molding [4,5]. These three processes are described in this section.

10.4.1 DIAPHRAGM FORMING

In diaphragm forming [18], the blank is placed between two very thin flexible sheets or diaphragms that can plastically deform as pressure is applied during the forming process. The diaphragms, not the blank placed between the diaphragms, are clamped around the edges of the mold using a clamping frame (Figure 10.18), and the air within the diaphragms is evacuated. The entire assembly is then placed inside an autoclave, and the autoclave temperature is raised to the forming temperature. After the forming temperature is reached, pressure inside the autoclave is increased. As the pressure increases and the diaphragms start to deform, the air between the mold surface and the bottom diaphragm escapes through the vents provided in the mold. It is also possible to apply vacuum on the bottom side of the diaphragms to facilitate their deformation. After the forming and consolidation are complete, the temperature is slowly reduced to room temperature, the pressure is released, and the

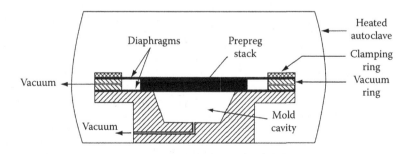

FIGURE 10.18 Diaphragm forming process.

part is removed. Figure 10.19 shows the temperature and pressure profiles used in a diaphragm forming process. The process shown in this figure is isothermal and has a relatively long cycle time. Preheating the autoclave to the forming temperature before placing the mold assembly inside can reduce the cycle time.

Since the blank is not clamped, it can slide within the diaphragms and deform with the diaphragms to the shape of the mold surface. The friction between the blank surfaces and the diaphragms create biaxial tensile forces on the material, which prevent wrinkle formation due to out-of-plane buckling. The thin layers of liquid polymer between the blank and the diaphragm surfaces influence the friction and affect the amount of deformation, fiber orientation, and thickness variation in the formed part.

The diaphragm material can be either metallic, such a superplastic aluminum alloy, or polymeric, such as a thermoplastic polyimide. Silicone rubber has also been used as a diaphragm in some applications. One advantage of silicon rubber is that it is reusable, since it does not retain the deformed shape after diaphragm forming. The main requirement for the diaphragm material is that it should be highly stretchable, is able to withstand high temperatures, and will not rupture during stretching. The deformation characteristics of the diaphragm depend on the modulus of the diaphragm material selected and its thickness.

The process parameters for diaphragm forming are the forming temperature, heating rate, forming and consolidation time, pressure, pressure application rate, and diaphragm material. Table 10.3 lists the process parameters found most suitable for diaphragm forming of an AS4 carbon fiber/PEEK composite [19]. The minimum forming temperature was 380°C, which is 30°C higher than the crystalline melting point of PEEK. The minimum pressure was 10 bar. Below this pressure, the consolidation was found to be poor and the interlaminar shear strength was low. The forming and consolidation time was 5 minutes. At longer times, significant thickness variation was observed due to matrix flowing out of the blank.

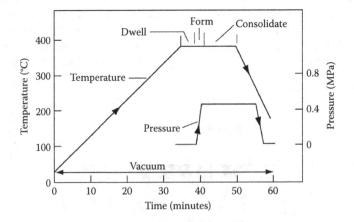

FIGURE 10.19 Temperature and pressure profiles used in diaphragm forming of a carbon fiber/PEEK composite.

TABLE 10.3
Diaphragm Forming Parameters for AS4 Carbon Fiber/APC-2 PEEK

Process Parameter	Influence on Quality and Performance	Minimum Recommended Value
Forming temperature	Consolidation quality, dimensional stability, mechanical properties	380°C
Heating rate	Does not have much influence	7°C/min
Forming and consolidation time	Consolidation quality, thickness variation	5 min
Maximum applied pressure	Consolidation quality, mechanical properties	10 bar
Pressure application rate	Out-of-plane buckling, wrinkling, thickness variation	5 bar/min
Cooling rate	Degree of crystallinity, mechanical properties	10°C/min
Diaphragm material	Surface finish	Polyimide

Source: Reprinted from *Composites: Part A*, 33, S. G. Pantelakis and E. A. Baxevani, Optimization of the diaphragm forming process with regard to product quality and cost, 459–470, Copyright (2002), with permission from Elsevier.

10.4.2 THERMO-STAMPING

Since thermoplastic matrix composites can be shaped and formed by the application of heat and pressure, some of the forming processes used in the metals and plastics industries can be applied to them with some modifications. Among these processes are thermo-stamping, hydroforming, and thermoforming (Figure 10.20). These are highly efficient processes for converting flat sheets to complex shapes at relatively fast production speeds. To apply these processes to thermoplastic matrix composites, the prepreg stack (either unconsolidated or consolidated) is heated outside the forming press to a suitable forming temperature so that stack can be deformed under pressure without tearing. It is then placed in the mold, shaped by applying either pressure

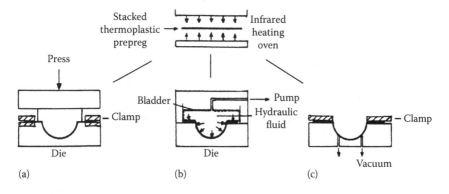

FIGURE 10.20 Forming processes for thermoplastic matrix composites (a) thermo-stamping, (b) hydroforming, and (c) thermoforming.

or vacuum, and cooled to produce the composite part. The thermo-stamping process is described in greater detail in this section.

Thermo-stamping is a matched metal die forming process similar to the warm stamping process used for sheet metals. In this process, the blank is clamped around its edges in a blank holder and heated to the forming temperature outside the mold. Heating is done either in a hot air-circulating oven or in an infrared oven. The blank holder serves several functions [10]: (1) it prevents shrinking of the blank during preheating, (2) it maintains fiber orientation in the heat-softened blank as it is transferred from the oven to the mold, (3) it allows rapid transfer of the preheated blank to the mold without significant heat loss, and (4) it suppresses wrinkle formation during press forming. For clamping to be effective, the clamping force in the blank holder must create suitable in-plane tensile stresses in the blank.

After the blank holder is placed on the die or the lower half of mold, the punch, which is the upper half of the mold, is quickly brought in contact with the top surface of the blank and is moved downward at a controlled speed. The mold is maintained at a much lower temperature than the blank temperature to allow rapid cooling and fast processing. The blank is drawn into the die as it slides out of the blank holder (Figure 10.21).

During the forming phase, there is a considerable amount of frictional resistance at the interface between the blank and the mold surfaces. Since the frictional force acts opposite to the forming force, it has a large influence on the deformation of the blank. Due to the presence of thin layers of liquid polymer on the heated blank surfaces, the principal mode of frictional resistance is hydrodynamic friction (instead of dry friction) [12]. The parameters that control the frictional resistance are the interface temperature, normal pressure at the interface, sliding velocity, fiber orientation, and mold surface roughness. If a mold release agent is used on the mold surfaces to ease the part removal, its presence will also affect the frictional resistance and, therefore, the deformation process. Friction experiments conducted with APC-2

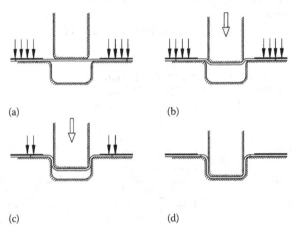

(a)

(b)

(c)

(d)

FIGURE 10.21 Stages of blank deformation during thermostamping in matched metal dies: (a) punch making contact with the heated blank, (b) punch pressing on the blank, (c) blank being drawn into the die, and (d) at the completion of forming.

(a unidirectional carbon fiber-reinforced PEEK composite) have shown that under isothermal conditions, the coefficient of friction between a 0° ply of APC-2 and steel foil (representing a steel mold surface) increases with increasing sliding velocity as well as temperature (Figure 10.22), but decreases with increasing normal load. The coefficient of friction also decreases as the fiber orientation angle with respect to the sliding direction is increased.

Defects in thermo-stamped parts include out-of-plane wrinkling, fiber rupture, ply separation, and nonuniform thickness distribution. Tension on the blank as it is clamped in the blank holder can considerably reduce wrinkling. However, if the clamping force is too high, the blank may tear along the rim. Thus, a suitable clamping force is essential to form a good-quality part without any wrinkles or fiber tear. The clamping force depends on both reinforcement architecture (e.g., unidirectional or fabric) and forming ratio (i.e., ratio of blank area and the projected area of the die cavity) [20]. A low forming speed also reduces the possibility of wrinkling by generating lower resistance to interply and intra-ply shearing, thus allowing the blank to deform more easily. Thickness distribution is affected by both forming speed and reinforcement architecture. A more uniform thickness distribution is obtained if a low forming speed is used. The thickness variation is greater in unidirectional fiber-reinforced parts than in fabric-reinforced parts. The reason for this is the transverse flow of fiber and matrix in the unidirectional architecture. If it is a laminated structure, the stacking sequence has a strong influence on the formability of thermo-stamped parts. For example, out-of-plane wrinkles and ply separation are observed in quasi-isotropic 0/90/45/−45 laminates, but not in 0/90 laminates [21].

The thermo-stamping process parameters include the preheating temperature, stamping time, stamping speed, and pressure. The acceptable processing window is a combination of these parameters that produce parts without fiber buckling and breakage, excessive resin migration, and intense thinning. Friedrich et al. [20]

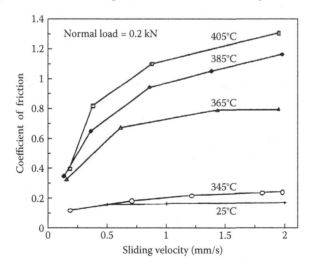

FIGURE 10.22 Effect of sliding velocity and temperature on the coefficient of friction between APC-2 and release-agent treated steel foils.

reported that the temperature range for successful thermo-stamping of an L-shaped part was within the small range above the melting point (T_m) for semicrystalline polymers, such as PP and PEEK. For PEI, which is an amorphous polymer, the optimum temperature was 60°C above its T_g. Stamping pressure and velocity were dependent on the volume fraction of fibers and the melt viscosity of the polymer. At low stamping velocities and pressures, the fibers were likely to buckle or break at the bend region. At high stamping velocities or at too high a pressure, on the other hand, there was excessive resin migration and thinning. The excessive resin migration resulted in resin being squeezed out of the outer layers. The optimum processing conditions as reported by Friedrich et al. are listed in Table 10.4.

A study conducted to make a direct comparison of diaphragm forming and thermo-stamping showed that the surface finish of thermo-stamped parts are better than diaphragm-formed parts [13]. In both cases, a spherical dome-shaped part with a flange was formed. Since diaphragm forming exerts lower constraints on material movement, the thickness variation is lower in diaphragm-formed parts. Another difference between the two forming processes is in the mode of deformation. In diaphragm forming, the outer edges of the blank are pressed onto the die flanges before the central area is drawn into the cavity. In addition, tensile stresses are created in the blank as the diaphragms are stretched. The combination of the two prevents out-of-plane wrinkling in diaphragm formed parts even in the absence of any clamping force on the blank [20].

The main deformation mechanism in both processes is interply slip that allows relative movement to adjacent plies to conform to the die shape. Intra-ply slip and interply rotation needed for 3D shape formation are not observed in thermo-stamping, but they occur in diaphragm forming. Thus, even though diaphragm forming is a much slower forming process than thermo-stamping, it is capable of producing more complex 3D shapes. Thermo-stamping, on the other hand, is more suitable for 2D and simple 3D shapes.

TABLE 10.4
Optimum Thermo-Stamping Conditions for Several Thermoplastic Matrix Composites

Composite	Fiber Volume Fraction (%)	Processing Temperature (°C)	Stamping Time (s)	Stamping Velocity (mm/min)	Stamping Pressure (MPa)
Carbon fiber/PP	20	180	>15	200–500	1.5–8
Glass fiber/PP	33	180	>15	200–500	2–9
Carbon fiber/PEEK	60	380	>15	200–500	4–9.5
Glass fabric/PEI	50	280	>30	300–700	5–9

Source: Reprinted from *Composite Sheet Forming*, K. Friedrich, M. Hou, and J. Krebs, Thermoforming of continuous fibre/thermoplastic composite sheets, Copyright (1997), with permission from Elsevier.

10.4.3   COMPRESSION MOLDING

Compression molding (also known as flow molding) is used for manufacturing continuous mat-reinforced thermoplastic composite parts. One such material is glass mat thermoplastics (GMTs) which contain 20–40 vol.% of randomly oriented continuous E-glass fibers in a PP matrix. The fibers are dispersed in a swirl pattern, and the sheet thickness is usually between 3 and 4 mm. GMT sheets containing unidirectional continuous E-glass fibers as well as E-glass fabric are also made. Another polymer used as matrix material in GMT is polyethylene terephthalate. Melt impregnation is the common method of production for GMT sheets. It involves passing several layers of glass fiber mat through a preheating oven and sandwiching them between extruded layers of the liquid polymer matrix. The sandwich is then passed through a series of calendaring rollers and a heated double-belted press for consolidation.

In the compression molding process for GMT parts, blanks cut from GMT sheets are stacked and preheated in an infrared oven or a hot air convection oven to a temperature above the melting temperature of the thermoplastic matrix. If the matrix is PP, the preheating temperature is between 200°C and 220°C. As the matrix starts to melt, the GMT layers in the stack tend to open up or *loft* considerably, in some cases, to two to three times the original sheet thickness. The heated stack of blanks is quickly transferred to a molding press, where it is placed in the lower half of an open mold. Both halves of the mold are usually preheated to a temperature higher than the room temperature, but much below the solidification temperature of the matrix. The press is closed at a high speed (typically 400–600 mm/s), and then the upper mold half is moved downward to press on the stack at a lower speed (typically 20–30 mm/s). As the pressure on the stack builds up to 100–200 bar, the material starts to flow outward to fill the cavity between the upper and lower molds. During flow, the material starts to cool and its viscosity starts to increase, and therefore, it is important that the cavity be filled before the viscosity becomes too high. The mold is kept closed under pressure until the average temperature of the material becomes equal to the mold temperature. The upper half of the mold is then opened, and the compression-molded GMT part is pushed out of the lower half of the mold using ejector pins. Typical compression molding conditions for PP matrix GMT are listed in Table 10.5. The cycle time for compression molding GMT parts is between 30 and 60 s compared to 1–5 minutes for SMC parts. Because of relatively high production rate and low material cost, PP matrix GMT has found several applications in automobiles, such as bumper beams, seat backs, and inner door panels.

One of the advantages of compression molding is that it can produce near-net shape parts with complex geometry and varying thickness. For simple shapes in which the final part thickness is close to the original sheet thickness, thermo-stamping is a less expensive and faster process. It involves heating the GMT blank in an infrared oven to 20–60°C above the melting temperature of the polymer matrix and quickly transferring it to a matched mold held at room temperature. The part shape is formed as the mold is closed, and pressure is applied on the heated blank. Very little flow of material takes place during thermo-stamping. For PP matrix GMT, the recommended blank temperature prior to placing it in the mold is 200°C, the mold closing speed is 750–1500 mm/min, and the molding pressure is 100–140 bar.

TABLE 10.5

Compression Molding Conditions for Glass Mat Thermoplastics with PP Matrix

Blank surface temperature (°C)	195–220
Blank heating time (s)	180–240 in infrared oven
	240–480 in hot air convection oven
Mold surface temperature (°C)	25–80
Mold closing speed (mm/s)	400–600
Pressing speed (mm/s)	20–30
Molding pressure (bar)	100–200
Total closing time (s)	<5
Cooling time in the mold (s)	4–5 per mm of part thickness
Total cycle time (s)	45–90
Shrinkage after molding (%)	0.2–0.5

REFERENCES

1. A. G. Gibson and J.-A. Månson, Impregnation technology for thermoplastic matrix composites, *Composites Manufacturing*, Vol. 3, No. 4, pp. 223–233, 1992.
2. J. Mack and R. Schledjewski, Filament winding process in thermoplastics, in *Manufacturing Techniques for Polymer Matrix Composites (PMCs)* (S. G. Advani and K.-T. Hsiao, eds.), Woodhead, Cambridge, UK, pp. 182–208, 2012.
3. P. Mitschang and M. Christmann, Continuous fiber reinforced profiles in polymer matrix composites, in *Manufacturing Techniques for Polymer Matrix Composites (PMCs)* (S. G. Advani and K.-T. Hsiao, eds.), Woodhead, Cambridge, UK, pp. 209–242, 2012.
4. R. Brooks, Forming technology for thermoplastic composites, in *Composites Forming Technology* (A. C. Long, ed.), Woodhead, Cambridge, UK, pp. 256–276, 2007.
5. A. B. Strong, *High Performance and Engineering Thermoplastic Composites*, Technomic, Lancaster, PA, 1993.
6. D. M. Bigg, Processing characteristics of thermoplastic sheet composites, *International Polymer Processing*, Vol. 2, pp. 172–185, 1992.
7. J. D. Muzzy and A. O. Kays, Thermoplastic vs. thermosetting structural composites, *Polymer Composites*, Vol. 5, No. 3, pp. 169–172, 1984.
8. A. C. Loos and M.-C. Li, Consolidation during thermoplastic composite processing, in *Processing of Composites* (R. S. Davé and A. C. Loos, eds.), Hanser, Munich, pp. 209–283, 2000.
9. J. D. Muzzy and J. S. Colton, The processing science of thermoplastic composites, in *Advanced Composites Manufacturing* (T. G. Gutowski, ed.), John Wiley & Sons, New York, 1997.
10. M. D. Wakeman and J. A. E. Manson, Composites manufacturing—Thermoplastics, in *Design and Manufacture of Textile Composites* (A. C. Long, ed.), Woodhead, Cambridge, UK, pp. 197–241, 2005.
11. J. C. Barnes and F. N. Cogswell, Transverse flow processes in continuous fibre-reinforced thermoplastic composites, *Composites*, Vol. 20, pp. 38–42, 1989.
12. A. M. Murtagh and P. J. Mallon, Characterization of shearing and frictional behavior during sheet forming, in *Composite Sheet Forming* (D. Bhattacharyya, ed.), Elsevier Science, Amsterdam, pp. 163–216, 1997.

13. J. Krebs, K. Friedrich, and D. Bhattacharyya, A direct comparison of matched-die versus diaphragm forming, *Composites: Part A*, Vol. 29A, pp. 183–188, 1998.
14. G. W. Ehrenstein, *Polymeric Materials*, Hanser, Munich, 2001.
15. J.-A. E. Manson, T. L. Schneider, and J. C. Seferis, Press-forming of continuous-fiber-reinforced thermoplastic composites, *Polymer Composites*, Vol. 11, No. 2, pp. 114–120, 1990.
16. N. Zahlan and J. M. O'Neill, Design and fabrication of composite components: The spring-forward phenomenon, *Composites*, Vol. 20, No. 1, pp. 77–81, 1989.
17. A. Lystrup and T. A. Andersen, Autoclave consolidation of fibre composites with a high temperature thermoplastic matrix, *Journal of Materials Processing Technology*, Vol. 77, pp. 80–85, 1998.
18. P. J. Mallon, C. M. O'Bradaigh, and R. B. Pipes, Polymeric diaphragm forming of complex-curvature thermoplastic composite parts, *Composites*, Vol. 20, No. 1, pp. 48–56, 1989.
19. S. G. Pantelakis and E. A. Baxevani, Optimization of the diaphragm forming process with regard to product quality and cost, *Composites: Part A*, Vol. 33, pp. 459–470, 2002.
20. K. Friedrich, M. Hou, and J. Krebs, Thermoforming of continuous fibre/thermoplastic composite sheets, in *Composite Sheet Forming* (D. Bhattacharyya ed.), Elsevier Science, Amsterdam, pp. 91–262, 1997.
21. P. De Luca, P. Lefébure, and A. K. Pickett, Numerical and experimental investigation of some press forming parameters of two fibre reinforced thermoplastics: APC2-AS4 and PEI-CETEX, *Composites: Part A*, Vol. 29A, pp. 101–110, 1998.

11 Joining and Repair

The previous chapters in the book describe the processes for manufacturing PMC parts. In order to build a complete structure, the composite parts have to be joined either with other composite parts or to parts of other materials. The number of joints depends on the number of parts in the structure, whereas the joining method primarily depends on the materials and the load transfer requirement between the parts being joined. For example, a large commercial aircraft contains two to three million different aluminum and composite parts and almost an equal number of joints. Most of these joints are made of rivets, pins, and other threaded fasteners. The body and chassis of automobiles today contain several hundred different steel parts, and most of these parts are joined by using resistance spot welds.

The primary purpose of a joint is to transfer loads from one member to another member in a structure. The load transfer takes place through shear, tension, or a combination of shear and tension. Joints are usually the weakest areas in any structure, and service failure of many load-bearing structures often initiates at the joints. The structural performance of a joint depends on the joining method, joint design, and joint quality. The joint design has a critical significance in PMCs, since they do not possess the forgiving characteristics of ductile metals, namely, their capacity to redistribute local high stresses by yielding.

This chapter describes the two most common joining methods used in PMCs, namely, mechanical fastening and adhesive bonding. Joining of thermoplastic matrix composites using fusion bonding is also described. Finally, repairing techniques of service-damaged PMCs is addressed at the end of the chapter.

Mechanical joints are created by fastening the substrates with bolts, rivets, or pins; bonded joints use an adhesive interlayer between the substrates (commonly called the adherends). Both types of joints can be used to join composites with composites and composites with other materials, most notably metals. The emphasis in this chapter is on the processing aspect of these two types of joining methods for PMC parts. When used for joining composites with metals, additional steps may be necessary; for example, corrosion prevention measures for the metal substrate. There are several good references [1–3] that provide information on stress analysis of joints in composite materials, and further details for joint design and performance evaluation are available in these references.

11.1 JOINING IN COMPOSITES

For thermoset matrix composites, the basic joining methods are mechanical fastening and adhesive bonding (Figure 11.1). A combination of these two joining methods is also used. Mechanical fastening and adhesive bonding are also used with thermoplastic matrix composites. While thermoset matrix composites cannot be welded together, several different welding methods are available and used for thermoplastic

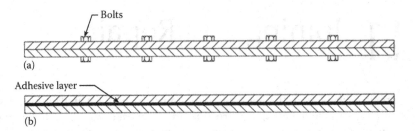

FIGURE 11.1   (a) Mechanically fastened and (b) adhesively bonded laminates.

matrix composites. Two of these welding methods are vibration welding and ultrasonic welding. They are described later in the chapter.

The advantages and disadvantages of mechanical joints and adhesive bonded joints are listed in the following:

1. Mechanical joints using fasteners
   a. Permit quick and repeated disassembly for repairs or replacements without destroying or damaging the substrates
   b. Require little or no surface preparation
   c. Are easy to inspect for joint quality
   d. Require machining or punching of holes in the laminates, which may interrupt fiber continuity, create localized damage around the holes, and reduce the strength of the substrate laminates
   e. Create highly localized stress concentrations around the joints that may cause failure initiation in the substrates
   f. Add weight to the structure
   g. May create potential corrosion problems in unprotected fasteners, for example, in an aluminum fastener if it is used for joining carbon fiber/epoxy laminates
   h. Prone to fretting
   i. May not be seal-proof, and therefore, water ingression into the laminates may take place
2. Bonded joints using adhesives
   a. Distribute the joint load over a larger surface area than mechanical joints
   b. Improve the overall stiffness of the structure by providing a continuous joining between the substrates (Figure 11.1b)
   c. Do not require holes; therefore, there are no stress concentrations in the substrates
   d. Add very little weight to the structure
   e. Provide continuous sealing against water ingression
   f. Are difficult to disassemble without either destroying or damaging the substrates
   g. May be affected by service temperature, humidity, and other environmental conditions due to their influence on adhesive properties

    h. Require careful surface preparation for good bonding and high joint strength

    i. Are difficult to inspect for joint quality

We will now consider each of these joining methods separately.

11.2 MECHANICAL FASTENING

11.2.1 Mechanical Fasteners

The mechanical fasteners used with PMCs fall into three categories: bolts, rivets, and pins. The materials for the fasteners are steels, aluminum alloys, and titanium alloys. The primary function of a fastener is to hold two or more parts together (usually under a clamping load), transfer service load between the fastened parts, and prevent any relative motion between them. Since in many applications, the joints may experience high fatigue loads and vibration, the fastener must be resistant to both fatigue failure and clamping load loss due to vibration. For fasteners used with PMCs, there are other requirements that must also be considered.

1. Fastener material: In selecting the fastener material for PMC parts, the two most important considerations are (1) its design allowable strength and (2) its resistance to galvanic corrosion. The difference in the coefficients of thermal expansion of the fastener material and the composite must also be considered. If the joint is exposed to large temperature variations during the service operation, the difference in thermal expansions or contractions may change the clamping load between the joined parts.

   Corrosion is not usually a problem with most fastener materials in contact with either glass or Kevlar fiber composites. However, if aluminum fasteners are used with carbon fiber composites, they tend to corrode unless they are separated from the composite parts with a corrosion-resistant material or coating. This is because of the large difference in galvanic corrosion potentials between carbon fibers and aluminum. Similar corrosion problems will occur if steel is used as the fastener material instead of aluminum. Since galvanic corrosion potential between carbon fibers and titanium is relatively small, titanium alloys, such as Ti-6Al-4V, are often selected as the fastener material with carbon fiber composites. The other fastener materials compatible with carbon fiber composites are superalloys (e.g., A286), multiphase alloys (e.g., MP35N and MP159), and nickel alloys (e.g., alloy 718).

2. Fastener design: The fasteners commonly used in the aircraft industry are lock bolts, which consist of a pin with a threaded end and a threaded collar (Figure 11.2). The shank of the pin can be either plain or threaded, and its head can be either a protruding type or countersinking type. They are installed by swaging the collar that locks it into the grooves of the threaded end of the pin.

   In Hi-Lok® bolts, a popular aircraft lockbolt, there is a hex key in the threaded end to react to the torque applied during collar installation.

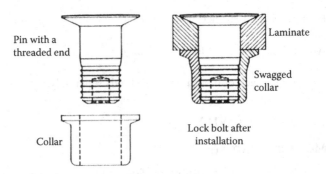

FIGURE 11.2 Lock bolt with a countersinking head.

The collar includes a frangible pintail that reacts to the axial load during the swaging operation. When the swaging load reaches a predetermined limit, the pintail breaks away at the breakneck groove.

There are many other types of mechanical fasteners that are used in the aircraft industry. They include threaded pins with mating nuts or self-locking collars and rivets with or without internal threads in the shank. Outside the aircraft industry, standard bolts, nuts, and washers are commonly used as fasteners for composites.

Since clamping can cause a high compressive force on the composite, fasteners used for composites should have large heads for distributing the clamping load over a large surface area and reducing the possibility of crushing of the composite under the fastener head. The threaded portion of the bolt or pin should not make contact with the hole surface to avoid damaging the composite during installation or service load application. This can be done by selecting appropriate clearance between the hole and the fastener diameters, centering the fastener during installation, and, if needed, using a sleeve to separate the threads in the fastener from the hole surface. To prevent fretting, fasteners can be bonded by using an adhesive between the fastener head and the composite surface.

3. Tightening torque: When a fastener is tightened using a torque wrench, a tensile preload is created in the fastener and a compressive clamping load is created in the joined parts. In general, the joint failure load increases with increasing tightening torque until a limit torque value is reached. Above the limit torque, the joint load does not appreciably increase. If the compressive clamping load generated by the tightening torque is too high, it can cause damage in the composite, such as matrix cracking, fiber breakage, crushing, and subsurface delamination.

4. Fastener placement and spacing: The mode of joint failure and the load-carrying capacity of mechanically fastened joints strongly depend on the fastener placement and location.

The important mechanical joint design parameters are e/d ratio, w/d ratio, and d/h ratio, where e is the edge distance, d is the hole diameter, and w and h are the width and thickness of the composite part, respectively

(Figure 11.3). In a multifastener configuration, spacing between the holes and their spatial arrangement in the composite part are also important, since they determine the load sharing arrangement between the fasteners being used. If the fasteners are spaced too close to each other, the increase in stress concentration at the fastener holes can cause early failure. If they are spaced too far apart, the joint efficiency becomes low.

The failure modes observed in the joint area of PMCs are schematically shown in Figure 11.4. They include (1) net tension, (2) bearing damage, (3) shear out, (4) tension and cleavage, and (5) cleavage. Combinations of these basic failure modes are also possible. In addition, there may also be fastener failure by fracture or by excessive bending. In general, failure by shear-out occurs at low e/d ratios and failure by net tension occurs at low w/d ratios. If the laminate contains nearly all 0° fibers, cleavage failure is also possible. Unless the e/d and w/d ratios are very large, the full bearing strength is

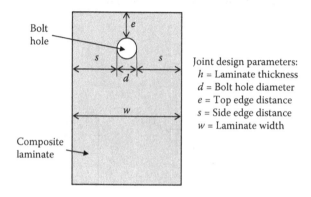

Joint design parameters:
h = Laminate thickness
d = Bolt hole diameter
e = Top edge distance
s = Side edge distance
w = Laminate width

FIGURE 11.3   Joint design parameters for mechanically fastened joints.

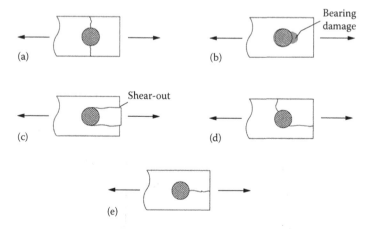

FIGURE 11.4   Basic failure modes in mechanically fastened composites: (a) net-tension failure (b) bearing damage, (c) shear-out failure, (d) tensile and cleavage failure, and (e) cleavage failure.

seldom achieved. In general, shear-out and net tension failures are avoided if e/d is >3 and w/d is >6. The edge distance needed to reduce shear-out failure can be reduced by increasing the laminate thickness at the edge or by inserting metal shims between various composite layers near the bolted area.

The joint design parameters are usually determined by conducting pin-bearing tests. In pin-bearing tests, a pin is inserted through the hole at one end of the composite, which is then loaded in tension until failure at the hole occurs (Figure 11.5). It is observed in these tests that the pin-bearing strength increases with both e/d and w/d ratios (Figure 11.6). Shear-out occurs at low e/d ratios and bearing failure occurs at high e/d ratios. Similarly, net tension failure occurs at low w/d ratios and bearing failure occurs at high w/d ratios. The e/d and w/d ratios, at which the failure mode changes from shear-out or net-tension failure to bearing damage, increase with increasing clamping torque. The use of a washer increases the joint strength, since it increases the clamping area and provides a lateral restraint. But it is observed that for a constant clamping

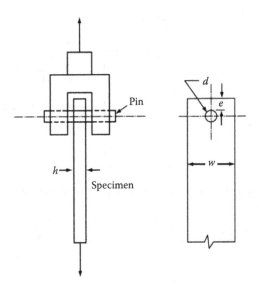

FIGURE 11.5   Pin-bearing test.

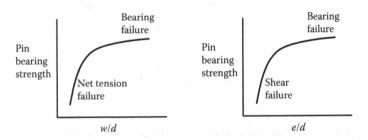

FIGURE 11.6   Pin-bearing strength as a function of w/d and e/d ratios.

torque, the joint strength decreases with increasing washer diameter, since it reduces the clamping pressure. Washers with sharp edges should not be used since they can create damage in the surface of the composite part.

11.2.2 Holes for Mechanical Fastening

Holes for mechanical fastening in a composite part can be either machined in a post-molding operation or formed in place during the molding operation. Machining is preferred, since molded-in holes may be surrounded by misoriented fibers, resin-rich areas, or knit lines. Drilling is the most common method of machining holes in a cured laminate; however, unless the proper drill, fixture, and cutting speed are used, the material around the drilled hole (particularly on the exit side of the drilled hole) may be damaged. High-speed water jets and lasers can produce cleaner holes with little or no damage, but they are more expensive than drilling.

11.2.2.1 Drilling

The most common tool for drilling is a conical point twist drill with a preferred point angle of 135°, helix angle between 10° and 36°, and clearance angle between 6° and 16° (Figure 11.7). The drill materials are solid carbides, carbide-coated steels, diamond-coated steels, or zirconium nitride-coated steels. The reason for selecting these hard or hard surface-coated materials instead of standard high-speed steel (HSS) is the high abrasiveness of glass and carbon fibers that causes the HSS tool to quickly wear out, therefore requiring frequent tool change. Kevlar fibers are not as abrasive, but it is difficult to drill clean holes in Kevlar fiber composites.

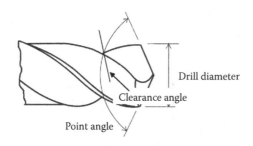

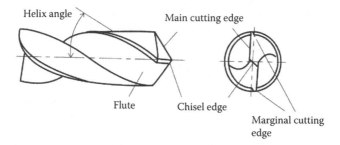

FIGURE 11.7 Drill design parameters.

This is because Kevlar fibers, when cut, tend to fray or shred. Specially designed drills are used for Kevlar fiber composites.

The drilling process parameters that affect the quality of drilled holes in the composite are the cutting speed and feed of the rotating drill. Cutting speeds in the range of 2,000–20,000 revolutions per minute and slow feed rates are recommended. A small pilot hole is often drilled, which is then enlarged to the final hole diameter and finished with a reamer. Cutting fluids are not commonly used for drilling composite parts. It is good practice to use vacuum to remove carbon dusts generated during drilling holes in carbon fiber composite parts, since airborne carbon dust particles may contaminate and short-circuit the electric motors of machineries present nearby.

Drilling of holes in PMCs can create several defects that can not only influence the quality and appearance of the drilled holes, but also affect the joint strength [3,4]:

1. Delamination or separation of two or more layers in a composite laminate can occur at the entrance as well as exit sides of the hole. The entry-side delamination is called the peel-up and the exit-side delamination is called the push-out (Figure 11.8).

   The peel-up delamination occurs as the drill reaches the interfaces between the top two or three plies of the laminate. The delamination starts at the periphery of the cutting edges of the drill, and much like the upward chip movement in metals, the delaminated area tends to move up, giving rise to the peel-up phenomenon. In general, the peeling effect increases with increasing helix angle. The size of delamination increases with increasing cutting speed and feed of the drill.

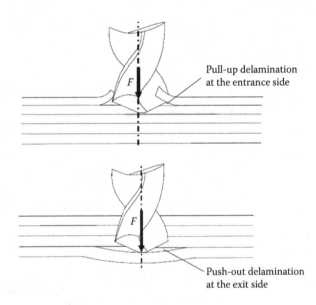

FIGURE 11.8 Delaminations caused by drilling of a composite laminate (force F in the figure represents the thrust force exerted by the drill).

The push-out delamination is attributed to high thrust force produced by the chisel edge of the drill. If the laminate is unsupported at the exit end, the thrust force tends to bend the bottom plies as it tries to break out of the laminate. If the thrust force is greater than a critical value, delamination occurs at the interfaces between the bottom two or three plies. The thrust force increases with increasing feed rate, but is not much influenced by the cutting speed. It also depends on the tool geometry, tool material, and physical condition of the tool. A worn out tool creates a higher thrust force than a fresh tool. The critical thrust force depends on the delamination energy and modulus of the laminates and the uncut depth under the tool.

The push-out delamination can be reduced by clamping a support board on the back side of the laminate. When a support board is used, the drill exits through the support board instead of through the laminate. It also provides a more uniform support on the back side of the laminate. After drilling the hole, the support board is discarded and a new support board is used for drilling the next hole.

2. Damage due to hole drilling: Drilling a hole in a composite part can damage the surface of the hole as well as the surface of the part adjacent to the hole. The hole surface damage can take the forms of surface roughness, chipped hole edges, exposed broken fibers, fiber pullout, and matrix cracks.

3. Matrix overheating: Since the thermal conductivity of PMCs is low, heat generated by drilling is not efficiently dissipated and the temperature of the composite material close to the hole increases. If the temperature rise is very high, it can cause considerable softening of the matrix in the composite. If the matrix is a thermoplastic polymer, it can start to melt. A thermoset polymer matrix may not melt, but can burn and char if the temperature increase is very high.

11.2.2.2 Water Jet Hole Machining

In water jet hole machining, a stream of water mixed with granular silicate or similar abrasive particles is injected at a very high pressure (3800–6200 bars) and high speed (3500 km/h) onto the composite surface using a 0.254 mm diameter nozzle. The water jet quickly erodes the matrix and shears off the fibers as it cuts the hole in the composite part. Water jet hole machining can be used in parts that are up to 150 mm thick. It does not generate any heat or create any dust. It also does not cause any delamination in the composite part.

11.2.3 Joint Strength Improvement

The strength of mechanical joints can be significantly improved by relieving the stress concentrations surrounding the joint. The following are a couple of the methods used for relieving stress concentration:

1. Softening strips of lower modulus material are used in the bolt bearing area. For example, strips made of E-glass fiber plies can be used to replace some of the carbon fiber plies aligned with the loading direction in a carbon fiber-reinforced laminate.

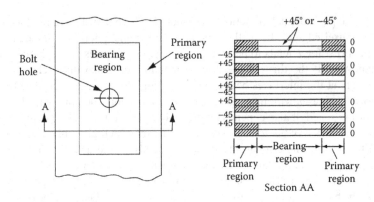

FIGURE 11.9 Laminate tailoring method for improving mechanically fastened joint strength of composite laminates.

2. The laminate tailoring method [5] divides the bolt-hole area into a primary region and a bearing region. This is demonstrated in Figure 11.9 for a $[0_2/\pm45]_{2s}$ laminate. In the bearing region surrounding the bolt-hole, the 0° plies in the laminate are replaced by ±45° plies. Thus, the lower modulus-bearing region is bounded on both sides by the primary region of $[0_2/\pm45]$ laminate containing 20–60% 0° plies. The majority of the axial load in the joint is carried by the high-modulus primary region, which is free of fastener holes. The combination of low axial stress in the bearing region and the relatively low notch sensitivity of the [±45] laminate delays the onset of the net tension failure commonly observed in nontailored $[0_2/\pm45]_s$ laminates.

11.3 ADHESIVE BONDING

11.3.1 ADHESIVES

Adhesives used for joining structural composites are either epoxies, polyurethanes, or acrylics. Their general characteristics are listed in Table 11.1. Each of these adhesives can be formulated to produce a wide range of properties [6]. Epoxies are the most commonly used adhesive in the aircraft industry. They are selected for their high tensile strength, temperature resistance, and fatigue durability. They have the ability to bond well to a variety of materials, such as composite to composite and composite to metals. Their impact resistance is low, but it can be significantly improved by formulating them with elastomeric tougheners. There is a variety of hybrid epoxies that provide improved properties, such as improved heat resistance or improved toughness at low temperatures. Among them are epoxy–acrylic, epoxy–cyanate ester, epoxy–urethane, and epoxy–polyamide.

Even though epoxies are also used in the automotive industry, polyurethanes and acrylics are preferred, since they require less bonding time than epoxies. Polyurethanes bond to a variety of materials and they provide strong impact and shock resistance to the joint, but they have lower tensile strength and temperature

TABLE 11.1

Characteristics of Adhesives Used for Joining PMCs

Adhesive	Characteristics
Epoxy	• Two-component or single-component formulation • Forms strong, durable bonds with many substrates • Requires heat curing • Refrigerated storage required • Low temperature/high strain rate, which can cause brittle failure
Acrylic	• Needs a surface activator • Can be used with less than ideal surface preparation (for example, without degreasing) • Relatively fast cure time (5–15 minutes) • Viscosity control can be difficult • Safety and handling issues due to toxic monomer (requires good venting)
Urethane	• Two-component or single-component formulation • Forms strong, durable bonds • Relatively high flexibility and excellent elongation to failure • Moisture resistance may be poor • Temperature limitation is approximately 100°C

resistance than epoxies. Acrylics are noted for their ability to bond to most surfaces with little or no surface preparation. Their mechanical properties are as good as or better than many epoxies, but they have a much higher curing shrinkage, which is a source of high residual stresses in the adhesive.

Adhesives are available in liquid, paste, or film form. The liquid and paste forms can be either a two-part system or a single-part system. The two-part system is composed of two reactive chemicals that are mixed before it is applied on the substrates to be joined. The curing reaction, which may require elevated temperatures, transforms the mixture into a solid adhesive over a period of time. The single-part system can be of two types: (1) the two chemicals are premixed and the mixture remains in a partially cured, nonreactive state until it is dispensed on the surfaces to be joined and cured at elevated temperatures and (2) a single chemical that require an external energy source to initiate the curing reaction, such as moisture for some polyurethanes or UV for some acrylics. The film adhesive is also a single-part premixed system, but it is either in a semisolid or in a solid form and is usually supported on a thin carrier made of a nylon or polyester fabric. The carrier film helps in controlling the bondline thickness and provides toughness to the adhesive joint. Film adhesives are preferred for many aerospace composite joints, because they are easier to handle and can be applied in uniform thickness. After the film adhesive is applied between the surfaces to be joined, it is cured at an elevated temperature and under pressure. Typical specifications of an epoxy film adhesive are listed in Table 11.2.

11.3.2 Adhesive Selection

Adhesive selection depends on a number of factors, such as the substrates to be bonded, adhesive application method, surface preparation requirement, adhesive bonding time,

TABLE 11.2
Typical Specifications of an Epoxy Film Adhesive

Type	Epoxy film
Thickness	0.127 mm
Width	900 mm
Weight	140 g/m^2
Scrim	Nonwoven polyester
Volatiles	Less than 1%
Shelf life	1 year at −18°C
Out-time	30 days at 30°C and 50% relative humidity (RH)
Gel time	1 hour when heated from 50°C at 2°C/min
Cure cycle	60–120 minutes at 177°C and180–700 kPa

operating environment (temperature, humidity, chemical, etc.), stress levels, and cost. Adhesive bonding time depends on the adhesive type, curing mechanism, and curing condition. In applications requiring a very rapid cure, a fast curing two-part toughened acrylic may be preferred over an epoxy that usually takes a longer time to cure. However, a slower curing adhesive may be selected if the bond area is large, since in such a case, the rapid curing system may need a very fast mixing and dispensing system; otherwise, the adhesive mix will start to gel and increase in viscosity before it is properly spread over the entire joining surface. Film adhesives are selected over liquid or paste adhesives for applications requiring precise control of adhesive thickness and positioning. Since they will not start to cure until the appropriate pressure and temperature are applied, they can be disassembled and repositioned if any adjustments are needed to be made during the assembly process.

Adhesive properties and their variation due to moisture absorption, temperature, etc., are also important considerations in adhesive selection. The following adhesive properties should be considered in the selection of adhesive for structural joint applications.

1. Shear and tensile properties (strength, modulus, and strain-to-failure) and their variation with environment, such as temperature and humidity
2. Fracture toughness, which is usually determined by conducting a double cantilever beam test
3. Coefficient of thermal expansion (relative to the coefficients of thermal expansion of the substrates, since it is one of the factors in controlling the thermal residual stresses in the cured adhesive)
4. Creep (which depends on both time and temperature) particularly for long-term applications under sustained loads
5. Chemical resistance, if the adhesive is exposed to chemicals that can alter its properties

The important characteristics of a good adhesive are high shear and tensile strengths but low shear and tensile moduli. Flexible adhesives with low modulus make the stress distributions in the adhesive more uniform and reduce the maximum stresses that occur at the ends of the bondline. High ductility and fracture toughness for the adhesive are important properties to consider if the substrates have dissimilar stiffness values or if the joint is subjected to impact loads.

11.3.3 ADHESIVE JOINING PROCESS

The joining process with epoxy adhesives involves the following steps. Similar steps are also needed for other types of adhesives.

1. Surface preparation: Proper substrate surface preparation is needed to assure high-quality bonded joints. It starts with lightly sanding or grit blasting the substrate surface to make it rougher. After removing the dirt and abraded surface particles, solvent cleaning (using acetone or other chemicals) is used to make the surface oil and grease free. If there is a peel ply on the surface, it must be removed before starting the surface preparation. If one of the substrates is a metal, its surface should be cleaned of any corrosion film using chemical etching or anodizing. It may also be necessary to apply a corrosion-inhibiting primer on the metal surface.

   The next step in surface preparation may include the application of a chemical coupling agent, such as a silane, that improves bonding of the adhesive to the substrate surface and reduces the possibility of adhesive failure at the interface between the two.

2. Adhesive preparation: Adhesive preparation for a two-part epoxy adhesive includes two steps: metering the right amounts of the adhesive ingredients and mixing them thoroughly before dispensing the mix on the substrate surfaces. In a two-part epoxy system, the epoxy resin is mixed with the hardener in the ratio usually provided by the adhesive supplier. Thorough mixing requires the use of a high-shear blender with rotating blades. The chemical reaction between the resin and the hardener during mixing may increase the temperature of the mix, which needs to be controlled; otherwise, there may be early gelling, rapid increase in viscosity, and reduction in its working life. Mixing often produces air bubbles in the adhesive mix, and if they are not removed before dispensing, they may appear as voids in the cured adhesive.

   In a one-part epoxy system, such as in a film adhesive, the epoxy and the hardener are premixed, but the chemical reaction between them is slowed down by storing the mixture at a subzero temperature in a freezer. For such a system, it is important to completely thaw the adhesive and bring it to room temperature before removing it from the protective storage bag; otherwise, it will start to absorb moisture from the surrounding atmosphere, which may cause porosity in the adhesive after curing. It is also important to check the viscosity of the adhesive before applying it on the substrates. If during storage the viscosity has increased to a high level, the adhesive may not spread uniformly as it is being applied on the substrate surface and the possibility of air entrapment and void formation will increase.

3. Adhesive application: Before dispensing the metered and mixed adhesive formulation on the substrate surfaces, it is important to make sure that the substrates are properly fixtured for good fit and adhesive thickness control. Adhesive thickness is one of the key parameters in the design of adhesive joints. If it is too thick, it will not only take a longer time to cure, but will also increase the possibility of forming voids and porosity. If it is too thin, the joint strength will be low. It is also important to keep the same thickness throughout the entire length. This can be accomplished by mixing glass beads with the adhesive or using a scrim cloth or a nonwoven mat.

4. Adhesive curing: Epoxy adhesives require elevated temperatures and several hours for proper curing. The cure cycle is usually provided by the adhesive supplier. For many epoxy adhesive systems, the temperature is usually increased in two steps. In the first step, as the temperature is increased to a precure temperature and held constant for a short time, the viscosity of the adhesive mix decreases which allows it to flow and wet-out the substrate surfaces. In the second step, the temperature is increased to the cure temperature and held constant for a longer time. Pressure is applied to achieve a good wet-out and remove entrapped air and volatiles. The cure cycle ends with slowly lowering the temperature to room temperature. Postcuring may be used to achieve full cure and improved mechanical properties.

11.3.4 BONDED JOINT DESIGN

The simplest adhesively bonded joint is a single-lap joint (Figure 11.10a) in which load transfer between the substrates takes place through a distribution of shear stresses in the adhesive. However, since the loads applied at the substrate ends are off-centered, a bending moment is generated at the adhesive joint. The bending action sets up a distribution of peel stresses that are directed normal to the substrate surfaces. Both shear and peel stress distributions exhibit high values near the lap

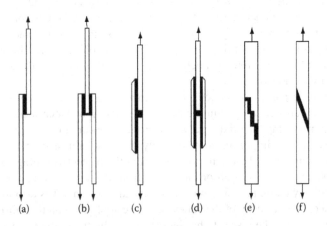

FIGURE 11.10 Basic bonded joint designs: (a) single-lap joint, (b) double-lap joint, (c) single- and (d) double-strap joints, (e) stepped lap joint, and (f) scarf joint.

ends of the adhesive layer and relatively low values over much of the lap length. The double-lap joint, shown in Figure 11.10b, eliminates much of the bending action and the resulting peel stresses present in the single-lap joint. Since the average shear stress in the adhesive is also reduced by nearly one-half, a double-lap joint has a higher joint strength than a single-lap joint. The use of a bonded strap on either side or on each side of the substrates (Figure 11.10c) also improves the joint strength compared to single-lap joints. Stepped lap (Figure 11.10d) and scarf joints (Figure 11.10e) can potentially produce very high joint strengths; however, in practice, the difficulty of machining the steps or steep scarf angles often overshadows their advantages. If a stepped lap joint is used, it may often be easier and less expensive to lay up the steps prior to curing the laminate. This eliminates the machining operation and prevents damage to the fibers. Depending on the part design and complexity, other joint designs are possible. Some examples are shown in Figure 11.11.

Bonded joint strength depends not only on the adhesive strength, but also on the substrate stiffness and joint design, both of which influence the stress distributions in the adhesive and adhesive–substrate interfaces. For example, in a single-lap joint, increasing the substrate stiffness by increasing its thickness reduces the bending action and, therefore, reduces the maximum peel stress. The maximum shear stress is also reduced. While both help in improving the joint strength, the substrate weight is increased due to increased thickness.

The joint design parameters for single-lap bonded joints are the adhesive thickness (h), lap length (L), and substrate end design (Figure 11.12). In order to achieve good bonding and good joint strength, a minimum adhesive thickness is required. As the adhesive thickness is increased above the minimum thickness, the joint strength becomes lower [7,8]. Increasing the lap length results in higher joint strength, but at high lap lengths, the increase in joint strength is relatively small. The ratio of lap length L to substrate thickness h significantly improves the joint strength at small L/h ratios. At high L/h ratios, the improvement is marginal (Figure 11.13). Tapering the substrate ends (Figure 11.14) reduces the high normal stresses at these locations.

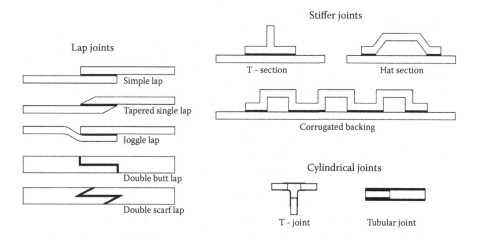

FIGURE 11.11 Examples of several different bonded joint designs.

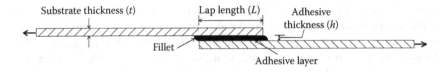

FIGURE 11.12 Joint design parameters in a single-lap bonded joint.

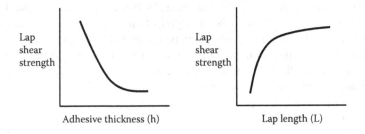

FIGURE 11.13 Joint strength of single-lap bonded joints as a function of adhesive thickness and lap length.

FIGURE 11.14 Tapering of substrate ends to improve joint strength.

In many instances, when the pressure is applied during bonding, small amounts of adhesive are squeezed out which form a fillet at the substrate ends. Studies have shown that fillets tend to reduce stress concentration at the joint ends and improve the joint strength.

11.3.5 QUALITY OF BONDED JOINTS

The quality and strength of bonded joints depend on the number and types of defects in the bondline. These defects are produced during the bonding operation. Some of these defects are often difficult to detect even with nondestructive inspection, such as ultrasonic C-scan, and cannot be corrected without destroying the substrates. This is one of the reasons for adding mechanical fasteners to adhesive joints in many primary aircraft structures so that the combined joint will be more reliable. In these hybrid joints, mechanical fasteners share the load with the bonded joint and act as the secondary load path after the bonded joint has either weakened or failed.

Various defects observed in adhesive joints are listed in the following:

1. Voids are caused by entrapped air bubbles generated during mixing or volatiles generated during the curing reaction. A cluster of small voids in close
   proximity of each other and located in a small area or spread over a small
   length of the bondline is called porosity. Voids and porosity occur if the

pressure applied during the bonding operation is insufficient, the bondline thickness is too high, or the adhesive viscosity increases too rapidly before the entrapped air or the volatiles are expelled. Voids cause stress concentration and are often the source of crack initiation within the adhesive.

2. Unbonds are areas where the adhesive has not bonded to one or both substrates. The reasons for unbonds are poor adhesion due to inadequate surface preparation, presence of contaminants on the surface, poor wettability, and insufficient pressure application. High adhesive viscosity and premature curing before complete surface coverage by the adhesive may also contribute to unbonds.

3. Weak bonds are caused by inadequate or uneven curing of the bondline. Incorrect formulation of the adhesive, poor mixing, presence of contamination, and uneven heating are generally the reasons for weak bonds. If the proper curing cycle is not followed, the adhesive will be undercured and its properties will be lower than the design properties. If the curing temperature is too high, it can cause thermal degradation of the adhesive and turn it into a brittle material. Poor surface preparation can also cause weak bonds. Weak bonds are the most difficult defect to detect even by using nondestructive inspection.

4. Cracks formed within the adhesive or at the adhesive–substrate interface are associated with curing shrinkage and residual thermal stresses arising in the bondline after curing.

5. There is a lack of fillet at the ends of the bondline.

Three types of failure modes are observed in bonded joints in PMCs (Figure 11.15). They are (1) adhesive failure, which occurs at the interface between the adhesive and the substrate; (2) cohesive failure, which occurs in the adhesive layer; and (3) a combination of adhesive and cohesive failures. In addition, there may be substrate failure, primarily by interlaminar fracture or delamination between the first and second layer of the composite laminate adjacent to the bondline. In general, adhesive failure occurs if there are unbonds or weak bonds, whereas cohesive failure occurs if there are voids and porosity, cracks, contaminations, and poorly cured areas in the adhesive layer. A combination of adhesive and cohesive failures is also very common. In such cases, failure may start in one mode (for example, cohesive failure mode)

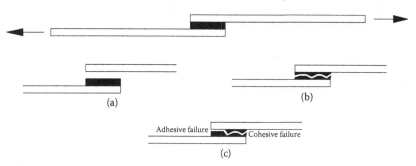

FIGURE 11.15 Basic failure modes in bonded joints (a) adhesive failure, (b) cohesive failure, and (c) mixed failure.

and then change to the second mode (adhesive failure mode). The failure modes may alternate several times before the complete failure of the bondline.

11.4 COCURING

Cocuring is a process in which curing and bonding are simultaneously achieved in the same curing cycle. The polymer matrix at the interface between the two surfaces being joined acts as the adhesive, and therefore, no additional adhesive is needed. It reduces not only the material cost by eliminating the adhesive, but also the processing cost since no surface preparation is needed and the adhesive cure time is also eliminated. One concern in cocuring is the difference in the coefficients of thermal expansion of the mold and the parts to be cocured. If the difference is large, it is possible that there will be high stresses in the joining area at the heating or cooling stages that may cause early failure in the cocured area.

A variation of cocuring is called *cobonding* in which one of the two composite parts in the assembly is fully cured and the other part is uncured. The two parts are bonded as the latter is cured using the appropriate curing cycle. In this process, careful surface preparation of the previously cured part is required, and additional adhesive may have to be added at the interface.

11.5 JOINING OF THERMOPLASTIC MATRIX COMPOSITES

Thermoplastic matrix composites can be joined with each other using mechanical fasteners, adhesive bonding, and fusion bonding. Mechanical fastening and adhesive bonding are described in Sections 11.2 and 11.3; however, in general, adhesive bonding is more difficult with thermoplastic matrix composites due to their low surface energy. Because of this, adhesive bonding of many thermoplastic matrix composites needs an additional step of increasing their surface energies using corona discharge, plasma treatment, etc.

Unlike the thermoset matrix composites, *fusion bonding*, also called *fusion welding*, can be used with thermoplastic matrix composites, since the matrix in these composites will melt and form a liquid pool when heated to a sufficiently high temperature (which is usually greater than $(T_m + 30°C)$ for semicrystalline polymers and $(T_g + 100°C)$ for amorphous polymers) and solidify when cooled to room temperature. Two principal advantages of fusion bonding over mechanical fastening and adhesive bonding are (1) relatively fast cycle time and (2) no additional material is needed. Like adhesive bonding, fusion bonding does not require any holes and creates a permanent joint, but unlike adhesive bonding, it does not require much surface preparation.

Fusion bonding of thermoplastic polymers involves the application of both heat and pressure in a localized area, called the *weld zone*. Heat is needed to increase the temperature of the thermoplastic polymers on both sides of the weld zone to their respective melting range, and pressure is needed to consolidate and join them together. Consolidation and joining occur due to the intermixing and diffusion of polymer molecules across the interface, which also create molecular bridging between the liquid pools in the weld zone (Figure 11.16). Upon cooling, solidification causes the development of joint strength in the weld zone, which can be close to the strength of the substrates.

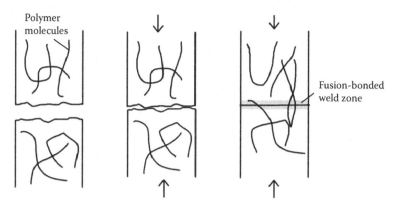

Polymer
molecules

Fusion-bonded
weld zone

FIGURE 11.16   Polymer molecules in a fusion bonded weld zone.

TABLE 11.3
**Fusion Bonding Techniques for Thermoplastics
and Thermoplastic Matrix Composites**

Thermal	Mechanical	Electromagnetic
• Hot gas welding	• Vibration welding	• Resistance welding
• Hot plate welding	• Spin welding	• Induction welding
• Extrusion welding	• Ultrasonic welding	• Dielectric welding
• Infrared welding	• Friction stir welding	• Microwave welding
• Laser welding		

The quality of fusion bonded joints depends on three factors—temperature, pressure, and consolidation time. The presence of voids (due to air entrapment) and inadequate fusion (for example, due to uneven heating or pressure distribution) will also affect their quality. If the fiber volume fraction in the thermoplastic matrix composite substrates is high, there may not be enough polymer available for melting, and therefore, the joint may not be properly consolidated. With semicrystalline thermoplastics, the development of crystalline structure during the cooling stage is also an important factor in determining the joint quality.

Several different fusion bonding techniques (Table 11.3) are used in the thermoplastics industry [9,10]. The difference in these techniques is mainly in the difference in the heat generation method. The most common fusion bonding techniques for thermoplastic matrix composites are vibration welding and ultrasonic welding. They are described in the following.

11.5.1   VIBRATION WELDING

In *vibration welding*, also called linear friction welding, the surfaces of the two parts to be joined are held in contact under pressure (typically 1–5 MPa), and while one part is fixed, the other part is vibrated parallel to the interface (Figure 11.17).

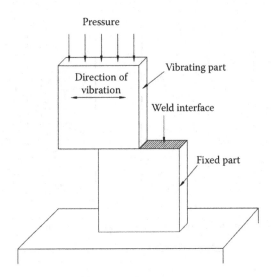

FIGURE 11.17 Vibration welding process.

The frequency of vibration is typically in the range of 100–200 Hz, and the amplitude of vibration is between 1 and 5 mm. As the vibration is continued, the friction-generated heat raises the temperature of the interface of the two parts and a thin layer of liquid polymers is created at the interface. Thereafter, the mechanism of heat generation changes from friction to viscous heat dissipation in the liquid pool. After the vibration is stopped, the liquid polymer pool cools down to room temperature while the parts are still being held under pressure.

Since during vibration welding, the applied pressure acts normal to the interface, the liquid polymer pool at the interface will have a tendency to flow laterally outward, which will cause the upper and lower parts to come closer to each other. The distance by which the upper part moves closer to the lower part is called weld penetration. The joint strength increases with increasing weld penetration until a threshold value is reached, above which weld penetration has a relatively small influence on the joint strength. The lateral flow of the liquid polymer pool may also cause extrusion of small amounts of liquid polymer on both sides of the interface, which will appear as flash on two sides of the joined parts.

Vibration welding is commonly used for fusion bonding semicrystalline polymers such as PP and PA-6. It also works well with short glass fiber-reinforced thermoplastics in which these polymers are used as the matrix material. One problem of using this technique with continuous fiber reinforced composites is the possibility of fiber movement and displacement, which may cause fiber misorientation in the weld zone.

Spin welding is a variation of vibration welding in which one part is rotated relative to the second part at their interface using a spinning motion while they are held in contact with each other under pressure. The surface speed of rotation is between 1 and 20 m/s. It can be applied to both solid and hollow parts with round cross-sections. Friction-generated heat at the interface of the rotating parts increases the local temperature and causes melting of the polymer surfaces in contact.

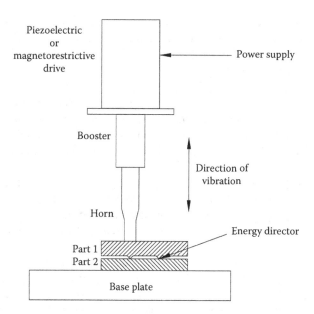

FIGURE 11.18 Ultrasonic welding process.

11.5.2 ULTRASONIC WELDING

In *ultrasonic welding*, the parts to be joined are clamped together under pressure and then subjected to very low amplitude oscillations at ultrasonic frequencies normal to the interface using an acoustic tool, called a horn (Figure 11.18). The frequency of oscillation is in the range of 20–40 kHz and the amplitude is between 10 and 100 μm. Heat is generated by a combination of surface friction and viscous heat dissipation in the material. The efficiency of ultrasonic welding is improved if the vibratory energy is transmitted over small contact areas instead of the entire interface. This is accomplished by molding small protrusions on one or both surfaces of contact. These protrusions, usually triangular in shape (Figure 11.18), are called energy directors. As the horn is pressed against the top part and the oscillation is started, melting of the polymers first occurs at the energy directors and then it spreads laterally. Without the energy directors, surface contacts at the interface occur only at the surface asperities, and since they are not uniformly distributed, heat generation also becomes nonuniform, which then results in nonuniform welding.

Ultrasonic welding is used for joining a wide variety of thermoplastics; however, they are generally more suitable for amorphous polymers than for semicrystalline polymers which require higher amplitudes of oscillation to generate sufficient heat for them to melt.

11.6 REPAIR

As the use of PMCs increases in aircrafts, automobiles, and many other applications, the need for developing effective repair technology has also increased. Replacement of composite parts that contain manufacturing defects or are damaged during transport, assembly, and service operations can be very expensive, particularly if they are made of carbon fiber composites. Table 11.4 lists the service-induced damages typically encountered in PMC aircraft parts [11]. Depending on their severity, they can be divided into three categories:

1. Allowable damage: Upon inspection and structural testing, some damages may be considered either cosmetic or not severe. If they do not affect the structural integrity or affect the functioning of the part, they will be categorized as allowable damage; however, they may still require sealing or minor patching.
2. Repairable damage: Some of the other damages, such as delaminated or disbonded layers and matrix cracks, may be severe enough to reduce the strength of the damaged part below the design allowable or, if continued in service, can cause further degradation in strength. However, these damages may be repairable.
3. Nonrepairable damage: Some damages, such as through-thickness penetration, may be very severe, and repair may be either too difficult to perform or not economical or may not restore the part to its original strength. Such damages are considered nonrepairable, and the damaged part must then be discarded and replaced.

The assessment of damages on composite parts starts with a visual inspection of the damaged area. Some damages, such as a scorch, a dent, or an abrasion, are easily visible. If the damage is not visible or barely visible, its presence

TABLE 11.4
Service-Induced Damages in PMC Aircraft Parts

Defect	Typical Causes
Cuts, scratches, abrasions, paint chips	• Mishandling
Penetrations	• Mishandling
	• Bird strike
	• Collision with service vehicle
Abrasion	• Rain/grit erosion
Delaminations	• Low energy impact: tool drops, hailstones, runway debris
	• Freeze/thaw contraction/expansion
	• Aerodynamic peeling
	• Overload, e.g., during assembly or removal
Disbonds	• Degradation of metallic interfaces
	• Failure of adhesive joints
Hole elongation and damage	• Bearing failure in a mechanical joint
	• High tightening torque causing surface damage
Dents	• Crushed core in sandwich constructions due to impact
Edge damage	• Poor fit
Burn marks	• Lightning strike
	• Excess heat exposure

may be detected by tapping the composite part surface with a coin or a light metal hammer and listening to the sound. The damaged area will produce a dull sound compared to the undamaged area which will produce a sharp, ringing sound. By tapping repeatedly, the boundary of the damaged area can be roughly mapped. But the determination of the severity of the damage requires nondestructive inspection and evaluation (NDI/NDE) that can identify its nature, size, and location under the surface. The most common NDI/NDE method used in the aerospace industry is ultrasonic C-scan, which has become a very useful tool for detecting delaminations and internal matrix cracks. Some minor damages, such as minor delaminations close to the surface, can be repaired by directly injecting a compatible liquid resin into the delaminated area and curing it in situ. If the damage is considered repairable after nondestructive inspection, the next steps are (1) removal of the damaged area, (2) preparation of the repair area, (3) patch repair of the damaged area, and (4) verification and evaluation of the completed repair. The goal of the patch repair is to restore the load path interruption caused by the damage and return the structure to its original strength and/or stiffness with minimum additional weight.

There are four main patch repair techniques [11,12]: (1) one-sided or double-sided lap-bonded patching, (2) scarf-bonded patching (Figure 11.19a and b), (3) bolted patching (Figure 11.19c), and (4) a combination of bonded and bolted patching. Bonded patches are preferred over bolted patches, since they do not require hole drilling. However, bonded patches require extensive surface preparation and drying. Lap-bonded patches are used in thin laminates (typically 2 mm thickness). Scarf-bonded patches are used for thick laminates. Since they produce a repair that is flush or nearly flush with the surface, there will be little or no protrusions, and therefore, the surface characteristics may not change much. The scarf-bonded repair can be

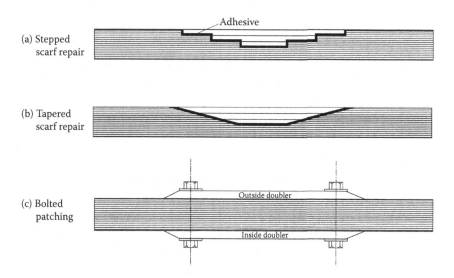

FIGURE 11.19 Damage repair techniques for composite laminates: (a) lap-bonded patching, (b) scarf-bonded patching, and (c) bolted patching.

stepped or tapered. The scarf angle is usually in the range of 3–5° to maintain an acceptable level of shear stress in the bonded joint. However, to machine such a small taper, a large quantity of material has to be removed, some of which may be from outside the damaged area. Furthermore, if a scarf repair is used, the patch layers must match the layers in the original laminate in terms of fiber orientation angle. Bolted patching is sometimes used for the repair of thick composite laminates. It is limited to applications where the strain levels are low. In general, bolted repairs produce half the strength of bonded repairs.

To start the repair process by patch bonding, the damaged area is removed by carefully machining out the damaged layers using a high-speed milling cutter or a router with carbide or diamond-coated tools. After the machining operation, the removed area is cleaned of any debris or dust and then properly prepared for adhesive bonding. Since moisture can cause problems by vaporizing and forming voids in the adhesive, it is also important to dry the laminate in the repair area.

There are three different methods of applying the repair patch: (1) the patch can be a precured laminate, called a doubler, which is bonded to the removed area using an adhesive; (2) the patch can be made of a stack of prepreg layers, which are cocured with an adhesive; and (3) the patch can be made using a wet layup process and then cured using the vacuum bag technique. Both film adhesive and two-part paste adhesives are used for bonding the repair patch. Film adhesives are more convenient than film adhesives, since they do not require weighing and mixing of resin and hardener.

REFERENCES

1. R. B. Heslehurst, *Design and Analysis of Structural Joints with Composite Materials*, DEStech, Lancaster, PA, 2013.
2. S. D. Thoppul, J. Finegan, and R. F. Gibson, Mechanics of mechanically fastened joints in polymer-matrix composite structures—A review, *Composites Science and Technology*, Vol. 69, pp. 301–329, 2009.
3. V. Krishnaraj, R. Zitoune, and J. P. Davim, *Drilling of Polymer-Matrix Composites*, Springer, Heidelberg, 2013.
4. E. Persson, I. Eriksson, and L. Zackrisson, Effects of hole machining defects on strength and fatigue life of composite laminates, *Composites: Part A*, Vol. 28A, pp. 141–151, 1997.
5. J. R. Eisenmann and J. L. Leonhardt, Improving composite bolted joint efficiency by laminate tailoring, in *Joining of Composite Materials*, ASTM STP 876, p. 238, ASTM, West Conshohocken, PA, 1985.
6. D. A. Dillard (ed.), *Advances in Structural Adhesive Bonding*, Woodhead, Cambridge, UK, 2010.
7. M. D. Banea and L. F. M. da Silva, Adhesively bonded joints in composite materials: An overview, *Journal of Materials: Design and Applications, Proceedings of the Institution of Mechanical Engineers*, Vol. 223, pp. 1–17, 2009.
8. M. Davis and D. Bond, Principles and practices of adhesive bonded structural joints and repairs, *International Journal of Adhesion and Adhesives*, Vol. 19, pp. 91–105, 1999.
9. C. Ageorges and L. Ye, *Fusion Bonding of Polymer Composites*, Springer, London, 2002.

10. A. Yousefpour, M. Hojjati, and J. Immarigeon, Fusion bonding/welding of thermoplastic composites, *Journal of Thermoplastic Composite Materials*, Vol. 17, pp. 303–341, 2004.
11. A. Baker, Joining and repair of aircraft composite structures, in *Composites Engineering Handbook* (P. K. Mallick, ed.), CRC Press, Boca Raton, FL,. pp. 671–776, 1997.
12. S.-H. Ahn and G. S. Springer, *Repair of Composite Laminates*, Technical Report No. DOT/FAA/AR-00/46, US Federal Aviation Administration, Washington, DC, 2000.

Appendix: Health and Safety Issues

When working with PMCs in the laboratory or on the production floor, proper care must be taken to handle fibers and polymers to avoid health and safety issues. Some of the resins, solvents, cleaning fluids, and even fibers can be hazardous if proper and timely precautions are not taken or if they are handled carelessly. It is important to read and follow the material safety data sheets commonly provided by the material suppliers. Some of the health and safety issues and protections against them are described in the following.

A.1 HEALTH AND SAFETY ISSUES

1. Fibers: Glass and carbon fibers have the potential to cause irritation of skin, eyes, nose, and throat. Since glass fiber diameter is greater than 6 μm and usually do not break into lengths smaller than their diameter, they are not respirable when airborne. Carbon fiber diameter is also greater than 6 μm, but they can splinter into smaller sizes, which are respirable. Fibers are not known to be toxic and carcinogenic. The sizing used on fiber surfaces may cause skin irritation or sensitization.

2. Resins, curing agents, and other ingredients in the resin formulation: In general, uncured epoxy, polyester, and vinyl ester prepolymers have low levels of toxicity. However, direct contact with them should be avoided to prevent skin sensitization and irritation. Isocyanates, one of the chemicals for producing polyurethane resins, are highly toxic and can cause health effects that can range from eye, skin, and respiratory track irritation to bronchitis and pulmonary edema. Thermoplastic resins in general are not known to be harmful, but molding them at high temperatures may generate gases that are irritating to eyes and inhalation.

   Curing agents or hardeners used for curing epoxy resins have high levels of toxicity and can cause severe irritation on both direct contact and inhalation. The aromatic amine hardeners are more toxic than amide hardeners and can cause more severe health hazards that can range from skin irritation to eye and liver damage.

   Styrene monomers used as a reactive diluent in the curing reaction of polyesters and vinyl esters are highly irritant to eyes and respiratory tract. Since styrene is highly volatile, it easily evaporates into the workplace environment and creates an odor that is readily recognizable. The effect of long-term exposure to styrene on humans has been widely studied and ranges from changes in blood cell chromosomes to brain damage.

TABLE A.1

Chemicals Commonly Used in the Composite Industry

Type	Examples	Health and Safety Issues
Ketones	• Acetone • Methyl ethyl ketone • Methyl isobutyl ketone	Eye, nose, and throat irritation; prolonged contact may cause dermatitis; in high concentrations, may cause headache, nausea, dizziness, etc.; volatile and flammable
Alcohols	• Methanol (methyl alcohol) • Ethanol (ethyl alcohol) • Isopropanol (isopropyl alcohol)	Do not usually present serious health hazards; can cause skin irritations; volatile and flammable
Chlorinated hydrocarbons	• Methylene chloride (dichloromethane) • 1,1,1-Trichloroethane • Trichloroethylene	Irritation of the eyes and upper respiratory tract; dizziness, nausea, etc.; dermatitis on prolonged skin contact; ability to depress the central nervous system; not particularly flammable (which is the reason for using them instead of ketones)

Styrene is classified as a possible carcinogen, although there is no proof that it can cause cancer. The permissible exposure limit of styrene varies from country to country. In the United Sates, it is regulated by the Occupational Safety and Health Administration (OSHA). The current OSHA standard is 100 parts of styrene per million (ppm) of air averaged over an 8-hour work shift, with a ceiling of 200 ppm and an acceptable peak of 600 ppm for 5 minutes in any 3-hour period of exposure.

3. Solvents: Many different types of chemicals and solvents are used for cleaning molds, tools, and equipment. Examples of these chemicals are given in Table A.1. Direct contact to these chemicals should be avoided. Prolonged inhalation of some of these chemicals can pose serious health hazards.

A.2 PERSONAL AND WORKPLACE PROTECTIONS

The following personal and workplace protections should be taken when working with PMCs:

1. Eye protection: Eye protection is required to prevent damage from splashes of chemicals and solvents when mixing and pouring them. Eye protection is also required to protect eyes from dusts generated during machining or cutting of composite parts. Wearing of safety glasses with side shields, goggles, or a face shield is highly recommended for eye protection.
2. Respiratory protection: Respiratory protection is needed if the dust and airborne solvent levels are high. Dusts are generated during machining and cutting or sawing composite parts. Grinding, routing, sanding, and polishing also create dusts. Much of the dusts generated by these processes are

fine particulates (3.5–5 μm in diameter) and respirable. Properly fitted dust masks provide the respiratory protection needed. The use of vacuum with a suitable high-efficiency particulate air (HEPA) filter is recommended near the source of dust. A downdraft table with good airflow can be used as the working surface for cutting, grinding, etc. A cutting saw with running water flow on its sides is also another way to remove dust while cutting composite parts.

3. Skin protection: Skin protection is needed if handling of fibers, resin, and other chemicals (such as curing agents) causes skin irritation and rashes. Fibers protruding out of machined, saw-cut, or fractured composite surfaces may splinter into the skin and cause lacerations. Proper gloves and protective clothing should be worn for skin protection.

4. Protection from burning: Many of the operations in composite processing, such as curing, forming, etc., are done at high temperatures. If proper protection is not taken, there may be mild to severe burning. Protection from burning requires the use of thermal gloves and careful handling of the hot parts after they are demolded.

5. Fire safety: Most solvents are volatile and flammable. They should be stored in a fire-proof cabinet and their containers must be kept closed when they are not in use.

6. Electrical safety: Dust from carbon fibers or carbon fiber composites can short-circuit electrical damage and electronic equipment problems due to its high electrical conductivity. Vacuum cleaners fitted with appropriate filters should be used to remove carbon fiber dust near its source. Nearby electrical or electronic equipment, switches, and plugs must be covered or sealed to protect them from being damaged.

7. Ventilation: The workplace should be fully ventilated to remove odor, evaporated chemicals, and airborne dust. In closed molding processes, such as bag molding, liquid molding processes, and compression molding, there is very little solvent evaporation, but in filament winding and pultrusion, the resin baths used for resin impregnation of fibers are usually open. In filament winding, the winding operation is done in the open air. Thus, in both pultrusion and filament winding, there will be significant solvent evaporation. For both these processes, local ventilation hoods at the sources of solvent evaporation must be installed. Places where prepregs are assembled or SMC sheet is manufactured must be well ventilated with good air flow.

Index

Page numbers followed by f and t indicate figures and tables, respectively.

Printed in the United States
by Baker & Taylor Publisher Services